GM Food on Trial

Genetics and Society

The books in this series, all based on original research, explore the social, economic and ethical consequences of the new genetic sciences. The series is based in the ESRC's Centre for Economic and Social Aspects of Genomics (CESAGen), the largest UK investment in social-science research on the implications of these innovations. With a mix of research monographs, edited collections, textbooks and a major new handbook, the series will be a major contribution to the social analysis of new agricultural and biomedical technologies.

Governing the Transatlantic Conflict over Agricultural Biotechnology
Contending Coalitions, Trade Liberalisation and Standard Setting
Joseph Murphy and Les Levidow

New Genetics, New Social Formations
Peter Glasner, Paul Atkinson and Helen Greenslade

New Genetics, New Identities
Paul Atkinson, Peter Glasner and Helen Greenslade

The GM Debate
Risk, Politics and Public Engagement
Tom Horlick-Jones, John Walls, Gene Rowe, Nick Pidgeon, Wouter Poortinga, Graham Murdock Tim O'Riordan

Growth Cultures
Life Sciences & Economic Development
Philip Cooke

Human Cloning in the Media
Joan Haran, Jenny Kitzinger, Maureen McNeil and Kate O'Riordan

Local Cells, Global Science
Embryonic Stem Cell Research in India
Aditya Bharadwaj and Peter Glasner

Handbook of Genetics and Society
Paul Atkinson, Peter Glasner and Margaret Lock

The Human Genome
Chamundeeswari Kuppuswamy

Debating Human Genetics
Contemporary Issues in Public Policy and Ethics
Alexandra Plows

Community Genetics and Genetic Alliances
Eugenics, Carrier Testing, and Networks of Risk
Aviad Raz

Genetic Testing
Accounts of Autonomy, Responsibility and Blame
Michael Arribas-Ayllon, Srikant Sarangi and Angus Clarke

Scientific, Clinical and Commercial Development of the Stem Cell
From Radiobiology to Regenerative Medicine
Alison Kraft

GM Food on Trial
Testing European Democracy
Les Levidow and Susan Carr

The Making of a Syndrome
The case of Rett Syndrome
Katie Featherstone and Paul Atkinson

Barcoding Nature
Claire Waterton, Rebecca Ellis and Brian Wynne

Gender and Genetics
Towards a Sociological Account of Prenatal Screening
Kate Reed

Neurogenetic Diagnoses, the Power of Hope, and the Limits of Today's Medicine
Carole H. Browner and H. Mabel Preloran

GM Food on Trial

Testing European Democracy

Les Levidow and Susan Carr

First published 2010
by Routledge
711 Third Ave, New York, NY 10017

Simultaneously published in the UK
by Routledge
2 Park Square, Milton Park, Abingdon, Oxon OX14 4RN

Routledge is an imprint of the Taylor & Francis Group, an informa business

First published in paperback 2012

Typeset in Sabon by IBT Global.

Library of Congress Cataloging in Publication Data

Levidow, Les.
GM food on trial : testing European democracy / by Les Levidow and Susan Carr.
p. cm.—(Genetics and society)
Includes bibliographical references and index.
1. Genetically modified foods—Government policy—Europe. 2. Genetically modified foods—Law and legislation—Europe. 3. Agricultural biotechnology—Government policy—Europe. 4. Agricultural industries—Government policy—Europe. I. Carr, Susan, 1943– II. Title.
TP248.65.F66L48 2009
363.19'2094—dc22
2009018971

ISBN10: 0-415-95541-6 (hbk)
ISBN10: 0-203-86669-X (ebk)

ISBN13: 978-0-415-95541-6 (hbk)
ISBN13: 978-0-203-86669-6 (ebk)
ISBN13: 978-0-415-65501-9 (pbk)

Contents

Tables

Figures

Acknowledgements

For research material from several EU member states, we would like to thank our partners in three research projects.

'GMO Releases: Managing Uncertainty about Biosafety', project during 1994–95
'Safety Regulation of Transgenic Crops: Completing the Internal Market?', project during 1997–99
'Precautionary Expertise for GM Crops', project during 2002–2004.

We also thank the European Commission for funding the three projects, as well as Bernhard Zechendorf for his assistance at the Bio-documentation Centre, DG Research. Also thanks to our departmental colleague David Wield for support through those projects.

The book frequently cites our partners' articles, which were originally published in journal special issues as follows:

Levidow, L. and Carr, S. (eds) (1996) 'Biotechnology Risk Regulation in Europe', special issue of *Science and Public Policy* 23(3): 133–200.
Levidow, L. and Carr, S. (eds) (2000) 'Precautionary Regulation: GM Crops in the European Union', special issue of the *Journal of Risk Research* 3(3): 187–285.
Levidow, L. and Carr, S. (eds) (2005) 'Precautionary Expertise for EU Agbiotech Regulation', special issue of *Science and Public Policy* 32(4): 257–324.

Some chapters draw upon our earlier articles, as follows:

Chapter 2

Levidow, L. (1996) 'Simulating Mother Nature, industrializing agriculture', in G. Robertson et al. (eds), *FutureNatural: Nature, Science, Culture,* London: Routledge, pp. 55–71.
Levidow, L., Søgaard, V. and Carr, S. (2002) 'Agricultural PSREs in Western Europe: research priorities in conflict', *Science and Public Policy* 29(4): 287–95.

Chapter 3

Levidow, L. (2000) 'Pollution metaphors in the UK biotechnology controversy', *Science as Culture* 9(3): 325–51.

Chapter 4

Levidow, L. (2007) 'European public participation as risk governance: enhancing democratic accountability for agbiotech policy?', *East Asian Science, Technology and Society (EASTS): An International Journal* 1(1): 19–51.

Chapter 5

Levidow, L., Carr, S. and Wield, D. (2005) 'EU regulation of agri-biotechnology: precautionary links between science, expertise and policy', *Science and Public Policy* 32(4): 261–76.

Levidow, L. and Carr, S. (2007) 'Europeanising advisory expertise: the role of "independent, objective and transparent" scientific advice in agri-biotech regulation', *Environment & Planning C: Government & Politics* 25: 880–95.

Chapter 6

Levidow, L. and Carr, S. (2007) 'GM crops on trial: technological development as a real-world experiment', *Futures* 39(4): 408–31.

Chapter 7

Levidow, L. and Bijman, J. (2002) 'Farm inputs under pressure from the European food industry', *Food Policy* 27(1): 31–45.

Chapter 8

Levidow, L. and Boschert, K. (2008) 'Coexistence or contradiction? GM crops versus alternative agricultures in Europe', *Geoforum* 39(1): 174–90.

For helpful comments on drafts of this book, we would like to thank the following readers: Andrew Barry, Nigel Clark, Jim Dratwa, Maria Paola Ferretti, Frank Fischer, Eric Gall, Peter Glasner, Anne Gray, Benjamin Holtzman, Sheila Jasanoff, Steve McGiffen, Éric Montpetit, Natasha Posner, Mike Saward, Rachel Schurman, Frans Seifert, András Székács, Bronislaw Szerszynski, Helge Torgerson.

Introduction

1 WHAT HAS BEEN ON TRIAL?

Since the late 1990s, public controversy has kept agricultural biotechnology on the defensive, especially in Europe. As some commentators have noted, 'GM food is on trial'. In ordinary conversation, 'GM food' as come to mean far more than food products derived from GM crops. Much more than agbiotech has been kept 'on trial'—a phrase whose meaning has likewise expanded.

Even naming agbiotech has been difficult and contentious. In the 1980s its US promoters chose 'genetically engineered organisms' to name their project of precisely, safely redesigning nature. European promoters anticipated difficulty with public attitudes. They soon chose instead 'genetically modified organisms' (GMOs or GM plants) as a more subtle, less hubristic name; likewise OGM in Latin languages. Opposition activists devised ominous names such as 'GM pollution' or *Gen-Müll* ('genetic garbage') for putting agbiotech on trial.

In the mid-1990s 'trial' generally meant a field experiment for testing how a GM crop performed outside the laboratory. These field trials were meant to demonstrate that GM crops were being kept safe, while also testing them for evidence of safety if grown commercially. Often disputes arose over whether the experiments had adequate control measures to protect the environment, a rigorous design for detecting any risks, and a legitimate societal purpose. Safety experiments were turned into symbolic tests of government responsibility.

Experimental trials have led to judicial ones. Since the late 1990s, some field experiments have been sabotaged by activists—at first covertly, and later overtly as political theatre for the mass media. Opponents asserted their own democratic legitimacy in the name of protecting the environment from agbiotech. These attacks led to prosecutions of activists, who used the opportunity to put the state symbolically on trial for failure to protect society.

In France such multiple trials resulted from actions of the Left-wing farmers' union, the Confédération Paysanne. In January 1998 approximately a hundred members forced entry into a Novartis warehouse and destroyed a stock of Bt 176 GM seed. When three activists were prosecuted, they turned their prosecution into a symbolic trial of agbiotech in the mass media. As a defence argument, they denounced Bt 176 GM maize as 'the very symbol of a type of agriculture and society that we reject'.

In 1999, moreover, Confédération Paysanne activists symbolically dismantled a McDonald's restaurant under construction in the town of Millau. The attack came as a response to a WTO ruling in the dispute over hormone-treated beef, involving an EU ban on US beef exports. In 1998 the WTO Appellate Body had ruled in favour of the US government and then authorised higher tariff barriers on some European food exports, as compensation for the USA's economic loss. The US tariff list targeted specialty products such as Roquefort cheese, produced by French *paysans*. When retaliating against McDonald's, activists linked hamburgers with beef hormones and GM food—as types of *malbouffe* or junk food. When prosecuted in court, the defendants put 'globalisation' and government policy symbolically on trial, while attracting large protests in the town and nationwide attention in the mass media (see Chapter 3).

Also in the late 1990s the European food market underwent great conflict over GM labelling. In 1996–97 the first US shipments containing GM soya and maize arrived at European ports, where symbolic protest generated Europe-wide publicity about GM grain entering processed food without any labelling. Agbiotech companies were accused of 'force-feeding us GM food'. Governments and companies were put on the defensive for denying consumers 'the freedom to choose' non-GM versus GM food. NGOs carried out surveillance of food products for GM 'contamination', thus testing claims about the ingredients and putting companies on trial for deception. Companies accommodated the demand for GM labelling but soon decided instead to exclude GM ingredients, thus imposing a commercial boycott. Agbiotech had been promoted as essential for economic competitiveness; now tested by market forces, however, GM grain became a commercial liability (see Chapter 7).

As those different examples illustrate, the European controversy had surprising dynamics, with continuously expanding trials, defendants and arenas—what was put on trial, how, where and by whom. Each episode involved multiple overlapping trials, both formal and informal. In formal trials, an official body examines evidence in order to reach a judgement, e.g. in judicial, regulatory, expert, experimental and commercial arenas. Through parallel informal trials, meanings are contested through discourses of blame, irresponsibility, threats, etc. Informal trials have driven, shaped and permeated formal ones.

Through a pervasive informal trial, agbiotech has been symbolised by contending metaphors of beneficent promise versus ominous threat. Agbiotech was promoted as an eco-efficient tool which can safely intensify high-yield agriculture, thus providing common societal benefits. In the mid-1990s, when agricultural supply and pharmaceutical companies created mergers in the name of a 'Life Sciences' strategy, beneficent natural qualities were attributed to agbiotech. Yet it was stigmatised by critics through epithets such as superweeds, Frankenstein food and mad soya; the latter drew analogies to the mad cow pandemic and its origins in government irresponsibility.

Agbiotech critics were accused of exploiting irrational fears, as well as unfairly turning the technology into a proxy for extraneous issues, such as sustainable agriculture and globalisation. Yet agbiotech itself was being promoted in such terms. Harmonised European rules were officially justified by reference to objective imperatives—the EU internal market, global economic competitiveness and WTO treaty obligations. The free market would adjudicate the societal need for such products, whose eco-efficient benefits would make agriculture more sustainable.

This EU policy framework was attacked for favouring private interests over the public good. Safety claims were criticised for scientific ignorance and political bias. Broad patent rights for 'biotechnological inventions' were denounced as biopiracy. Agbiotech was turned into a symbol of a political-economic project and its promoters.

Through multiple expanding, overlapping trials, the issues became polarised. The term 'trial' gave pejorative meanings to agbiotech by putting its industrial and government promoters on the defensive. They were accused of acting irresponsibly and concealing political agendas behind expert claims. Such accusations were taken up widely by civil society groups, far beyond the NGOs that were campaigning against agbiotech.

As the controversy intensified, in 1999 several member states collectively announced their refusal to consider any more approvals of new GM products. As a precondition for resuming the EU-wide approval procedure, they demanded precautionary changes in EU rules, especially more stringent criteria for risk assessment and likewise for 'GM' product labelling. With these demands, a minority of member states created a procedural impasse, soon known as the *de facto* EU Council moratorium. In dispute were the rules for putting GM products on trial and thus for reaching legitimate decisions.

Governments have also undergone judicial trials. NGOs have brought court challenges against regulatory decisions to authorise GM products. Conversely, judicial procedures have tested government restrictions and regulatory blockages. When the WTO put the EU on trial for a *de facto* moratorium and for national bans on GM products, the Commission defended precautionary approaches that it had previously opposed. These formal trials offered an extra arena and stimulus for wider symbolic trials.

In all those ways, the term 'trial' acquired broader meanings. The many types of trial have included the following:

- Examining evidence in order to reach a judgement, e.g. on safety, risk, guilt, etc.
- Testing the performance of institutions, especially regarding irresponsibility and unaccountability.
- Competing for authority to set policy agendas and devise strategies that could address the societal conflicts.
- Testing institutional endurance and adaptation in pursuit of democratic legitimacy.

Those types illustrate four standard definitions of the term 'trial' (see box below). Often many types overlap in the same case or episode.

Multiple trials have also tested the democratic legitimacy of government decisions, at EU and national level. Agbiotech was turned into a symbol and test of the EU's democratic deficit. In this book, the phrase 'European democracy' denotes that contest of meaning and legitimacy—rather than any ideal form of democracy.

The European agbiotech conflict has a broader relevance, which also has been contentious. We hear recurrent appeals 'to avoid another GM' controversy over other innovations such as nanotechnology. In such appeals, the agbiotech conflict is often attributed to deficient public understanding, inadequate public consultation and poor communication—mainly about a technology. As this book argues, however, agbiotech acquired its public meanings through wider policy agendas—European integration, regulatory harmonisation, 'objective' expert advice, the internal market, trade liberalisation and the marketisation of public goods. Given these meanings, the agbiotech case has a broader relevance beyond technological controversies.

Starting from popular phrases about 'GM on trial', this book analyses the dynamics of European integration around agbiotech policy, spanning approximately a decade from the mid-1990s onwards. The book discusses questions falling into four categories—trial roles, government policies, democratic legitimacy and broader lessons learned—regarding the European agbiotech controversy.

Trial

• noun: (1) a formal examination of evidence in order to decide guilt in a case of criminal or civil proceedings; (2) a test of performance, qualities, or suitability; (3) an event in which horses or dogs compete or perform; (4) something that tests a person's endurance or forbearance.

on trial (1) being tried in a court of law; (2) undergoing tests or scrutiny; (3) trial and error: the process of experimenting with various methods until one finds the most successful.

—*Compact Oxford English Dictionary* (online)

Trial roles and types

- Who and what have been put on trial?
- What accusations have been made against the defendants?
- What types of trial have they undergone?

Government policies on trial

- What policies have been on trial?
- How and why have they changed since the late 1990s?
- How have these policies promoted or marginalised potential European futures?

Democratic legitimacy on trial

- What difficulties arose for the accountability of representative democracy?
- How has this case tested European integration, its democratic legitimacy, its deficits and possible remedies?

Broader lessons learned

- How can 'another GM' be avoided—or perhaps created?
- What broader lessons can be learned from this case?

2 MULTIPLE, OVERLAPPING TRIALS: CHAPTERS AS MINI-STORIES

The rest of this Introduction summarises how the 'GM on trial' story is told in instalments. Chapter 1 surveys analytical perspectives central to the book, while Chapters 2 and 3 set the scene for the Europe-wide conflict over agbiotech. Chapters 4–8 each provide a mini-story on a specific policy area. The Conclusion answers the questions posed above.

Chapter 1 Analytical Perspectives

Journalistic accounts, as well as some academic accounts, have emphasised how the European controversy became polarised between agbiotech proponents versus critics. This book looks more closely at state actors, whose policies have shifted somewhat—from initially accommodating agbiotech proponents, towards mediating the societal conflict or even supporting policies which block agbiotech. These dynamics can be analysed by drawing on various analytical perspectives, which are surveyed in four main sections:

i. Understanding the EU agbiotech conflict: briefly reviews relevant points in five previous books on the EU agbiotech conflict, as an entry point into key issues that will be taken up in this book.
ii. Diagnosing the EU's democratic deficit: explains how the internal market project generates legitimacy problems and examines various diagnoses of them.
iii. Governing societal conflict over technology: looks at ways to analyse the co-production of technology with socio-natural order, consequent societal conflicts and their management.
iv. Contending discursive frames: examines how societal problems are framed by policy coalitions for promoting specific futures.

The final section sketches our tripartite model of discursive frames. Agbiotech promoters elaborate a neoliberal eco-efficiency frame, while opponents elaborate various framings of threat and alternatives. Through managerialist framings, state bodies set and implement policies, while potentially accommodating or marginalising elements of the other two frames. These three policy frames have taken analogous forms across the various policy issues and multi-level trials (as shown in Table 1.1). This tripartite model is elaborated with further tables in relevant chapters.

Chapter 2 Making Europe Safe for Agbiotech, Generating Dissent

From an early stage, agbiotech served a larger political-economic project. Transatlantic agrochemical companies designed GM crops for further industrialising agriculture. GM crops featured single-gene traits for an agri-industrial model of more efficient production of standard commodities. As its global pioneer, the USA adopted neoliberal policies that could facilitate the commercial success of agbiotech. These policies included broader patent rights, their extension to public-sector bodies, R&D incentives for 'value-added genetics', no GM labelling, 'product-based regulation' based on existing legislation and global trade liberalisation.

In the EU, agbiotech was adopted as a symbol of European integration and societal progress, linked with the US political-economic model as an inevitable future for Europe. The technology was promoted as an objective imperative—essential for higher productivity, eco-efficiency gains, economic competitiveness in the global commodity market and thus survival of the European agri-food sector. EU policies also facilitated broader patent rights and marketisation of research, thus blurring the distinction between public and private sectors.

EU regulatory frameworks too were shaped by a neoliberal agenda. In the early 1990s special regulation of agbiotech was put on trial for deterring investment in European R&D, as grounds to streamline the procedures. Expert safety assessments served to minimise regulatory burdens in the mid-1990s; field experiments were cited as a basis for product safety and thus commercial authorisation. No special labelling was initially required for GM products, on grounds that such rules would lack any scientific basis and would unfairly impede the internal market. The public was modelled as unwitting consumers of GM food, thus supporting a presumably beneficial technology. The EU internal market was meant to enhance competitive advantage in the global economy. In all these ways, new policies and institutions were nominally making GM products safe for Europe, while also making Europe safe for agbiotech to achieve commercial success.

That agenda provoked conflict in every policy area, so the US neoliberal framework could not be simply replicated in the EU; rather, it underwent a difficult adaptation. The conflict can be analysed in terms of three different policy frames (corresponding to the tripartite model introduced in Chapter 1). Agbiotech proponents elaborated a *neoliberal eco-efficiency frame,* while critics counterposed an *apocalyptic frame,* whose accusations put agbiotech on trial. To deal with these antagonistic forces, state authorities elaborated a distinctive *managerialist frame* for selectively channelling, incorporating or marginalising dissent. Through interactions among those frames in every policy area, the EU policy system generated a distinctive, unstable neoliberal approach that increasingly diverged from the US model.

Chapter 3 Opening Up Risks, Disputing Un/sustainable Agriculture

Through European protest, agbiotech was turned into a symbol of multiple anxieties: the food chain, agri-industrial methods, their inherent hazards, global market competition, state irresponsibility and political unaccountability through globalisation. Agbiotech was stigmatised along several lines—as dangerous products, as unsustainable agriculture and even as pollutants that were contaminating democracy as well as the environment. In 1996–97 the first US shipments of GM soya and maize were turned into symbols of such threats. By linking these threats, activists catalysed a broad opposition movement. Similar issues were taken up across the conventional boundaries of NGOs, especially by environmentalist, consumer and some farmer groups.

Controversy extended to more arenas—from the cultural, to the judicial, regulatory, expert, experimental and commercial—where agbiotech could be put on trial. In regulatory arenas, normally favouring official scientific expertise, this was attacked for optimistic assumptions and reliance on inadequate scientific methods, thus opening up issues of risk and precaution. In judicial arenas, legal expertise competed with or drew upon other expertise; government decisions to authorise GM crops were put on trial, both symbolically and formally. When activists were prosecuted for sabotaging field trials, they used the opportunity to accuse the state of failure to protect society from agbiotech. Pro-biotech policies came under challenge by uninvited forms of public participation, as well as in official procedures.

Risk issues were linked with sustainable agriculture in contending ways. From an eco-efficiency frame, agbiotech was promoted as a way to achieve a sustainable, high-yield intensification of agriculture; this policy gained support from organisations of agri-industrial farmers and many European authorities. Here eco-efficiency meant an input-output efficiency of resource usage for producing standard global commodities—a framework at odds with agri-ecological meanings of sustainable agriculture. From the apocalyptic frame of opponents, GM crops meant unsustainable agriculture, threatening the European countryside, which was valued as an aesthetic landscape, a wildlife habitat, for its local heritage, for the stewardship role played by farmers, for farmers' economic independence and for 'quality' alternatives to industrial agriculture. Through this frame, organisations of less-intensive, organic and small-scale farmers made alliances with other opponents of agbiotech.

As the controversy deepened in the late 1990s, the defendant on trial was expanded—from product safety, to biotech companies, their innovation trajectory, regulatory decision-making, expert advisors, government policy and its democratic legitimacy. Activists expanded their accusations, which were taken up by many groups in civil society. Acting as a pervasive mobile jury, they raised questions about whose control and interests were driving the technological development. As agbiotech was put on trial as unsustainable agriculture, institutions faced greater pressure to test claims for agri-environmental improvement as well as safety.

Chapter 4 Channelling Participation, Testing Public Representations

Since the 1980s, various state bodies in Europe have sponsored participatory technology assessment (TA) of agbiotech. In responding to or anticipating societal conflict, participatory TA exercises were promoted for diverse aims: to democratise technology, to educate the public, to counter 'extreme' views, to gauge public attitudes, to guide institutional reforms and/or to manage societal conflicts. Such aims informed and shaped each exercise. As a distinct arena, state-sponsored participatory TA has some common features across different national contexts, especially where EU member states have had strong commitments to agbiotech.

In these TA exercises, individuals were selected for a deliberative process aimed at seeking to reach a group view. Participants discussed the normative, value-laden basis of expert claims, thus developing a lay expertise. They went beyond simply questioning experts. Some participants questioned whether agbiotech would provide a means for sustainable agriculture and a beneficent control over the agri-food chain. Some also raised the need for alternative options, thus implying broader citizenship roles. However, these questions were generally channelled and reduced into proposals for regulatory control measures. In these exercises, agbiotech was largely spared the deeper trial being undergone in the wider public debate.

Models of public representation and citizenship were also being tested in these exercises. Their design and management structured relations between expert and lay roles. Participants could question, evaluate, anticipate or even simulate expertise. However, such discussion generally focused on appropriate regulatory and expert advisory arrangements for GM products, perhaps as a means to influence official agendas by accepting their limits.

In some cases the exercises stimulated, complemented or reinforced policy changes towards greater state accountability for regulatory criteria, but not accountability for state commitments to agbiotech. Despite some aspirations to democratise technology, democracy was generally 'biotechnologised' by internalising policy assumptions about agbiotech as progress. State-sponsored participatory TA marginalised choices about innovation pathways, which implicitly became non-choices.

Chapter 5 Regulating Risk, Testing EU Reforms

Safety claims for GM crops became more contentious from the mid-1990s onwards. Official risk assessments accepted intensive monoculture as a baseline for accepting undesirable effects from GM crops, e.g. a genetic treadmill supplementing the pesticide treadmill. These agri-environmental standards were challenged by some member states. In response to those disagreements and wider societal conflicts, as expressed in the 1999 *de facto* moratorium, EU agbiotech legislation was revised to encompass a broader range of potential risks and uncertainties. In parallel the Commission made EU-level changes for agri-food regulation in general. The earlier policy of 'risk-based regulation' was recast as 'science-based regulation', equating science with 'objective' risk assessment by official EU expertise. A new European Food Safety Authority (EFSA) was meant to provide cognitive authority for expert advice, towards harmonising risk assessments and even precaution, as if EFSA could provide a policy-free basis for EU regulatory decisions.

These reforms generated conflicts which tested official claims for expert objectivity and thus 'science-based regulation'. When GM products were assessed under the new EU regulatory procedures between 2003 and 2005, EFSA's GMO Panel generally advised that each GM product was as safe as its conventional non-GM counterpart. By contrast, more and more member states criticised the available scientific information as inadequate. In parallel, agbiotech critics challenged optimistic assumptions in safety claims. Moreover, expert disagreements highlighted precautionary and policy issues in risk assessment; any science/policy boundary became contentious and unstable, thus undermining EFSA's cognitive authority as definitive scientific expertise. The EU's revised regulatory procedures turned out to extend the earlier conflicts from the late 1990s, rather than harmonising risk assessment.

In this multi-level trial, the defendant was expanded from safety claims, to 'objective' expert advice, to the EU regulatory procedure and its legitimacy. When the Commission approved GM products by citing EFSA's advice, with little support from member states, some criticised the Commission for a 'democratic deficit'. When it symbolically put national bans on trial in the EU regulatory committee, member states voted overwhelmingly against the Commission proposals, thus defending national sovereignty. Facing a legitimacy crisis in early 2006, the Commission's policy shifted, now asking EFSA to address the demands of member states for more transparent, rigorous risk assessments. Previously the Commission had depended upon EFSA to judge whether member states were guilty of unjustified dissent, but now EFSA too was being judged and put on the defensive.

In the same period, when the WTO agbiotech dispute put the EU literally on trial for blocking GM products, the procedure generated multiple trials. To demonstrate that the EU had no moratorium, the European Commission sought to approve new GM products and lift national bans on ones already approved. It attempted to put recalcitrant member states on the defensive but gained little support from other member states. At the same time, the Commission's statements to the WTO emphasised scientific unknowns, inadequate information in risk assessment and thus weaknesses in safety claims. Citing those arguments, NGOs defended regulatory blockages and even denounced the Commission for duplicity, given its demands on member states to approve GM products.

Chapter 6 Scaling Up GM Crops, Testing Commercial Operators

From the mid-1990s onwards, agbiotech opponents framed agbiotech risks in successively broader ways. Their discourses emphasised three ominous metaphors: 'superweeds' leading to a genetic treadmill, thus aggravating the familiar pesticide treadmill; broad-spectrum herbicides inflicting 'sterility' upon farmland biodiversity; and pollen flow 'contaminating' non-GM crops. These metaphors gave moral meanings to the agricultural environment, thus expanding the charge-sheet of suspected crimes for which GM products should be kept symbolically on trial. As a way forward through official procedures, the commercial stage was anticipated or designed as a 'real-world' experiment.

Elaborating a managerialist frame, regulatory authorities translated the three ominous metaphors into measurable effects to be anticipated, managed and prevented. These broader accounts of unacceptable effects meant greater uncertainty about whether GM crops could cause harm when scaled up to the commercial stage. As critics warned against 'uncontrollable risks' of GM crops, some national regulators proposed extra control and monitoring measures. These were aimed mainly at reassuring or marginalising critics, yet the outcomes were not entirely predictable. Such measures were anticipating or designing the commercial stage as a 'real-world' experiment.

This would test commercial operators (seed breeders, farmers, grain traders and processors), whose behaviour might cause harmful effects, including simply the spread of GM material. Agbiotech critics cast doubt on whether preventive measures would be reliable, realistic or morally responsible.

In the UK, farm-scale trials turned agricultural fields into a social laboratory for testing farmer behaviour and its effects. The trials examined whether broad-spectrum herbicides on GM crops would cause relatively more harm than conventional crop management to farmland biodiversity. The trials were originally intended to simulate farmers' likely cultivation practices and thus predict their environmental effects, as a basis to make a regulatory decision on commercial use. After the trial results turned out favourably for GM maize, however, those cultivation practices were made prescriptive, as rules to be enforced on farmers. UK authorities also required agbiotech companies to compensate non-GM farmers for any economic loss from GM contamination. Bayer declined the prospect of a real-life experiment with such regulatory burdens on its GM maize, so commercialisation did not go ahead. The trial results were unfavourable for GM herbicide-tolerant oilseed rape, which could not gain government support for commercial authorisation.

EU-wide regulatory harmonisation also became more contentious for new GM products. Some member states cited environmental uncertainties that could not be resolved before the commercial stage. On these grounds, they requested extra measures—scientific information, monitoring requirements or control measures to prevent potential harm—as a prerequisite for commercial authorisation. In planning a scale-up of GM crops, experimental designs underwent tensions between predicting, testing and/or prescribing operator behaviour. Within and among EU institutions, conflicts continued over a legitimate basis for scaling up GM crops for commercial use, foreseen as a further experimental stage.

Chapter 7 Labelling GM Products, Testing Free Choice

Early EU agbiotech policy granted safety approval for GM soya and maize, with no requirement for a special label on derived food products. GM grain was invisibly mixed in agri-food chains and processed food, so the public was cast as unwitting (and perhaps supportive) consumers of GM technology. Protesters warned that consumers would be human guinea pigs, being 'force-fed GM food'. By default of any clear means to act as citizens, many consumers 'voted' against GM food through their purchases and protest.

Actual or potential labelling rules were put 'on trial' along with agbiotech. In the overall debate on GM labelling, consumer choice was framed in contending ways. From a pro-agbiotech standpoint, any GM label should be based only on scientific information relevant to consumer safety. From an anti-biotech standpoint, environmental NGOs demanded comprehensive labelling of all food from GM sources, as a democratic right and defence against economic globalisation. Consumer NGOs wanted GM products likewise to be fully labelled, so that everyone would have the right to know their origin and could choose products accordingly. Governments and companies were put on the defensive for denying consumers 'the freedom to choose' non-GM versus GM food. As an extra pressure for GM labelling, NGOs carried out surveillance of food products for GM 'contamination', thus testing claims about the ingredients and putting companies on trial for possible deception.

In response to those pressures, EU labelling rules redefined what a 'GM' product is, according to detectability criteria that became successively broader in three stages during 1997–2000. Such rules faced multiple accusations concerning the following issues: whether the criteria neglected or adequately identified GM 'contamination'; whether they accommodated or denied 'free consumer choice'; and whether they would stabilise or undermine the EU internal market for processed food. By the late 1990s European retail chains excluded GM grain altogether from their own-brand products, rather than apply a GM label. This commercial boycott in turn deterred farmers from cultivating GM crops.

In 1999 several EU member states demanded full labelling and traceability, i.e. for all food and feed products of GM techniques, even those such as oils where no GM material was detectible. Such demands cited precautionary reasons: consumers and regulators must have adequate information to deal with uncertain risks. When the European Commission drafted new regulations accordingly, its proposals were debated along several lines: whether comprehensive GM labelling criteria would enhance or deny meaningful consumer choice, whether traceability requirements could be reliably implemented and whether such rules would intensify the transatlantic trade dispute. When the new GM labelling and traceability rules were finally enacted in 2003, they complemented a particular social and natural order: agbiotech products could be kept 'on trial' for uncertain risks in both the regulatory and commercial arenas.

Chapter 8 Segregating GM Crops, Contesting Future Agricultures

Since the late 1990s, the inadvertent mixing of GM material with non-GM products has posed an economic problem: such admixture could lower the price of non-GM grain or food, so commercial operators could lose income. Agbiotech opponents turned this admixture problem into an extra risk issue: GM crops would spread 'genetic pollution', thus endangering the environment and non-GM crops. In 2001 the European Commission developed a policy for the 'coexistence' of GM with conventional and organic crops: farmers would have a free choice among those options, protected through segregation measures as appropriate. Admixture was relegated to a mere economic problem, separate from the environmental issues of risk regulation. Coexistence rules were left to the discretion of member states, within limits set by the Commission—allowing the free circulation of GM products.

Commission policy mandated that coexistence rules must 'allow market forces to operate freely', yet any such rules would shape markets in ways favouring some freedoms over others. Commission policy linked product safety, EU expert authority for risk assessment, a European internal market for officially safe products, and segregation measures to keep GM material below the 0.9% labelling threshold in non-GM products. This policy framework depended upon clear, stable distinctions—between un/approved GM crops, environmental/economic harm, dis/proportionate economic burdens on GM farmers, un/free market choices, etc.

Those official distinctions were blurred by some proposals for segregation measures to minimise any GM material in non-GM products, regardless of any plausible economic loss from admixture. Uncertain environmental risks were cited as precautionary grounds for stringent segregation measures. Moreover, 'quality' alternatives were being promoted by some member states and the Assembly of European Regions. To protect those alternatives, some authorities devised stringent rules which would effectively preclude GM crops—implicitly in the name of 'coexistence', or explicitly as 'GM-free zones'.

Proposed segregation rules were put on the defensive from two opposite standpoints. According to the Commission, 'coexistence' rules could impose extra burdens on GM farmers only to the extent necessary to ensure an internal market; on these grounds, it rejected some regional draft rules for imposing a disproportionate burden that could deter or preclude GM crops. This policy was denounced by agbiotech critics as coercion—for legalising 'GM contamination', endangering non-GM crops and so precluding alternatives. Thus agbiotech was further put 'on trial', along with the Commission's coexistence policy, for denying a free choice of non-GM and quality products. Segregation rules expressed a wider European contest over possible future agricultures.

Chapter 9 Testing European Democracy

As this book shows, the European agbiotech controversy continuously expanded trials, defendants and arenas—what was put on trial, how, where and by whom. Through multiple overlapping trials, GM products were turned into symbols of various threats to the public good. These trials tested and undermined the official promotion of agbiotech as an objective imperative. Europe was told that it had no choice but to accept agbiotech, yet this imperative was turned into a test of democratic accountability for societal choices.

Sources of the controversy lay in early EU policies. Since the 1980s agbiotech has been promoted as a clean eco-efficient technology, symbolising societal progress and European integration. To make Europe 'safe' for the agbiotech innovation trajectory, the EU policy framework adapted the US neoliberal model in co-producing agbiotech with specific models of the natural and social order. As a precisely improved form of nature, GM crops were 'biotechnological inventions' warranting patent rights. These safe products would protect European agriculture from the dual threats of wild unruly Nature and global economic competition. The agricultural environment was modelled as an efficient factory for standard global commodities; safety claims accepted the normal hazards of intensive monoculture. The public was modelled as unwitting consumers of GM products, effectively supporting agbiotech development.

Those policies were promoted by invoking globalisation, economic competitiveness, treaty obligations, beneficent biotechnological properties and 'risk-based regulation'. Through these supposedly objective imperatives, government decisions could pre-empt societal choices. A neoliberal agenda could be depoliticised, while using agbiotech as its instrument.

Consequently, agbiotech became vulnerable to attack along multiple lines. GM products were reaching the commercial stage in a period when food hazards were widely attributed to agri-industrial methods, deregulatory policies and official expert ignorance. So attacks on agbiotech for 'tampering with our food' resonated greatly with public anxieties. Since the late 1990s 'GM food' has been widely stigmatised—as a multiple threat to health, agriculture, the environment, the public good and democratic sovereignty. Activists' accusations were taken up by large environmentalist and farmers' organisations, as well as by wider groups in civil society, acting as a pervasive mobile 'jury'. Through uninvited forms of public participation, as well as in official procedures, such groups have sought to hold governments accountable for their policies.

These conflicts have been played out through overlapping trials in judicial, regulatory, expert, experimental and commercial arenas. When activists were prosecuted for sabotage, they used the opportunity to put government and companies symbolically on trial. Beyond tests of product safety, agricultural fields were turned into laboratories for testing human behaviours which matter for potential effects of GM crops. When put on trial by NGO surveillance and market forces, agbiotech was forced out of the food market as a commercial liability: by 1999 supermarket chains had excluded GM grain from their own-brand products.

By then, GM products were being widely opposed and blocked throughout Europe. State bodies faced stronger pressure to justify agbiotech development as a desirable choice for the public good—or else to favour alternatives. Through various reforms, the EU has sought to govern societal conflicts over its deep commitment to agbiotech, thus reducing democracy to governance. EU institutions were created or adapted to channel the societal conflict into regulatory procedures and official expertise. The reforms too were put on trial. 'Science-based regulation' depended on credibly 'objective, independent' EU-level expert advice, but this official status was disputed. Calls for greater precaution served to open up official assumptions about the expert basis of rules for the internal market.

In opposing agbiotech, moreover, civil society groups generated alternative social identities. 'GM-free zones' became a widespread European slogan, implicitly meaning EU-free zones, while also promoting 'quality' alternatives to agri-industrial systems. These alternatives have arisen largely from outside EU institutions and along lines antagonistic to the internal market project, at least regarding agbiotech.

In all those ways, agbiotech became a focus for contesting claims about European 'democracy'. It became a test case of the EU's democratic deficit—its sources, diagnoses and possible remedies. Agbiotech was turned into legitimacy problem for representative democracy and its public accountability. Agbiotech acquired its public meanings through wider policy frameworks—regulatory harmonisation, 'objective' expert advice, the EU internal market, trade liberalisation and marketisation of public goods. All these policies became defendants in 'GM on trial'.

In drawing broader lessons, commentators generally assume that the European agbiotech controversy was about a technology. As this book shows, however, it was about how an innovation modelled the social and natural order, and how state bodies promoted that order as an objective imperative. These aspects can illuminate prospects for avoiding 'another GM' controversy—or perhaps prospects for creating one.

1 Analytical Perspectives

INTRODUCTION

Starting from popular phrases about 'GM on trial', the Introduction posed several questions about the European agbiotech controversy as a multi-level, multi-arena trial. To answer those questions, this book tells the story by drawing upon many analytical perspectives, especially interdisciplinary approaches, which are surveyed here. This chapter has the following sections:

1. Understanding the EU agbiotech conflict: how five previous books on the EU agbiotech conflict serve as an entry point into key issues that will be taken up in this book.
2. Diagnosing the EU's democratic deficit: how the internal market project generates legitimacy problems for European integration, with various diagnoses of a democratic deficit.
3. Governing societal conflict over technology: how technological designs are co-produced as socio-natural order, how this generates societal conflicts and how these are managed.
4. Contending discursive frames: how societal problems are framed by policy coalitions to promote specific futures.

1 UNDERSTANDING THE EUROPEAN AGBIOTECH CONTROVERSY: PREVIOUS ANALYSES

Many other writers have analysed the European agbiotech controversy and government responses. This book draws upon and extends insights from five such books. Their titles and key concepts provide the subheadings of this section, which identify issues to be taken up later. From issues raised in the five books, the rest of this chapter elaborates analytical perspectives in three broad categories: diagnosing the EU's democratic deficit, governing conflict over technology and framing issues to promote specific futures.

1.1 Molecular Politics: Ecological Modernisation

In his book *Molecular Politics,* Herbert Gottweis (1998) analyses the discursive framing of biotech policy and public debates, from the 1980s through the early 1990s. In European policy discussions, a biotechnological vision was elaborated as a solution to several problems; societal problems were attributed to genetic deficiencies (Gottweis, 1998: 228). European companies could not adequately compete in an increasingly global market, so they had to be converted or incorporated into competitive multinational companies. For this economic aim, along with a more efficient use of resources, an essential tool would be the application of modern biotechnology to European agro-food industries, according to official policy (ibid.: 170).

As environmental risks were disputed by proponents and critics, mainstream modernisation discourses incorporated ecological concepts, argues Gottweis. Ecological modernisation shaped new cognitive frames for risk management:

> . . . the new ecological discourse constituted not only 'a way to solve environmental problems', but also a social technology that attempted to create a particular social order. The ecology discourse had introduced concepts that made 'environmental problems' stable and manageable. (ibid.: 234)

These discourses informed changes in conceptual tools which 'constituted important reference points for the shaping of a policy counter-narrative' on agbiotech, argues Gottweis (ibid.: 235). Early EC policy came under public challenge, often from new actors bringing an ecological discourse.

> That [earlier regulatory] system was increasingly perceived as beset by 'regulatory gaps' and inconsistencies. But these regulatory loopholes were not the 'cause' of the emergence of a new regulatory discourse . . . [Rather] the changed discursive constellation had given rise to new actors and institutions, and they got a voice in the regulatory debates. (ibid.: 318–19: also 265).

As agbiotech framings became polarised in national debates, policymaking broadened the scope of issues which could be legitimately deliberated within regulatory practices, he argues. Policymakers sought to maintain or gain hegemony 'by re-absorbing discourses of polarity into a system of "legitimate differences" and by defining the locations where differences can be articulated', thus potentially absorbing critics' demands. The boundary between experts and non-experts became more permeable and negotiable. Controversial socio-economic issues about agbiotech were transformed and displaced onto ecological risk assessment; by contrast, the technological trajectory could not be legitimately debated, as industry was still officially accepted as a site of progress (Gottweis, 1998: 319–21). In these ways, societal conflict was potentially reduced to regulatory policymaking, though NGOs rejected such efforts (ibid.: 337).

Building on his insights, our book extends the story of those strategic displacements and boundary shifts. Conflicts over agbiotech were channelled into 'legitimate' regulatory issues and procedures—which became even more contentious, while hardly marginalising dissent. Although Gottweis associates ecological modernisation with critical discourses, our account emphasises its links with agbiotech promotion and regulation. Beyond ecological modernisation, agbiotech critics increasingly elaborated discourses antagonistic to agri-industrial efficiency, while promoting alternative agricultures and societal visions, especially by the late 1990s (beyond the scope of his book).

1.2 Designs on Nature: Democratic Accountability

In her book *Designs on Nature,* Sheila Jasanoff analyses biotechnology policy to inform theoretical and policy issues of democratic accountability. As biotech innovation devised novel 'designs on nature', its governance challenged some assumptions of liberal democracy. Official practices threw into question whether 'citizens have the capacity to participate meaningfully in decisions that seem increasingly to call for specialized knowledge and expertise'. Through various expert bodies and procedures, democratic control over biotech 'was sometimes set aside in favour of other culturally sanctioned notions about what makes the exercise of power legitimate' (Jasanoff, 2005a: 272, 287).

To analyse such legitimation strategies, she examines biotech policy in the US, the EU and some EU member states. 'Democratic engagement with biotechnology was shaped and constrained by national approaches to representation, participation, and deliberation that selectively delimited who spoke for people and issues, how those issues were framed, and how far they were actively reflected upon in official processes of policymaking' (Jasanoff, 2005a: 287). In each country, she argues, institutional boundary work demarcated some issues as being separate from ordinary politics. Such issues were rendered invisible or were designated as appropriate only for expert analysis. Building on that analysis, this book shows how such an expert/lay boundary was constructed and contested.

Her book also analyses how biotechnology policy shapes the EU, as well as vice versa:

> The issue of legitimacy relates back in turn to the kind of union Europe hopes to be, and indeed is permitted to be. (Jasanoff, 2005a: 70; cf. Schmitter, 2001).
>
> Seemingly technical questions about how the EU should promote biotechnology in the member states turned out . . . to be unanswerable

> without also taking on board the deeper question of what kind of European Union there should be. Complicating the analyst's task is the ambiguity of Europe itself as a geographical, political and cultural construct. . . .
>
> Europe has to be invoked as an agent with fixed attributes and capacity to act, even though its actions are the very means by which new European identities are being constituted. (Jasanoff, 2005a: 69)

Within the EU agenda of regulatory harmonisation, agbiotech regulation has undergone a tension between two political models: one seeking to eliminate national regulatory divergences, the other protecting national values that generate them, she argues (ibid.: 71).

Building on those insights, our book shows how agbiotech became a difficult test case for European integration. EU procedures had pervasive tensions between smoothing away national differences and accommodating them; moreover, such differences proliferated and even circulated around Europe, thus expressing societal conflict over agbiotech.

1.3 The Politics of GM Food: Discourse-Frame Analysis

In his book on *The Politics of GM Food,* David Toke (2004) develops a discourse-frame analysis to compare the policy systems of the US, the UK and the EU in the agbiotech sector. He draws upon analytical perspectives of Hajer and Foucault, especially for drawing links between cultural attitudes and political systems in each jurisdiction. On this basis, he explains regulatory differences and changes: jurisdictional differences in scientific risk assessment 'are related to different cultural attitudes' towards agbiotech. National policies claim a scientific basis, which in turn reinforces a given policy. Distinctive regulatory discourses are translated by scientists or expert advisors into 'factual' terms, he argues.

In Toke's view, a government may need to accommodate public concerns which eventually arise, so it would be better to anticipate them in advance. Yet an institutional path dependency—whereby past decisions constrain future ones—often makes a policy framework insensitive or inflexible to such changes in context, he argues (ibid.: 205–8). In his diagnosis, 'science-based regulation' readily becomes a constraint on policymaking. Starting from similar questions, our book develops a different analysis: that institutional rigidity (or path dependency) was less a problem than science/policy boundary-setting. We show how regulatory science both stimulated and followed policy change, thus flexibly translating some public concerns into expert advice.

The EU agbiotech case 'reveals the extent to which scientific truths are culturally bound', Toke argues (ibid.: 187). The BSE (or 'mad cow') crisis has been often blamed for anti-biotech attitudes and regulatory obstacles. However, it is more 'plausible to argue that the BSE crisis reinforced an existing discourse of mistrust of food quality and of regulators', especially in the UK, he argues. It is difficult to distinguish between the BSE crisis providing a new metaphor in anti-biotech rhetoric and actually causing changes in public attitudes, he says (ibid.). In a similar vein, our book shows how a specific discursive strategy turned 'mad cow' into a potent metaphor mobilising public distrust in regulatory arrangements by linking several policy issues.

In the agbiotech issues, the European Parliament has generally supported the pro-environmentalist stance of some member states, often against the Commission's more permissive stance (Toke, 2004: 53). This differs from political science perspectives which have seen the Parliament as generally supporting the Commission's pro-environmentalist stance against some member states which have more lax policies (Judge, 1993: 199). Why are the dynamics in this case different from those in other regulatory sectors? According to Toke, the Commission has sought to soften public demands for restrictions on GM food and crops, in order 'to pursue its normative role of producing harmonious trade relations, especially in pursuit of what the Commission still regards as an area of technical progress'. In response, the European Parliament influenced legislative outcomes along more pro-environmentalist lines, he argues (Toke, 2004: 188).

Building on that insight, our book shows how tensions between the Commission and the European Parliament resulted from public protest putting agbiotech promoters onto the defensive. Parliament allied with member states promoting more cautious or stringent approaches, often contrary to key aspects of the Commission's legislative proposals. Some disputed issues were 'environmental' only in the indirect sense of the word; Parliamentary amendments concerned, for example, regulatory procedures or GM labelling criteria. Beyond their sharp disagreements, the Parliament and Commission often converged on legislative changes that accommodated dissent within more stringent regulatory measures, while also potentially facilitating approval of GM products.

1.4 Genes, Trade and Regulation: Conflicts Among Interest Groups

In his book *Genes, Trade and Regulation,* Thomas Bernauer (2003) analyses agbiotech regulation in the US and EU, partly as a basis for explaining the transatlantic trade conflict. He questions explanations which cite differences in 'regulatory culture' or which blame stereotypical characteristics such as technophobia and protectionism. Instead he focuses on two dynamics: the struggle of interest groups for political and market influence; and the interactions between political forces seeking to influence regulatory frameworks. In relation to these processes, he states:

> In the European Union, both processes have worked in ways that have driven agbiotech regulation towards greater stringency. In

> the United States, they have worked in ways that have sustained agri-biotech-promoting regulation (Bernauer, 2003: 10)

Civil society organisations found greater opportunities to block agbiotech or to gain more stringent regulatory criteria in Europe, thus leading to a US-EU trade conflict, Bernauer argues. Moreover, they used the trade conflict to intensify domestic political conflict, stimulating the EU to make significant regulatory changes. Bernauer diagnoses these changes as a policy problem: that ever more complex, stringent, costly regulations are insufficiently backed by robust institutional structures for implementing them; consequently, the trade conflict may reduce investment in agri-biotechnology. As a remedy, he proposes more centralised forms of governance in food safety to increase consumer confidence (Bernauer, 2003).

By contrast, our book argues that the conflict originated in contending societal visions, irreducible to 'interest groups', i.e. conflicting interests . Although economic interests have been at stake, they were recast in response to political protest and commercial difficulties. Societal conflicts have been aggravated by moves towards centralised regulation—which therefore could not be a solution for the EU's problems, contrary to Bernauer's account.

1.5 Governing the Transatlantic Conflict: Contending Policy Networks

In their book, *Governing the Transatlantic Conflict over Agricultural Biotechnology,* Murphy and Levidow (2006) challenge the stereotype of a 'US versus EU' conflict over agbiotech. In their analysis, three contending policy networks have operated across the Atlantic, generating policy conflicts over agbiotech in both the USA and EU. The Transatlantic Business Dialogue promoted regulatory harmonisation for trade liberalisation of a benign technology and its safe products. The Transatlantic Consumer Dialogue promoted the right to know what food is being sold and the right to choose safe food, based on precautionary risk assessment. And the Transatlantic Environmental Dialogue demanded prior proof of safety for agbiotech products.

According to their book, those dynamics began in the mid-1990s. The Transatlantic Business Dialogue shaped the European Commission's New Transatlantic Agenda (NTA); this discourse-coalition identified 'barriers to transatlantic trade' as the main problem for EU and US policymakers. Their problem-definition drove the Commission's support for regulatory harmonisation, both within the EU and across the Atlantic, towards commercial approval of GM products.

The TABD-NTA agenda created the context in which mainstream NGOs could generate a broad popular movement against agbiotech, argue Murphy and Levidow. In the mid-1990s civil society protests began to challenge the neoliberal 'free trade' agenda, through movements widely called 'anti-capitalist' or 'anti-globalisation', especially in Europe. They attacked the NTA for favouring industry influence, for 'levelling down' standards and for jeopardising democratic sovereignty. Often led by environmental activists, anti-agbiotech movements linked several threats: trade liberalisation policies, industrial agriculture, economic globalisation, corporate power and unknown risks of GM products. By drawing analogies to BSE, GM products were framed as a threat of intensive agriculture, generating health hazards which elude the available scientific knowledge and official expert advice (ibid.).

As intense public controversy erupted over agbiotech in the late 1990s, the EU faced a legitimacy crisis, and US grain exporters faced European blockages. Critical voices gained a greater hearing in transatlantic and EU fora. Consumer groups elaborated a discourse of restoring public confidence through regulatory reform and precautionary expertise. Through this understanding of a common policy problem, policymakers developed a transatlantic governance process for managing societal conflict by incorporating or marginalising dissent (ibid.). By analogy, our book analyses a European governance strategy for channelling societal conflict into common problems of public confidence and product safety.

2 DIAGNOSING THE EU'S DEMOCRATIC DEFICIT

'Democracy' has been a long-standing theme of European integration, though with major changes in emphasis. In the 1980s this term referred to the accession of former dictatorships in southern Europe—Spain, Portugal and Greece. By the mid-1990s the term likewise referred to the accession of countries formerly in the Soviet bloc. This discourse took for granted the EU as a central instrument of democratisation.

At the same time, however, democracy was being highlighted as a core problem of European integration itself. EU decision procedures were being criticised for a 'legitimation gap', later called a 'democratic deficit'. This problem has had various diagnoses, each corresponding to institutional changes being proposed or implemented, as surveyed in this section.

For an ideal-abstract concept such as democracy, many commentaries follow an idealist method. This evaluates institutional arrangements according to first principles—e.g. freedom (versus coercion), pluralism, inclusiveness, self-government, etc. Ideal-moral meanings are often connoted by the term 'democracy' or by the epithet 'undemocratic'. Such terms may also denote exemplary institutions or countries, e.g. political rule through a Parliamentary system, election rules, etc. Connotative and denotative meanings are often conflated, thus leaving conceptual ambiguities. When those first principles are further examined, moreover, they appear more complex and even mutually inconsistent (Gupta, 2007: 22–23, 50).

Given the fluidity and contextual ambiguity of 'democracy' concepts, a realist method avoids any single definition (ibid.: 20). Instead, such a method analyses institutional practices and stakeholder demands regarding democracy, including the normative accounts of various actors. Such a method does not depend upon a particular normative account of democracy.

As a source of problems for democratic legitimacy, transnational relations have fragmented, shifted and blurred authority for decisions. Specialist territorialisations, and growing interdependence between them, have generated a 'radical dispersal of the public sphere'. Consequently, 'the regimes that are known as democratic are more properly differentiated and limited autocratic systems—that is, in traditional terminology, liberal oligarchies' (Zolo, 1992: 114, 181). Conflicts continue over the geopolitical locus and appropriate form of democracy:

> . . . boundaries within which democratic practice can be contained are simply not delineable any longer—they slip away between fragmentation and interdependence . . . this situation, while apparently undermining the operation of democracy, also raises demands for democracy. (Gupta, 2007: 27)

Indeed, geopolitical shifts in state authority lead stakeholders to attack decisions as 'undemocratic', while also demanding democratic accountability—often from contradictory normative standpoints. These global dynamics are exemplified by European integration, especially in its neoliberal form since the 1990s.

2.1 EU's Democratic Deficit: What Sources and Remedies?

From the start of the European Economic Community in the 1950s, its project has sought to overcome internal trade barriers to capital investment, goods and labour, so that these could freely circulate among member states. This process has generated conflicts of legitimacy, for several reasons. Given that the EU is not a state, it cannot provide democratic accountability in ways analogous to a national government, though European integration did not initially depend upon explicit claims for democratic status. Within its free-trade area, the EU could be somewhat protected from external economic forces, but by the 1980s the 'internal market' project became linked politically with global trade liberalisation. This agenda was represented as an objective imperative rather than as a political choice, thus aggravating the EU's democratic deficit.

Although legitimacy problems have arisen too for member states, the EU has had a somewhat open-ended character, with a fluid institutional structure. Consequently, 'the citizen is required to identify with something that is indefinite and changeable' (Lord, 2001: 168). Any institutional change shapes the 'EU' that needs to be legitimised.

> . . . the EU is an unprecedented experiment in the peaceful and voluntary creation of a large-scale polity out of previously independent ones. It is, therefore, singularly difficult for its citizens/subjects/victims/beneficiaries to compare this object politique non-identifié with anything they have experienced before. . . . the problem is compounded by the simultaneous need to legitimate not only what the unit should be—i.e., to define what 'Europe' is, but also the regime that should govern it, i.e., what its institutions should be. (Schmitter, 2001: 2)

National practices cannot provide the norms for democratic legitimation of the EU. 'Indeed, well-entrenched differences among its members may actually make it more difficult', e.g. by impeding efforts at legitimacy (ibid.). Given its treaty commitments to 'an ever closer union' through an internal market, the EU deals with national differences in ways which may limit or undermine its legitimacy.

Since the late 1980s, European integration has been driven more firmly by a neoliberal agenda. In the name of enhancing Europe's global economic competitiveness, this agenda has sought to eliminate constraints on trade and competition. Such proposals came especially from the European Round Table of Industrialists, founded in 1993 (vanApeldoorn, 2000, 2002). The ERT represented Europe's most transnational companies, many of them transatlantic. This transatlantic neoliberal agenda became more explicit in 1996 with the foundation of the Transatlantic Business Dialogue, whose main slogan proposed that all products should be 'Approved once, accepted everywhere'. This trade liberalisation agenda promoted a European economic space conducive to integrating capital investment and trade along transnational lines (Ziltener, 2004: 962–64), so that each company would become more transatlantic. Thus competition among transatlantic companies was represented as competition between the US and EU.

The European integration project exemplifies a global shift towards a 'competition state'. While its predecessor regime protected public goods and resources within a national territory, by contrast the competition state enhances domestic capacity for global competitive advantage. As Cerny argues, many collective goods are no longer controllable by the state—either because they have become transnational in structure or because they constitute private goods in a wider world marketplace. The 'competition state' extends marketisation in order to make domestic economic activities more competitive in international terms. Although some changes are called 'deregulation', this is generally a re-regulation, whereby market-controlling regulations are replaced by market-friendly ones (Cerny, 1999).

A more apt term would be 'market-creating' rules, i.e. rules forcibly marketising access to services. In the EU such liberalisation has been called 'negative integration', meaning the removal of institutions and rules that hinder the free movement of capital and goods. The European

Commission has prior authority to 'force member states to allow market competition in services that had hitherto been provided by state agencies' (Scharpf, 1999: 70).

Where key policies come under the control of politically non-accountable authorities, especially the European Commission, greater tensions arise between political democracy and the capitalist economy: 'Institutions of social protection and industrial citizenship, freely chosen only a few decades ago, are dismantled under the alleged compulsion of an economy that once was, but no longer seems to be, under the effective control of democratically accountable governments' (Scharpf, 1999: 27). Worse yet, this is happening with those governments' complicity, as they take responsibility for implementing EU policy. In these ways, the established mechanisms of democratic self-determination 'are being corrupted and hence delegitimised by exogenous influences. Or so it seems' (ibid.).

Many political theorists idealise EU institutions as a democratic space, while some also recognise the difficulties for democratic legitimacy. Habermas identifies a tension: the EU's liberalisation policies promote 'a forced desolidarisation of society', yet the European integration process creates new opportunities for citizens to create links across national borders. Therefore European democracy depends upon a cross-national 'common practice of opinion- and will-formation', whereby European citizens 'learn to recognise one another as members of a common political existence beyond national borders' (Habermas, 2001: 73, 99, 100). This positive scenario begs the question of how such cross-national links relate to EU institutions and bear upon their legitimacy.

2.2 European Integration via Regulatory Harmonisation

European integration was initially sought through negative integration. According to this perspective, national regulatory variations hinder trans-European mobility, including the circulation of labour, capital and goods, argues Andrew Barry. By eliminating these variations, Europe would be 'cleared for business'. In the neoliberal imagination,

> [regulatory harmonisation] would eradicate the striations of national-state capitalism and create the entirely smooth space of multinational capitalism. Against the background of the neoliberal project of the single-market programme, Europe was acted on as an economic space fractured by a vast array of legal, administrative, technical and linguistic barriers . . . Europe would be deterritorialised, transformed into a purely economic zone without any internal political boundaries of cultural identity. (Barry, 2001: 69–70)

Yet the EU could never become the pure smooth space of the neoliberal imagination, argues Barry. Such an outcome would require at least a mutual recognition of regulatory standards among member states, through EU competence for harmonisation. Such efforts have often become contentious, readily revealing differences in standards and practices across Europe. 'In practice, standardisation may produce new fractures and dislocations which may act as catalytic points for further political conflict' (ibid.: 63). Harmonisation can open up other spaces for political contestation of cultural values, which remain irreducible to politics in the conventional sense (ibid.: 72, 81, 82). Consequent legitimacy problems have been theorised as two types: input and output legitimacy.

Input legitimacy, also known as 'government by the people', depends upon an accountability process. Yet the EU lacks a pre-existing sense of collective identity, Europe-wide policy discourses, and a Europe-wide institutional infrastructure that could ensure the political accountability of office holders to a European constituency (Scharpf, 1999: 187). In analysing the EU's democratic deficit, accountability has been defined as democratic procedures where 'important public decisions on questions of law and policy depend, directly or indirectly, upon public opinion formally expressed by citizens of the community' (Weale, 1999: 14). However, that definition presupposes the prior existence of a distinctive European 'public opinion', 'citizenship' and 'community' relevant to a policy issue. Where there is no sense of common identity, many citizens question a system's right to make collectively binding decisions (Lord, 2001: 168).

Output legitimacy, also known as 'government for the people', means government capacities to solve collective problems in the common interest. This depends upon a shared perception of common interests, which may seem relatively more feasible than input legitimacy, especially for the EU. However, the internal market depends upon policy changes which become contentious at EU and national levels. Consequently, 'European policy must either avoid opposition by remaining below the threshold of political visibility, or it must search for conflict-minimizing solutions . . . ' (Scharpf, 1999: 23). As a common strategy, policy issues are technicised and relegated to specialist experts.

That relegation arises from a project designed to protect economic liberalism from democratic accountability. Through technocratic decision-making, the so-called 'free-market mechanism' is managed more competently through rule-based, extra-democratic systems. In this way, European integration has been designed to manage and contain the democratic aspirations of mass society. This managerialist agenda informs mainstream literature on the 'democratic deficit':

> On the whole, this literature is not about 'democracy' understood as the sovereignty of the people but rather about the lack of the 'legitimacy' of the institutions of the EU . . . In short, the indicated deficit . . . [concerns] the legitimation of the EU towards the demobilised and territorially regimented citizens. (Bonefeld, 2002: 138)

Much commentary on the EU's legitimacy problem takes for granted the neoliberal globalisation agenda—as progress, as an objective necessity or as both. By default, the mainstream debate has focused upon institutional structures. According to one diagnosis, a democracy dilemma arises because the EU cannot reproduce the electoral accountability of nation states:

> The present decision-making structures of the EU thus correspond less to a democratic deficit than to a fundamental 'democracy dilemma' . . . If the consequences of globalization necessitate further integration, but the standards of nation-state democracy are not simply transferable to Europe, which institutional shape should, and can, a future EU assume? (Decker, 2002: 258)

As a different diagnosis of the legitimacy problem, 'Regulators wield enormous power, yet they are neither elected nor directly responsible to elected officials', notes Majone (1996). This could be seen as a 'democratic deficit' if legitimacy depends solely upon electoral accountability (ibid.: 284–85). Yet it need not do, he argues. On the contrary, national governments interfere in regulatory issues for party-political reasons or for short-term electoral gain, thus intensifying distrust among governments and creating credibility problems for EU regulation. In his view, 'the root problem of regulatory legitimacy in Europe today is not an excess of independence but, on the contrary, the constant threat of politically motivated interference' (ibid.: 299–300). Thus he attributes EU-level difficulties to national electoral accountability—nearly the opposite of other diagnoses.

Towards a solution, Majone argues, the EU should be seen as a 'regulatory state'—i.e. as enacting rules and procedures which promote the internal market. This means a different basis for legitimacy than a nation state. By delegating policy to 'independent institutions', the EU 'would increase the credibility of domestic policies'. It would thereby facilitate 'the emergence of those networks of national and European regulators which alone hold the promise of resolving the dilemmas of regulatory federalism' (Majone, 1996: 282). In this way, the EU can enhance procedural legitimacy and accountability by stating clear reasons for decisions, rather than by enhancing public participation (ibid.: 291–94). His remedy begs many questions—in particular, how regulatory standards involve socio-cultural judgements, how an ''independent' body would address or disguise those judgements and how this already happens in current procedures.

As an alternative diagnosis, national differences can be seen as a resource for democratic accountability, rather than as an obstacle. Legitimate standard-setting depends upon deliberative procedures which evaluate and perhaps reconcile different national approaches as normative choices: 'Diversity is the strength of Europe, not its weakness or problem' (Renn, 1995: 154). From this perspective, diverse national approaches provide a basis for constructing European institutions in a more legitimate way.

Perhaps illustrating those divergent diagnoses, the European Environmental Agency (EEA) has undergone a tension between two models of knowledge production. The European Commission largely sees Europe as an emerging superstate needing harmonisation across diverse cultures, through objective information from expert sources such as the EEA, as if such information stood above cultural values. By contrast, the EEA sees Europe as a civil society imagining and shaping the EU along different potential pathways; for this scenario, objective information means relevant, timely knowledge for enabling citizens to evaluate uncertainty and contingency, from a standpoint independent of the Commission. Thus 'two very different versions of natural/social order co-exist uneasily within this new institution' (Waterton and Wynne, 2004: 105). These also imply different models for citizen roles and the EU's democratic legitimacy.

2.3 Democratic Legitimacy via Regulatory Standard-Setting?

For the European integration project, early strategists envisaged that national regulatory differences would be overcome or reconciled through a low-key standard-setting process. Rather than establish a new bureaucracy, European integration would facilitate new expert networks, develop European forms of knowledge and thus provide authority for European-wide policy. A decentralised, expert-based harmonisation would define the new Europe as well as integrate it. Named after a prime architect, the 'Monnet method' envisaged a low-politics process which would initially avoid the most contentious issues. A few hundred Commission staff would mobilise thousands of national experts to set technical standards, as a means to achieve an internal market. Under that subtle method, support for European solutions could eventually be extended to politically more sensitive areas (Weale, 1999: 44; Jordan, 2002: 5).

That method depended upon a separation of technical from policy issues. Monnet foresaw the EC as 'a depoliticized organization whose legitimacy would be based upon apparently objective technical criteria . . . ' These would be set by legal and economic experts, based on the French experience of a technocratic model. However, this modernist project proved to be impossible to sustain within European integration (Weale, 2000: 115).

As policy views were represented and disputed as technical issues, politicians became more dependent upon experts:

> Both supporters and opponents of particular regulatory measures usually cast their arguments in the language of 'regulatory science' rather than in the more traditional language of interest or class policies. Paradoxically, the very fact that the scientific basis is often uncertain and contestable tends to increase the role of experts at every stage of the regulatory process. (Majone, 1997: 157)

Indeed, EU-wide standard-setting involves a constant tension between politicisation and technicisation. Within regulatory committees, decision-making has featured conflicts of authority, especially over 'political' issues that may warrant a broader procedure beyond technical expertise. For example, regulatory decisions are often justified as 'adaptations to technical progress'. Yet 'the difficulty of distinguishing between political and technical questions also provides an opportunity to those who might wish to reduce political questions to technical ones' (Landfried, 1999: 181).

Trade barriers often resulted from member states devising their own product standards for health and safety, which were seen as potentially justified. To reconcile these national differences, the integration project was initially seen as a technical-administrative task of lowering trade barriers, but this strategy encountered great difficulties by the mid-1980s. Rather than seek uniform standards, 'the new approach to harmonization' had a more modest goal: mutual recognition of national standards, also known as negative integration (Pelkmans, 1987).

Although negative integration reduced national differences in some policy areas, especially by liberalising markets in services, this new approach did not work for product regulation, for several reasons. Mutual recognition could mean levelling down standards, so some EU member states blocked products from others which had less strict standards. These blockages often led to litigation at the European Court of Justice. Thus the internal market could be achieved only through regulatory harmonisation, also known as positive integration; this needed a deeper Community involvement in environmental, safety and health issues (Vos, 1997: 132, 138).

European integration through standard-setting posed at least two problems. The original EC Treaty required unanimity within the Council, which remained difficult to achieve. And EC law was made not directly by the people's representatives but rather by national representatives in the Council, with its Council proceedings kept secret and without Parliamentary debate. 'This democratic gap severely hampers public debate about possible measures to be taken' (van der Straaten, 1993: 72).

The 1987 Single European Act provided a partial remedy. The political negotiation process for new statutes was shifted from unanimity to a qualified-majority basis, involving a co-decision procedure between the Council and the European Parliament, while giving considerable power to the Commission. According to the new Article 100a, harmonisation measures for the internal market were to be based on 'a high level of protection' for human health and the environment. These changes aimed to remedy the democratic deficit in the Community's decision-making process (Vos, 1997: 133).

Often this process has resulted in relatively stringent standards for products, involving both input and output legitimacy. Through its Environment Committee, the Parliament has supported proposals for more rigorous standards through harmonised legislation; often this has meant allying with the Commission against some reluctant member states (Judge, 1993: 199). In setting standards for new products, output legitimacy too has been as effective at the EU level as at the national level. This can be true for product standards 'precisely because here national problem-solving capacity has not been destroyed by negative integration—and hence, is also less constrained by the economic pressures of regulatory competition' among EU member states (Scharpf, 1999: 107). Conversely, the Commission has been relatively more dependent upon member states for decisions on product approval than in other areas of trade liberalisation.

Standard-setting arrangements have featured a tension between two tendencies: transferring authority from national to EU-level expert bodies and the Commission (sometimes called centralisation), versus strengthening EU regulatory committees which represent member states. The role of those representative committees has remained essential for democratic legitimacy, argues Joerges: Assuming that 'the Europeanization of markets requires institutional structures which ensure both the effectiveness and the legitimacy of risk assessments, then we are bound to strive for institutional solutions which transcend the boundaries of our constitutional States without replacing these States with a Europeanized equivalent', i.e. without an EU superstate (Joerges, 1997: 221–22).

To facilitate Europe-wide standard-setting, advisory expertise too has undergone a Europeanisation process. The relevant expertise was available mainly at the national level. So in the 1990s the Commission sought to build up a European-wide scientific expertise acceptable to all participating experts. These EU-wide expert bodies encompass diverse traditions and views; likewise EU regulatory committees representing member states. The Commission has become dependent upon standard-setting in those bodies (Vos, 1997: 138–39).

Amid national conflicts over standard-setting, both within and among member states, EU regulatory committees have deliberated normative issues through the EU's formal comitology procedure, representing the member states. Their judgements link cognitive and normative issues in regulatory procedures (Joerges, 1999: 315, 319). This needs a deliberative style of problem-solving, as done in the comitology procedure, rather than a Weberian-type state bureaucracy simply implementing a prior legislative mandate (Joerges, 2006: 2, 7).

However, the comitology process per se may not provide democratic legitimacy for EU decisions, argues Joseph Weiler. On the one hand, a transnational deliberative procedure can expose national sectarian interests more readily than a national procedure could do. On the other hand, decisions favour some interests and norms, while excluding others. If proposals for 'transparency, openness and equal access' are fully realised, then they may impede EU decisions. On his reading, then, comitology 'is a microcosm of the problem of democracy, not a microcosm of the solution' (Weiler, 1999:

349). This tension may also apply the other way around, if deliberation is further limited by internal market imperatives.

2.4 Expertise Contested—and Democratised?

Across many political jurisdictions, regulatory standard-setting has featured disputes over the appropriate relation between science and policy, as well as the relation between expert and lay roles. Policy issues have often been represented or disguised as techno-scientific ones for delegation to expert advisors, thus limiting democratic accountability as well as political responsibility. In claims for the boundary between science and policy, relatively broader accounts of 'science' increase the scope for expert advice to influence, constrain or justify regulatory decisions (Jasanoff, 1987). Conversely, defining an issue as policy increases the scope for public involvement and state accountability.

Thus science/policy boundaries are drawn strategically, shaping institutional power in regulatory decision-making. Likewise expert/lay boundaries variously narrow or broaden entitlements to participate in such issues. Through a 'citizen science', opposition groups have challenged government claims for an expert consensus, while counterposing alternative solutions for societal problems (Irwin, 1995).

Science/policy boundaries have been deployed and challenged in a series of European scandals over food and medical safety. In particular the 1996 'mad cow' pandemic arose from the UK deregulation of animal feed processing; expert advice made over-optimistic policy assumptions about the prospects for the infectious agent being transmitted to humans and for preventive measures being implemented. The consequent scandal undermined official images of policy-neutral expertise at both national and EU levels. In the crisis, EU regulatory procedures were attacked for equating expert advice with 'science', as a means to pre-empt or conceal policy decisions. This legitimacy crisis stimulated a broad debate on citizenship and governance of techno-scientific issues. The European integration project faced difficulties in constructing or representing a citizenry compatible with expert-regulatory harmonisation for the internal market.

In its White Paper on European Governance, the Commission noted the EU's legitimacy problems and possible remedies: The EU 'will no longer be judged solely by its ability to remove barriers to trade or to complete an internal market; its legitimacy today depends on involvement and participation'. For legitimately governing the EU, the document emphasised the role of civil society—defined broadly to include: trade unions and employers' organisations ('social partners'), NGOs, professional associations, charities, grassroots organisations, organisations that involve citizens in local and municipal life, religious organisations, etc. (CEC, 2001a: 11).

However, the White Paper generally emphasised instruments for efficient planning, problem-solving and policymaking, especially for the internal market, rather than address the sources of the EU's legitimacy gap. Public involvement was promoted as an educational instrument rather than as a public means to test the validity of arguments. For example, greater transparency was needed to 'boost confidence in the way expert advice influences policy decisions' (ibid.: 5). As a result, 'improved participation is likely to create more confidence in the end result and in the institutions which deliver policy' (ibid.: 10).

Moreover, the White Paper framed the policy process as technocratic decision-making, for which civil society must be shaped, educated and incorporated. According to a critical analysis, quoting the White Paper:

> The authors put their trust in extended participation and active involvement of civil society—and 'with better involvement comes greater responsibility'. Another means proposed, thus, is partnership arrangements, which entail a commitment to additional consultations with civil society actors. The problem, here, is, on the one hand, the democratic danger of co-optation. . . . What is more: 'Participation is not about institutionalising protest. It is about more effective policy-shaping, based on early consultation and past experience.' Hence, there is the problem of domestication of civil society organizations. Civil society is not seen as an arena for voluntary action and for open and free public debate. The democratic division of labour between state and civil society is endangered when voluntary associations (NGOs) are used as merely instruments to implement policies more smoothly. (Eriksen, 2001: 62–63)

For decision-making, 'The technocratic vision has, for a long time now, dominated both the public and scholarly debate on the EC/EU: it is an élite game in the hands of economic interests and bureaucrats' (ibid.: 66). Thus the White Paper re-affirmed a citizenship model which underlies the EU's legitimacy problems, while suggesting ways to operationalise that model.

Within the wider debate on the democratic deficit, 'Science and Governance' has received special prominence as a legitimacy problem for government. As the Commission's White Paper noted, expert advice has often obscured political responsibility for decisions:

> It is often unclear who is actually deciding—experts or those with political authority. At the same time, a better-informed public increasingly questions the content and independence of the expert advice that is given. (CEC, 2001a: 19)

According to a supplement to the White Paper, experts are challenged by 'counter-experts':

> While being increasingly relied upon, however, expertise is also increasingly contested. . . . 'Traditional' science is confronted with the ethical, environmental, health, economic and social implications of its

> technological applications. Scientific expertise must therefore interact and at times conflict with other types of expertise. . . . (Liberatore, 2001: 6)
>
> Problems arise with regard to the credibility and legitimacy of expertise and its role in public policy and debate when it seems to be used mainly as a 'legitimating device' for decisions already made by politicians. The same applies when experts seem to replace political deliberation with 'technocratic' decisions apparently inaccessible and unaccountable to public scrutiny. (ibid.: 10)

In addition to democratising expertise, 'expertising democracy' can provide plural expert advice to democratic institutions and to a broader citizenry (Liberatore and Funtowicz, 2003: 147).

There have been proposals to enhance regulatory transparency, even to 'democratise expertise'—as in the title of the above report. A more subtle remedy would be to pluralise expertise. Advisory procedures need to develop 'an approach that makes apparent the possibility of unforeseen consequences, to make explicit the normative within the technical, and to acknowledge from the start the need for plural viewpoints and collective learning' (Nowotny, Scott and Gibbons, 2003: 153, 155). Knowledge can be tested in various ways—in partnership or in antagonism to those claiming cognitive authority.

For expert judgements in risk regulation, EU policy has given a high-profile role to the precautionary principle. Its meaning has been continuously disputed, despite efforts by the European Commission to formulate a 'common understanding'. Uncertainty about risks depends upon how questions are asked of science (Levidow, 2001; Levidow et al., 2005).

Precaution can highlight unknowns in risk assessment. There may be ambiguity in the choice of framing assumptions, e.g. about causal pathways of potential harm and thus the basis for generating evidence. Uncertainty may also involve epistemic ignorance—what we don't know that we don't know. Assessors make value choices about how to frame unknowns and indeterminacies, so expert judgements cannot provide a definitive 'sound science'. As an alternative approach, a deliberative process among stakeholders can highlight ambiguity and ignorance, so that framing assumptions can be made more transparent and deliberative: 'In acknowledging that the problems of scope, incommensurability and ignorance in risk assessment are otherwise intractable, active stakeholder engagement in the appraisal process becomes a matter of analytical rigour' (Stirling 1999: 20). By recognising and addressing unknowns, precaution can inform knowledge-generation. By considering a wider range of options, 'a precautionary process extends the knowledge base for risk appraisal' (Stirling, 2003: 53).

EC regulatory and judicial frameworks encompass scope for precaution, i.e. decisions citing scientific uncertainty, which may be indicated by expert disagreements. According to a Commission legal chief, 'The basic legal definition of scientific uncertainty reflects the potential for error inherent in science and scientific information.' In EC law, moreover, the precautionary principle plays normative roles, e.g. in placing the burden of evidence on the applicant for approval of a product (Christoforou, 2003: 208, 210).

3 GOVERNING CONFLICT OVER TECHNOLOGY

Technological controversy involves power struggles that arise over how to define the issues at stake, even over what is the technology. Proponents often draw distinctions between a technological device and its context or consequences, while critics emphasise links between those elements. Conflicts also arise over the extent and types of public participation in these issues. Such dynamics can be illuminated by critical perspectives on 'co-production' and 'governance'.

3.1 Co-production of Technology as Socio-Natural Order

'Co-production of technology, nature and society' provides concepts for analysing links between social and natural order. Science and technology can be understood as socio-technical hybrid constructs, ordering society in particular ways, while attributing that order to separate 'natural'characteristics (Jasanoff, 2004: 21). Within co-production perspectives, a constitutive approach emphasises how such order is constituted, created, and maintained—e.g. through new facts, things or concepts. According to one proponent, Society 'is no less constructed than Nature, since it is the dual result of one single stabilization process' (Latour, 1993: 94). In this process, people relegate part of their experience to an apparently immutable, objective reality. Constitutive approaches emphasise correspondences between knowledge and power, while giving little attention to moral or political conflicts that normally accompany social/natural demarcations (Jasanoff, 2004: 22–23).

By contrast to constitutive approaches, interactive approaches investigate conflicts around boundaries, e.g. between putatively un/changeable or natural/social realities. Interactive co-production emphasises contexts where these distinctions undergo challenge, amidst competing epistemologies. Through boundary-work, various practices make and stabilise those distinctions; for their authority, scientific claims depend upon specific changes in social order. This perspective can illuminate how technology provides a solution to a problem of socio-political order, while not taking for granted a particular form of political power or alliance as an explanation for such change (Jasanoff, 2004: 18–19, 30). Concepts of objectivity and expertise remain important for legitimising regimes as democratically accountable. Such concepts affect how research results are taken up in public realms, e.g. as persuasive, biased, inconclusive; how they are meant to 'solve' public problems; and how they are constructed to legitimise policy (ibid.: 34).

In the co-production of technology, nature and society, several instruments operate at the nexus of social and natural order. These instruments include the following.

- Making identities: Collective identities can help people to restore sense out of disorder, by putting things back into familiar places. These include characteristics such as European, professional, intelligent, etc. But these may also be contested or renegotiated in elaborating a different order.
- Making institutions: Institutionalised ways of knowing are reproduced in new contexts of disorder; they also serve as sites for testing or reaffirming a political culture. According to a model of market capitalism, for example, the human subject is able to form autonomous preferences, make rational choices and act freely upon the choices so made; any exceptions are interpreted as a market failure, rather than a problem with the model.
- Making discourses: Languages are produced or modified in ways which promote tacit models of nature, society, culture or humanity. For example, discourses may define the boundary between promising and fearsome aspects of a technology—e.g. links between 'un/natural' and 'un/safe' characteristics.
- Making representations: this includes means of representation in diverse communities of practice; models of human agency and behaviour; and the uptake of scientific representations by other social actors (Jasanoff, 2004: 38–41).

According to a political theorist of the EU, its input legitimacy depends upon similar elements—collective identities, institutional accountability and policy discourses—which have been lacking on a European scale (Scharpf, 1999: 187). A co-production perspective can illuminate strategies for establishing, destabilising and reshaping those instruments.

Technology promotion often appeals to 'natural' characteristics as threats or solutions. What we call 'nature' is a socio-cultural product, yet nature is discursively separated from society as two independent realms. This dichotomy provides a basis for supposedly objective imperatives, authorising specific societal agendas which in turn reproduce that form of nature (Castree, 2000). This can be illustrated by examples from agriculture.

The 18th century utilitarians represented the market as the natural regulator, complementing the natural liberty of the entrepreneur to trade without interference. Through metaphors of machine and market, this new discourse justified the Enclosures in transforming agricultural land into capital, along with new institutions to police dispossession from common lands. Such ideas undermined the earlier discourse of 'natural law', meaning the natural justice of the yeomanry living from their own labour on the land as a common societal resource (Williams, 1980: 79).

As an industrial project, agbiotech recasts natural resources in its own image of precise genetic changes. Organisms are seen as deficient factories needing efficiency improvements. Through genetic modification, they are made more orderly, controllable and thus predictable for efficient production. Thus nature is made safe for biotechnology, while displacing alternative concepts of natural resources (Sagoff, 1991).

More specifically, agbiotech invests crops with qualities of productive efficiency, as in 'smart seeds' or 'genetically improved crops'. This innovation extends a 'technological treadmill' on which farmers have been running for a long time, through earlier technologies likewise aiming to enhance output. The myth of the yeoman-farmer as a skilled craftsperson may persist, but the reality may be the 'propertied labourer', who follows contractual instructions from a corporation which already owns the crop in the field (Kloppenburg, 1988: 283).

Agbiotech policy illustrates how neoliberal agendas depend upon social power structures and a dominant culture. The technology is promoted as necessary to solve societal problems: 'Market liberalism and technocracy set the agenda, not democracy . . . the economism of globalisation discourse is combined with an authoritarian technological determinism' (Barben, 1998: 417).

3.2 Participation as Governance—Versus Democracy?

Governments have often relegated major technological decisions to administrators and expert advisors, thus undermining democratic accountability to competent citizens (as described above in section 2). 'Competence has long been recognised as a prerequisite for democratic citizenship'. What has changed over time is 'the understanding of what it means to be politically competent and, latterly, what counts at all as political'. The 20th century saw politics restructured—separated into a value-laden politics and technical-rational policy. 'As decision-making grew more technical and expert-driven, citizens were progressively distanced from the process of data-gathering and analysis that formed the backbone of administrative decisions' (Jasanoff, 2005b: 368).

Citizens' groups have expressed widespread scepticism towards science as a universal, objective knowledge. For example, farmworkers have criticised expert assumptions about real-world 'normal' conditions of pesticide usage (Irwin, 1995). In the last couple of decades, knowledgeable citizens have asserted a stronger voice in techno-scientific issues. Science and democratic politics are actively joined in three ways: in identity-making, in linking consumption and citizenship, and in politically relevant knowledge production (Jasanoff, 2004: 91).

Risk assessment has been widely promoted as a truly rational basis for policy decisions on techno-scientific issues, as well as an appropriate entry point for public participation in such issues, yet this focus has often created more social conflict. Such results have been theorised as a conflict between technical versus cultural rationalities, especially in environmental controversies. In an early US controversy over GM microbes, 'The political rhetoric of the popular culture stressed control over its environment', thus expressing a cultural rationality, which 'does not separate the context from the content of risk analysis'. By contrast, a technical rationality considers only measurable parameters (Krimsky and Plough, 1988: 107–8, 306).

When facing expert disagreements about technical matters, citizens rely more upon a socio-cultural assessment of the factors surrounding a decision.

> Citizens want to know how conclusions were reached, whose interests are at stake, if the process has a hidden agenda, who is responsible, what protection they have if something goes wrong, and so on . . . From the perspective of cultural rationality, to act otherwise would itself be irrational. (Fischer, 2005: 57)

Participatory processes often frame issues and model citizens in terms of risk issues. In efforts at participation, 'risk' is reified twice: by defining the universal public meaning of technological controversy as a 'risk' issue; and by selecting particular 'risk' definitions as natural, objective and universal. Citizens are modelled according to specific risk issues and definitions, while excluding other accounts (Wynne, 2005).

Legitimacy problems have resulted from governmental over-dependence upon science and its discursive equation with official expertise. In broad historical perspective, 'At stake here is the Enlightenment project, where objective science and representative democracy are combined to provide a new legitimation of the State', argue De Marchi and Ravetz (1999: 754). From the controversial cases of the BSE crisis and a GM maize product in the 1990s, they diagnose difficulties of governing potential hazards: even speculative hazards could undermine public trust in risk regulation. 'Here it is the uncertainties which dominate, and which require the reference to explicit values' (ibid.: 755). To deliberate these values, wider public participation has been advocated as an 'extended peer review' of official expert judgements. However, NGO involvement requires a somewhat 'self-contradictory balance between their functions as critics and as stakeholders', i.e. holding a stake in feasible outcomes (ibid.: 756).

Opening up technology choices means deliberating alternative accounts of the problem. According to an advocate of participatory TA: 'In practice, the relationship between representative democracy and participatory methods becomes most clear and complementary, when engagement is approached as a means to open up the range of possible decisions, rather than as a way to close this down'. In his view, such an evaluation makes 'more clear how to apportion responsibility and accountability in decision making' (Stirling, 2006: 5; likewise Stirling, 2005: 229).

EU discussions on governance, including 'science and governance', promote an apparent consensus on common problems to be addressed. In mainstream policy language, likewise governance is often understood as a co-operative means to deal with common problems (e.g. CEC, 2001a). In a similar vein, governance involves social institutions 'capable of resolving conflicts, facilitating cooperation, or, more generally, alleviating collective-action problems in a world of interdependent actors', according to a political scientist (Young, 1994: 15).

Such participation involves the premise 'that a problem is "common", in the sense that stakeholder advantage cannot be obtained—nor, often, defined—independently from collective reasoning'. Yet often such advantages are foreseen and pursued; some stakeholders pursue antagonistic agendas (Pellizoni, 2003). The conflict can be managed through 'governance', by selectively structuring participation around common problem-definitions, i.e. by highlighting some problems or agendas as common ones. According to critical perspectives on 'governance', this concept helps to contain or marginalise antagonistic agendas, while undermining representative democracy. Management and 'governance' presuppose pacified worlds in which common aims could be defined for the good of all (Pestre, 2008).

Governance strategies provide a 'discursive de-politicisation', effectively removing societal choices from the political agenda. 'The democratic public is dislodged from its position as (in principle) the ultimate judge and arbiter in the realm of "governing"; with governance, it is at best one among many stakeholders; it (the public) merits no privileged position' (Goven, 2006). Fundamental conflict can be displaced onto supposedly collective problems and solutions. Choices about societal futures can elude the formal accountability of representative democracy, even through participatory exercises.

For technology-related issues, 'Complex issues of governance are reduced to issues of scientific uncertainty . . . Thus the reduction is not merely one of the political to the scientific, but a further reduction of science to the paradigm of risk and its assessment and management'. Participation 'is essentially a managerial discourse, perhaps, even more narrowly, a crisis management discourse masquerading as a theory of democracy' (Rayner, 2003: 168–69). Thus 'governance' can play a managerial role, substituting for the accountability of representative democracy.

4 FRAMING ISSUES, PROMOTING FUTURES

Societal controversies can be analysed as contending discursive frames—as divergent ways to frame issues, to define the problems that need solutions and thus to promote divergent futures. The policy process can be seen as interacting frames and problem-definitions. According to Rein and Schön (1991: 263), 'Framing is a way of selecting, organizing, interpreting, and making sense of a complex reality to provide guideposts for knowing, analyzing, persuading and acting'. This section surveys perspectives on how such frames inform coalitions, while shaping group identities. Discursive frames contend for influence in various contexts; this section also surveys the 'arena' concept for analysing how each context favours different kinds of resources and discourses.

4.1 Contending Frames

In public debate, policy agendas are set by contending frames, each defining the societal problems that need solutions:

> Cognitive and normative frames . . . refer to coherent systems of normative and cognitive elements which define, in a given field, 'world views', mechanisms of identity formation, principles of action, as well as methodological prescriptions and practices for actors subscribing to the same frame. Generally speaking these frames constitute conceptual instruments, available for the analysis of changes in public policy and for the explanation of developments between public and private actors which come into play in a given field. (Surel, 2000: 496)

Each problem-definition implies future scenarios for linking nature, society and technology. Political power can be exercised through discursive accounts of reality, by promoting one future vision against others. New story-lines can bring together actors into coalitions. 'Political change may therefore well take place through the emergence of new story-lines that re-order understandings', argues Hajer (1995: 55–56). 'Interactive policymaking is now a practice within which people generate new identities' (Hajer and Wagenaar, 2003: 12).

When social movements seek to challenge or change government policy, they devise their own 'collective action frames'. These help to produce shared meanings of issues, to mobilise adherents and resources, and thus to use or create new political opportunities for influence. Resonance with public meanings depends less upon whether discursive claims have factual validity than upon whether they offer culturally believable diagnoses of societal problems (Benford and Snow, 2000). In these ways, opposition movements can mobilise effective resources beyond specialist expertise and stimulate policy change.

Environmental discourses have been widely analysed as issue-framing: 'the environment' in particular has become a terrain of contested social values. Although environmental threats are often attributed to nature, they are always framed by policy agendas, through story-lines which selectively problematise some aspects of physical and social reality. The narrative devices include images, causal models and metaphors. Environmental discourses define problems and structure reality so that some framings seem plausible, while others are foreclosed (Hajer, 1995).

Agricultural innovations have been long promoted and contested through divergent languages of nature. After the rise of agribusiness, especially the agrichemical-intensive methods after World War II, these methods came under widespread attack as 'unnatural', as well as dangerous for human health and the environment. In their strategic response, the agri-food industry emphasised deficiencies of nature: 'the modern-day Prometheans of the food industry spelled out the ways in which Mother Nature grossly neglected her human offspring, especially in the field, at the cash register and in the kitchen', noted a cultural analysis (Belasco, 1989: 117). Its technological innovations promised to compensate for those deficiencies but met sceptical responses.

Discursive frames can be exemplified by the controversy over bovine growth hormone, a product which was designed to increase milk production. Each side appealed to accounts of the natural, partly by naming the product in different ways. The euphemistic term 'bovine somatotropin' implied a natural basis for a benign productive efficiency. On the other side, the term 'hormone' implied an unnatural chemical, by analogy to other capital-intensive innovations which had already harmed environmental quality, small-scale farming and its independence (Buttel, 1998).

4.2 Sustaining What Development?

Since the 1990s, the idea of 'sustainable development' has become mainstream—and all the more contentious, especially in Europe. Some accounts promote technological innovation as eco-efficient improvements which provide economic advantage and minimise environmental pollution. Critics have questioned such claims, and even the eco-efficiency framework, while counterposing alternative development trajectories.

Such conflicts about sustainable development can be illuminated by theories of ecological modernisation (EM). These emphasise the potential for re-embedding an ecological dimension of economic practices within modernist institutions, by institutionalising ecology in production and consumption processes. Some propose government measures to stimulate self-regulation of industry, thus transferring responsibilities from the state to the market (Mol, 1996: 306).

While EM literature often takes for granted the claims for eco-efficient innovation, some perspectives analyse their strategic role. Such policy discourses promote specific technological changes as techno-fixes, in ways which constrain policy choices and pre-empt alternatives.

> [ecological modernisation] . . . uses the language of business and conceptualises environmental pollution as a matter of inefficiency, while operating within the boundaries of cost-effectiveness and administrative efficiency . . . [EM] is . . . basically a modernist and technocratic approach to the environment that suggests that there is a techno-institutionalist fix for the present problems. (Hajer, 1995: 31–32)

Such critical perspectives can help to illuminate tensions among divergent environmentalist approaches:

> . . . the late 1990s showed how citizens not so much opposed eco-modernist governmental policies but conceived of the environmental problem in different, more culturally loaded terms . . . Furthermore, governments could be seen to strengthen the ties between eco-modernist thinking and neo-liberal economic discourse. . . . (Hajer and Versteeg, 2005: 179)

Thus eco-modernist discourses promote specific policy agendas and cultural meanings which may underlie societal conflict.

European Commission policy has been interpreted as ecological modernisation, as in the following paraphrase: 'the market must ensure that environmentally friendly goods and services have a competitive advantage over those that cause pollution and waste' (Hanf, 1996: 210). According to a similar analysis of Commission strategy, a post-industrial economy depends on stringent environmental standards stimulating the capacity for high-quality, high-value products (Weale, 1992: 77, 1993: 210). Although such policy links may not be consistently realised in practice, this discourse can be understood as a coalition-building device, especially for reconciling powerful economic interests with environmental protection, argues Weale (1993: 213).

EU 'sustainable development' policy has accommodated the aim of 'completing the internal market'. Avoiding distortions of market competition has been the main aim, while also providing extra opportunities for EU-level legislation which may limit environmental degradation (Burchell and Lightfoot, 2001: 36). That linkage has generated divergent regulatory frames. According to a prevalent neoliberal view, different national standards impede trade and economic progress, so the Community should promote a mutual recognition of standards in order to complete the internal market, as already discussed above. According to another view, the internal market could bring products and environmental changes which are either positive or negative, so the Community should set standards which favour environmental improvement. The latter view appeared in some DG-Environment policy documents, but it was not easily adopted or implemented, for many reasons. DG-Environment had a weak role within the Commission; environmental issues were readily subordinated to imperatives of economic competitiveness (Weale and Williams, 1993).

EM perspectives have been cited in various ways to analyse the European agbiotech controversy. When many policymakers there opposed any special regulation of GMOs in the late 1980s, others advocated such EU legislation as necessary for environmental and trade reasons. EM concepts provided 'important reference points for the shaping of a policy counter-narrative', according to an early study (Gottweis, 1998: 235). By contrast, later studies have associated EM concepts with pro-biotech agendas. In the UK, New Labour supported GM crops as a means to maintain an economic competitive advantage, while enhancing sustainability through more intensive, eco-efficient agri-production (Barry and Paterson, 2004). Agbiotech supporters make claims for environmental advantages over conventional agriculture. Opposition groups rejected that framework along with industrial agriculture, while appealing to natural characteristics of alternatives, especially organic farming (Toke, 2002).

This book adapts EM perspectives for our overall tripartite model of three policy frames. Agbiotech has been promoted as 'sustainable agriculture', sometimes explicitly meaning 'eco-efficiency', narrowly understood as input-output efficiency of biophysical resources. Critics have counterposed different accounts of sustainable development. Through these frames, policy coalitions seek to undermine, accommodate or influence the others (see next sub-section and Table 1.1).

4.3 Tripartite Models of Contending Policy Frames

Public controversies have been generally stereotyped as 'two sides', especially in the case of contentious technologies. This stereotype often informs analytical accounts, thus assuming that the state simply takes one side against the other. To analyse its role, a more complex interactive model is needed.

Cultural Theory has elaborated a tripartite model of discursive frames, especially for technological-environmental controversies. Each case can be analysed as three main political cultures, each promoting its own 'myth of nature', in turn justifying preferred social relations. In particular, an 'individualist' culture frames nature as benign, providing a cornucopian source of societal benefits, while restoring global equilibrium from any perturbations. An 'egalitarian' culture frames nature as ephemeral, inherently vulnerable to harm from industrial activity. A 'hierarchist' culture frames nature as 'perverse/tolerant', able to correct any perturbations within finite limits, beyond which it undergoes serious harm; thus the limits must be anticipated and respected to maintain stability. Each frame justifies policies which would licence, constrain or block industrial activities, respectively (Schwarz and Thompson, 1990).

Although any typology over-simplifies reality, it can help provide valuable insights. For that reason a similar tripartite model is applied to the groups involved in the European agbiotech controversy. Analogous models are elaborated across the chapters, as summarised in Table 1.1 below.

The agbiotech controversy begins with three frames—eco-efficient, managerialist and apocalyptic—analogous to the individualist, hierarchist and egalitarian cultures (respectively) of Cultural Theory. Based on each frame, a discourse coalition promoted different links between the social and natural order. Over the timescale analysed in this book, these three coalitions underwent significant changes, encompassing more supporters and policy issues. When agbiotech risk issues were linked with 'sustainable agriculture', that discourse too had three divergent frames: high-yield intensification within a neoliberal frame; multifunctional agriculture within a managerialist frame; and quality agriculture within a community-territorial frame. Here our taxonomy draws upon another tri-partite model which was devised mainly for the global South (Woodhouse, 2000), as well as one oriented to Western societies (Dobson, 1996: 418).

By the start of the 21st century, those three frames had become linked with contending paradigms of rural development. In conflicts over future agriculture, 'rural space within Europe has become a "battlefield" of knowledge, authority and regulation', argue Marsden and Sonnino (2005:50). They theorise a competition among three paradigms. As the dominant one, complementing a neoliberal policy framework, the 'agri-industrial' paradigm globalises the production of standardised food commodities for international markets, as the framework for designing and commercialising GM crops. In the 'post-productivist' paradigm, rural spaces become consumption spaces for urban and ex-urban populations. In the 'agrarian-based rural development' model, agri-production is relocalised by embedding food chains in highly contested notions of place, nature and quality (ibid.). These paradigms promote divergent frames of the agricultural environment to be protected (Levidow, 2005).

However, the post-productivist paradigm does not operate as a distinctive frame. It can be integrated with agrarian-based rural development, e.g. through eco-tourism partly based on 'quality' agriculture, agri-ecology, etc. (Marsden and Smith, 2005: 441). Likewise it can be integrated with the agri-industrial paradigm, e.g. by shifting the least productive land into non-agricultural uses. To analyse the state's role in this conflict, an 'agricultural diversity' frame mediates conflicts between the agrarian-based rural development and agri-industrial paradigms; each has potential links with a post-productivist paradigm. For that reason, framings by state bodies are shown in the middle of Table 1.1. These three policy frames have taken analogous forms across the various policy issues and multi-level trials. The tripartite model is elaborated with a full table in relevant chapters.

Table 1.1 Discursive Frames across Agbiotech Policy Arenas

Actors/Coalitions	*Agbiotech Promoters*	*State Bodies (legislators & regulators)*	*Agbiotech Opponents*
Arenas and Issues			
innovation and regulation (Chapters 2, 5, 6)	eco-efficiency	managerialist, (precautionary regulation)	apocalyptic
risk and sustainable agriculture (Chapter 3)	sustainable intensification	multifunctional agriculture	quality agriculture
GM labelling (Chapter 7)	economic individualism	market completion	democratic rights
agricultural futures (Chapter 8)	agri-industrial	agricultural diversity	agrarian-based development

4.4 Arenas of Policy Conflict

'Risk' discourses play symbolic roles. When groups become involved in a controversy, 'risk' may not be the trigger but rather a symbolic meaning for societal decision-making processes and power structures. 'Such groups use the risk arena to mobilise social resources to influence policies in other arenas', e.g. to oppose big business or favour deregulation (Renn, 1992: 191).

As a metaphor in political science, the 'arena' concept describes the symbolic location of political actions which influence collective decisions and policies. As symbolic locations, arenas are neither geographical places nor organisational systems; rather, they describe actions of all relevant actors involved in a policy issue. Examples of such arenas include elections, markets and regulatory procedures. Different policy systems may interact in an arena but still preserve their autonomy (Hilgartner and Bosk, 1988).

The outcome of a policy conflict depends on interactions among groups seeking influence across all relevant arenas. Arena theory can help explain how social groups respond to and shape policy issues. Actors in a controversy can mobilise resources of several kinds—money in economic arenas, authority in political arenas, social influence in overall social systems, value commitments in cultural arenas and evidence in scientific arenas (Renn, 1992: 186). Each arena is characterised by dominant actors—e.g. scientists in the regulatory arena, consumers in an economic arena, citizens in the political arena, etc. Each actor may have different identities and roles across those arenas (Joly and Assouline, 2001: 23).

Scope for effective action depends on rules of access, which affect the arguments and resources which can be used effectively in each arena. Each arena is characterised by particular rules. These include formal rules that are coded and monitored by a rule enforcement agency, and informal rules that are learned and developed through interactions among the actors. Formal rule changes require institutional actions, while informal rules may change according to whether or not rule-bending is penalized. 'Several actors may join forces to change the rules even if they disagree on the substance of the issue' (Renn, 1992: 182). In some cases, 'actors may decide to ignore some of the rules if they feel that public support will not suffer and if the rule enforcement agency is not powerful enough to impose sanctions on actors who violate the rules' (ibid.: 184). That perspective can help to illuminate how actors use, bend, challenge and/or change the rules of access, both within and across arenas. By doing so, a discursive frame and its policy coalition can gain advantage over others.

In the European agbiotech controversy, opposition discourses gained influence in every arena by circumventing, bending or changing the rules (as this book will show). Opposition strategies mobilised cultural resources across all relevant arenas—judicial, regulatory, expert, experimental and commercial. Throughout these arenas, the controversy featured three contending policy frames. Agbiotech proponents elaborated a neoliberal eco-efficiency frame, while critics counterposed an apocalyptic frame, whose accusations put agbiotech on trial. To deal with these antagonistic forces, state authorities elaborated a distinctive managerialist frame for selectively channelling, incorporating or marginalising dissent. Through interactions among those frames in every policy area, the EU policy system shifted somewhat from a neoliberal eco-efficiency frame to a managerialist frame. Institutions underwent pressure to put pro-biotech claims on trial—and were themselves put on trial. This controversy in turn became a test case for the democratic legitimacy of EU policy.

2 Making Europe Safe for Agbiotech, Generating Dissent

INTRODUCTION

GM products were initially developed by transatlantic multinational companies in the early 1990s. Meanwhile they were buying up US seed companies—as a source of high-quality germplasm, and a means of marketing seeds to farmers. Commercialisation was anticipated first in the USA, where critics raised questions about the promised benefits, their societal distribution and alternatives. However, such concerns had no bearing on company or government decisions. The US government promoted agbiotech as technological, economic and societal progress.

By default of any effective means to question the innovation, US debate focused on possible risks. Questions were asked about whether or how GMOs could be made predictably safe for the environment. In the margins of this risk debate, a philosopher turned that predictive question into a normative issue. He analysed how organisms were being standardised for predictable, efficient agri-industrial uses through genetic modification, and thus how nature was being made safe for agbiotech. Agbiotech meant a normative shift in what counts as natural, beneficial, rational, etc. (Sagoff, 1991).

From a similar perspective, this chapter will explore the following questions about developments from the 1980s to the mid-1990s:

- How was agbiotech linked with specific forms of nature and social norms?
- How was Europe being made safe for the development and commercialisation of agbiotech?
- How did this framework model European integration—and generate political conflict?

The story begins with the US promotion of agbiotech, as a potential model that was eventually adapted in Europe. This adaptation emerged through much conflict in several policy areas—industrialising agriculture, extending patent rights and specifying agri-environmental criteria for risk regulation. Early dissent against this project generated three contending risk-frames (Table 2.1).

1 US AGBIOTECH AS CO-PRODUCTION

US metropolitan capital drove the industrialisation of rural America, especially through transport links linking grain production with urban mass markets. These production-consumption links dissolved boundaries between nature and technology, between rural and urban. Nature was put in its proper place, largely outside the agri-industrial system, e.g. constructed and idealised as 'wilderness' in the National Parks system. As the epicentre of the new system, Chicago could be ironically called *Nature's Metropolis* (Cronon, 1991). The agribusiness model linked technological innovation, productive efficiency, greater agri-input sales, dependence on pesticides for 'clean' fields, government subsidy for agricultural production, and trade liberalisation for commodity exports. Through the new WTO treaty in 1994, the US government further institutionalised trade liberalisation for agri-food products, which could thereby be treated as equivalent across national borders, beyond any cultural differences (McAfee, 2003).

As a new instrument for this agricultural model, agbiotech mined nature for its industrial utility and commercial value. In the 1930s molecular biology had already reconceptualised 'life' in physico-chemical terms. Through a computer metaphor, DNA became coded 'information' which could be freely transferred across the species barrier. 'As technology controlled by capital, it is a specific mode of the appropriation of living nature—literally capitalizing life' (Yoxen, 1981). A 'molecular vision of life' diagnosed societal problems as genetic deficiencies (Kay, 1992). Biotechnology R&D has been celebrated as 'value-added genetics', i.e. identifying genes which can enhance the market value of proprietary genes and novel seeds (Lawrence, 1988: 32). These metaphors complemented R&D priorities for further industrialising agriculture.

Multinational companies have sought to enhance the market for high-productivity agricultural inputs such as GM seeds, to drive down the price of bulk food commodities and to obtain multiple sources of inputs for the global food processing industry (McMichael, 1998). Such pressures were already aggravating US farmers' economic problems—for which GM seeds were supposed to provide a solution. Thus agbiotech extends a long history of technological treadmills through more capital-intensive inputs. GM crops attracted greatest interest from US farmers who use the most intensive cultivation practices for producing standard commodities.

In that context, GM crops were designed for an 'external' control of crop pests. Agbiotech R&D has identified and inserted extra genes to function in ways analogous to the chemical pesticides being replaced or supplemented. By contrast, an 'internal' control strategy would restructure agronomic-ecological relationships so as to minimise the need for pest-control agents; it would seek to manage pests rather than entirely eradicate them. With the former strategy, 'biotechnology products provide additional options that are not . . . being sought after by farmers but are consistent

Table 2.1 Three Contending Risk-Frames for GM Products in the Mid-1990s

Frame (discourse-coalition)	*Eco-efficiency*	*Managerialist*	*Apocalyptic*
Institutions	agbiotech business, e.g. SAGB/EuropaBio, EPP in European Parliament	DG-Environment and national regulatory authorities	environmental NGOs, Rainbow/Green MEPs, Coordination Paysanne
Agricultural problem	inefficient farm inputs and uncompetitive outputs	extensive cultivation methods which undermine the resource base for agriculture	intensive monoculture, farmer dependence on multinational companies, pesticide treadmill
Nature framed as	cornucopian potential to be reaped by simulating natural qualities, e.g. defences against external natural threats	environmental resources to be managed and protected from uncertain biophysical effects of a new technology	fragile resources under threat from uncontrollable, irreversible risks of industrial methods
GM crops framed as	safe tools for greater input-output eco-efficiency, with economic and environmental benefits	novel organisms, e.g. by analogy to exotic ones: potential hazards to be evaluated and managed	pollutants threatening the environment, democracy and societal values
Needs and benefits	Agbiotech offers additional opportunities for farmers to protect or improve their crops and to use natural resources more effectively.	Agbiotech can improve the economic viability of agri-food production in environmentally sustainable ways.	Europe does not need agbiotech, whose benefits go mainly to biotech companies.

Policies			
EU internal market	Common standards are needed to avoid trade barriers.	Common standards can be raised to stimulate beneficial innovation.	Common standards pose a threat of levelling down through trade liberalisation.
R&D funding priorities	Molecular-level knowledge can enhance economic competitiveness.	Molecular-level knowledge can enhance economic competitiveness, employment and the quality of life.	More extensive cultivation methods can avoid the hazards of intensive monoculture and enhance sustainable agriculture.
Patent rights and biopiracy	Living material warrants patents as a reward and incentive for innovation. Biopiracy = unauthorised use of bio-technological inventions.	Judgements must distinguish between discovery/invention and between un/ethical patents.	Patents would be unethical and privatise living material. Biopiracy = patents on life.
Risk legislation	Vertical, product-based legislation is necessary for non-discriminatory approaches to GMOs.	Horizontal, process-based legislation is appropriate framework for risk-based regulation.	Horizontal, process-based legislation is necessary for precautionary approach to unpredictable hazards.
Regulatory aims and effects	Over-regulation stigmatises a safe technology, impedes innovation and deters R&D investment in Europe.	Internal market should be based on a high level of protection for human health and environment. Avoid transboundary risks and regulatory differences across countries.	Strict regulation is needed to prevent known and unknown hazards.
Harm to be prevented	Prevent only harm to human health or the non-agricultural environment	Prevent harm to human health or non-agricultural environment (before the scope was broadened in late 1990s)	Prevent all potential effects including hazards of intensive monoculture, e.g. genetic treadmill

Note: All text here quotes or paraphrases public statements. The managerialist frame, placed in the middle, mediates conflicts between the other two, while elaborating its own distinctive perspective.

with the general needs and interests of specific agricultural subsectors', i.e. competitive pressures for greater productivity or cost reduction (Krimsky and Wrubel, 1996: 9).

To promote agbiotech, the US government provided extra policies: broader patent rights for turning natural resources into capital, incentives for research on GM techniques and a predictable basis for safety approval of GM products. In 1980 new laws increased the incentives for public-sector institutions to pursue proprietary aims. New patent legislation relaxed criteria for Federal approval of licensing agreements between universities and businesses. A revision of tax laws provided an extra incentive for private investment in universities. Moreover, investors could benefit financially from patents resulting from research funded by the Federal government (Krimsky, 1991: 66–67). Together these incentives were generating a 'university-industry complex' around agbiotech (Kenney, 1986). New arrangements provoked debate about 'conflicts of interest' between public and private sectors.

By the mid-1980s, 'public-sector' research was shifting its priorities towards patentable knowledge and artefacts, as well as imposing confidentiality rules (Kloppenburg, 1998: 232–34). Funded by government, universities and Land Grant Colleges went beyond a mere conflict of interest. They were redefining their interests along business lines, thus blurring the boundary between public and private. Genetic material formerly provided a common resource, while now molecular knowledge and techniques could be commoditised through broader patent rights; this shift was institutionalised through the US Patent Office and court decisions.

Through these policies, the public sector was being commercialised:

> Although cloaked in the discourse of 'national competitiveness', essentially this involves putting fundamental research in the public sector at the service of the applied, 'near-market' development work undertaken by industry. [Contractual arrangements] not only direct the research agenda by pre-empting research capacity but also spread the mantle of trade secrecy over its output . . .
>
> The balance of control over agri-biotechnology is being swung towards the corporate sector by policy decisions deliberated intended to weaken the bipolar institutional framework characteristic of conventional plant breeding in the United States. (Goodman and Redclift, 1991: 176, 180)

Long before agbiotech, the US agri-industrial system was generating public unease, far beyond issues such as pesticide hazards. Aims were in conflict: 'the public has begun to raise broad questions about the fundamental goals for the food and agricultural system, including such issues as equity, efficiency, resilience, flexibility, conservation and consistency with other objectives of US society' (Busch and Lacy, 1983: 211).

The agri-food industry turned those public concerns into a new business opportunity, especially by capitalising new images and simulations of nature—equated with beneficent, wholesome qualities. Advertisements emphasised the 'natural' status of food products and additives, while concealing their technological origins. R&D strategies have been promoted as simulating 'natural' origins, methods and characteristics of products—while often claiming proprietary rights for the human artifice involved. In the 1980s Monsanto celebrated genetic engineering as 'a natural science', this language appealing to the natural recombination of genetic material (Monsanto, 1984). When people question whether GM food is 'natural', therefore, they have taken their cue from industry messages about other food products. As a philosopher sardonically described this paradox, 'Only the most sophisticated technology will assure your product a clean, all-natural label. The natural is patentable' (Sagoff, 2001: 129).

In the USA, food products from GM crops have been an exception to special labelling or advertisements. According to the agbiotech industry, a 'GM' label would wrongly imply that GM techniques affect the safety of such food products (ibid.: 131). The US FDA readily accepted this rationale. When some state governments proposed to require a GM label, moreover, the FDA sought to prevent such a rule.

Non-labelling of GM food opened up a market opportunity for organic food, but its non-GM status came under attack. For a long time, agribusiness had attempted to incorporate organic farming into large-scale monocropping systems. To facilitate this process, in 1997 the USDA proposed a rule which would permit GM crops, sewage sludge and food irradiation within certified 'organic' production. The proposal was widely attacked—as a Trojan horse imposing an alien agenda on organic farming, and was defeated through a mass protest campaign (Rigby and Bown, 2007: 88–89).

Since the late 1980s, US environmental NGOs have questioned whether agbiotech would bring environmental improvements (Krimsky and Wrubel, 1996). In particular, they denounced herbicide-tolerant crops as a 'threat to sustainable agriculture'. From their critical standpoint, such crops would extend farmers' dependency upon herbicides, while potentially supplementing the familiar 'pesticide treadmill' with a 'genetic treadmill' (BWG, 1990). Environmental NGOs counterposed systems approaches, such as Integrated Pest Management. IPM 'could avoid the need for the majority of pesticides, now and into the future'—rather than fund GM research to develop 'better' pesticides (Mellon, 1991). From that perspective, they challenged GM crops for addressing the wrong problem: safely extending intensive monoculture.

US agricultural policy has promoted the intensive productivist model for which GM crops are designed. When the US government eventually took up sustainability discourses, these generally diagnosed agricultural problems as inadequate or unstable productivity. According to the Department of Agriculture, meeting the central challenge of sustainability 'is essential if USDA is to be a successful partner in sustaining the people and productivity associated with U.S. agriculture and natural resources' (USDA, 1996). As a solution, new inputs would exploit natural capital to enhance productivity for economic competitiveness.

US research policy likewise has promoted input-output efficiency as sustainability. This means reducing herbicide usage and selection pressure for herbicide-tolerant weeds, which could undermine the efficacy of GM herbicide-tolerant crops. Thus sustainability meant maximising the lifespan of tools for efficient agri-industrial production.

Within US policy, agbiotech is regulated under normal product legislation, within existing statutory-bureaucratic frameworks. 'Risk-based' policy discourse served to avoid any new legislation, to foreclose opportunities for judicial review and to limit any influence by the dissenting public (Jasanoff, 1995, 2005a). In particular, the Office of Science and Technology Policy issued a Coordinated Framework for biotechnology regulation. It set guidelines for classifying GMOs according to their product categories, for assigning each category to the relevant Federal agency. Under the Coordinated Framework, GMOs would be regulated (if at all) like their unmodified counterparts, under existing product legislation. As partners of industry, regulatory agencies symbolically normalised products as 'safe', thus complementing a monocultural system (Levidow and Carr, 2000).

Through the above policies, agbiotech was co-produced with specific forms of nature and society. US policy combined a globalisation discourse with an authoritarian technological determinism. As seen in the US case, a narrow definition of risks can justify 'a specifically neoliberal linkage between the freedom of science and the freedom of capital'. Within this framework, 'neoliberal strategy proves flexible and adaptable' (Barben, 1998: 410, 417).

2 AGBIOTECH FOR INDUSTRIALISING EUROPEAN AGRICULTURE?

In Europe the US model of intensive agri-industrial production was appropriated as an inevitable future. Agbiotech symbolised European progress through a clean, precise technology. In the 1980s the European Commission developed a narrative linking technology policy with economic, market and environmental objectives (see Table 2.2).

A biotechnological vision was promoted as an overall solution to several problems: European companies could not adequately compete in an increasingly global market, so they must be converted into competitive multinational companies. For this economic aim, an essential tool would be the application of modern biotechnology to European agro-food industries. According to a 1982 Commission policy document, our 'Biosociety' increasingly depends on sustainable processes and recyclable products,

Table 2.2 Commission Agbiotech Policy as Ecological Modernisation

Discursive framing	*Early 1990s*	*Since Late 1990s*
Innovation Policy Framework Programme (research funds)	Europe needs agbiotech for economic competitiveness, eco-efficiency, clean technology, etc. (CEC, 1993a) Prioritise 'pre-competitive' biotech research and (later) sectoral convergence with pharmaceuticals through 'Life Sciences'.	Europe needs pro-active policies to exploit biotech in a responsible manner, consistent with European values and standards (CEC, 2002b) Prioritise research on 'biotech for health', and novel crops for non-food uses.
Problem-closure of technological solutions	Agro-food industry is 'based on biotechnology'—hence necessity to accommodate market pressures for greater productivity and efficiency (CEC, 1993a)	GM crops can generate 'more sustainable agricultural practices'. Regulatory oversight expresses societal choices regarding agbiotech (CEC, 2002a, 2002b).
Regulatory aims and criteria	Scientific uncertainty, e.g. about GMOs upsetting 'the delicate balance existing in nature', justifies regulatory controls (CEC, 1993b). Regulatory burdens on applicants should be proportionate to the identified risks (CEC, 1994). GM crops must avoid harm to human health and the natural environment.	Through technical precaution, scientists evaluate the available knowledge and may request more data. Risk assessment should also include agri-environmental effects; it should explain uncertainties about any identified risk (EC, 2001).
Social accommodation (mainly by DG-Environment)	EU regulatory procedure includes opportunity for public comment on GM products (EC, 1990). Environmental NGOs attend discussions on risk-assessment methods and criteria. NGOs receive funds to hold conferences on regulatory issues.	EU regulatory procedure requires public consultation (EC, 2001). In deliberative fora, NGOs counterpose alternatives to agbiotech—choices which lay outside EC policies and regulatory procedures.

Note: This table shows temporal shifts relevant to later chapters of this book. The categories in the left-hand column are adapted from Hajer (1995: 266).

especially of agriculture; modern biotechnology can extend our control over such processes (CEC, 1983b). In this way, the Commission problem-diagnosis promoted a specific future scenario requiring biotechnology (Gottweis 1998: 170).

Policymakers associated biotech with productive efficiency for economic competitiveness. The Commission identified biotechnology as necessary to create wealth through more efficient innovation; its new language blurred any distinction between conventional and GM techniques. According to a 1993 White Paper of the Commission, 'biotechnology' has a direct impact on sectors which comprise 9% of value-added in the EU. The global revenues of 'the biotechnology industry' would reach 100bn ecu by the year 2000; 'perhaps only modern biotechnology has the potential to provide significant and viable thrusts. . . . ' (CEC, 1993a). Therefore its maximum exploitation must be facilitated by removing any obstacles.

Reinforcing a neoliberal policy framework, moreover, the 1993 White Paper counselled European adaptation to inexorable competitive pressures: 'The pressure of the market-place is spreading and growing, obliging businesses to exploit every opportunity available to increase productivity and efficiency' (CEC, 1993a: 92–93). This imperative was linked with innovations such as biotechnology: 'The European Union must harness these new technologies at the core of the knowledge-based economy' (ibid.: 7). The entire agro-food industry became discursively 'based on biotechnology', i.e. economically dependent upon the products of genetic modification, as essential tools for the future (ibid: 100–103). Biotech was promoted as a clean technology, epitomising 'the new development model':

> [This aims to] decouple future economic prosperity from environmental pollution and even to make the economic-ecological relationship a positive instead of a negative one. The key for doing this will ultimately lie in the creation of a new 'clean technology' base. (ibid.: 147)

In parallel, EC environmental policy likewise sought new technology which could provide efficiency gains while achieving environmental objectives (see Table 2.3). 'Sustainable development' was conceptualised in eco-efficiency terms, as theorised by ecological modernisation (see Chapter 1):

> Many of the new clean and low-waste technologies not only reduce pollution substantially, but also economise on the consumption of raw materials to such an extent that cost savings can more than offset initial higher investment costs and thereby reduce unit production costs. A case in point is represented in the development and use of new techniques in the field of genetic engineering and biotechnology; these offer considerable potential for useful applications in agriculture, food processing, chemicals, pharmaceuticals, environmental clean-up and the development of new material and energy sources. (CEC, 1993b: 28)

Table 2.3 Sustainable Development in EC Environmental Action Programmes (EAPs)

EC environmental policy has elaborated 'sustainable development' in terms of eco-efficiency (as theorised by ecological modernisation; see Chapter 1). Examples from Environmental Action Programmes:

3rd EAP

Environmental resources 'are the basis of—but also constitute the limits to—further economic and social development and the improvement of living conditions' EAP (CEC, 1983a: 3).

4th EAP

'. . . the protection of the environment can help to improve economic growth and facilitate job creation' (CEC, 1987: 4).

5th EAP

'. . . it is essential to view environmental quality and economic growth as mutually dependent' (CEC, 1993b: 28). 'The concept of shared responsibility requires a much more broadly based, active involvement of all economic players including public authorities, public and private enterprise in all its forms, and, above all, the general public, both as citizens and consumers' (ibid.: 26).

Review of 5th EAP

Aims to achieve 'sustainable industrial development, involving the formulation of the concept of ecoefficiency, and partnerships between governments and industry, using industry's capacity for innovation' (CEC, 1998a: 5).

6th EAP

'Business must operate in a more ecoefficient way, in other words producing the same or more products with less inputs and less waste, and consumption patterns have to be more sustainable' (CEC, 2001b: 3). Those perspectives informed EU research priorities. According to the Fourth Framework Programme: 'In particular, efforts will be made to identify the science and technology options with the most favourable impact on growth, competitiveness and job creation in Europe' (CEC, 1994). This agenda equated 'favourable' with 'competitive'. The Programme also mentioned environmental, health and ethics issues—mainly to accommodate the European Parliament.

Priority was given to genetic characteristics which could enhance productive efficiency. State-funded research was meant to be 'pre-competitive', i.e., developing knowledge that would facilitate competitive innovations, rather than create products. R&D funding on molecular-level knowledge was prioritised through special budgets—e.g. the Biotechnology Action Programme in the 1980s, BRIDGE in the early 1990s, Biotechnology in the mid-1990s, and the Life Sciences and Biotechnology programme in the late 1990s. 'Life Sciences' denoted strategies for industrial integration—between seeds and agrichemicals companies, as well as between agricultural supply and pharmaceutical companies. By merging companies, linking R&D programmes and focusing on the molecular level, synergies were expected in identifying valuable genes and chemicals.

Molecular-level information had already been prioritised for medical and agricultural research funded by the European Commission. According to an advisor to the Fourth Framework Programme, Europe must develop bio-informatic technology for processing gene sequences, and must achieve better 'understanding of how genes work'. With such knowledge, European policy could offer 'added value' to the life sciences through inter-European teams, research infrastructure and training (DG XII, 1994: 8–9). In supporting this research agenda, the European Parliament added a proviso: 'that human life is not a marketable commodity and that there can be no commercial competition in this sphere' (CEC, 1993c: 304). No such constraint was put on plant material, though patent rights remained contentious (see next section).

In the name of the 'knowledge economy', privatisation of natural resources was promoted by the 1987 Single European Act and subsequent Maastricht Treaty. Aiming 'to complete the internal market' by 1992, these treaties redefined both the state and the environment as instruments of marketisation. In that vein, market metaphors informed a new research policy. According to the European Commission, citing both treaties, public authorities 'must bring about the creation and maintenance of an overall economic "environment" and a respect for free competition, which is necessary so that firms can effectively develop supply policies' (CEC, 1992).

By 1990 Commission funds for biotech research became dependent upon industry partners committing resources to any project proposal. Research was given a clear economic function, with 'more careful attention to the long-term needs of industry', according to managers of the DG-Research Biotechnology Division (Magnien and de Nettancourt, 1993: 51). Here they reflect on their policy mission:

> [In the 1980s] the EEC engaged in a battle against time for reinforcing its industrial competitiveness, and research was clearly seen as one of the strategic ingredients of a recipe to rejuvenate industrial activities. . . . (ibid.: 50)
>
> Knowledge remains the principal and most limiting factor of innovation, as opposed to technology, which becomes more and more widely accessible. The most vital resource for the competitiveness of

> the biotechnology industry is the capacity to uncover the mechanisms of biological processes and figure out the blueprint of living matter. (ibid.: 53)

This research agenda conceptualised nature as an information machine whose deficiencies had to be corrected, as an essential means towards European industrial regeneration and competitive advantage.

Molecular knowledge was sought as a key to industrial competitiveness. Biotechnology was promoted as essential for making precise genetic changes which could safely protect crops and enhance agricultural production. This R&D agenda complemented the wider aim to 'industrialize agriculture', in the words of a lobby group for crop biotech (GIBiP, 1990).

Environmentalist groups attacked this entire agenda, especially for aiming 'to convert agriculture into a branch of industry', according to a Green MEP (Haerlin, 1990). Critics foresaw that agbiotech would undermine farmers' independence and livelihoods, while taking agriculture further along a misguided route. It was developing single-gene solutions for problems which derive from a monocultural farming system, designed for industrial models of efficiency (Genetics Forum, 1989). Moreover, novel crops could disadvantage farmers who use more traditional methods, or could displace substances hitherto imported from tropical countries, thus jeopardising rural livelihoods (e.g. Hobbelink, 1991).

Replying to such criticism, an industry lobby group invoked objective imperatives of market competition and even democracy:

> Let there be no illusions: as with any innovative technology, biotechnology will change economic and competitive conditions in the market. Indeed, economic renewal through innovation is the motor force of democratic societies. (SAGB, 1990: 15)

In that discourse, democratic powers are attributed to technological and economic progress; market forces become the natural regulator, thus naturalising any disruption to livelihoods (cf. Williams, 1980).

Indeed, the biotech industry anticipated and emphasised greater market pressures on agriculture, as an imperative and opportunity for a biotechnological solution. EC agricultural subsidies would be reduced and would lose their former link with production. In the view of many company managers, liberalisation of markets and reduction in CAP payments would continue, and this would be good for future product development, as an opportunity to sell inputs to farmers (Chataway et al., 2004: 1053).

Within such policy frames of agbiotech as progress, GM crops would protect Europe from economic insecurity and agricultural-environmental problems. Thus agbiotech promoted a specific problem-diagnosis of inefficient agriculture, for which precisely redesigned GM crops would provide the solution. As a potentially self-fulfilling prophecy, the eco-efficiency discourse-coalition could facilitate a co-production of agbiotech, nature and society.

3 NATURALISING AGBIOTECH, SIMULATING THE NATURAL ORDER

Agbiotech developments generated dissent from the start on both sides of the Atlantic. By contrast to the USA, in Europe views became more polarised between eco-efficiency versus apocalyptic frames, especially in Germany (see Table 2.1). In several countries, early opposition elaborated arguments that would later gain prominence.

In Europe the agbiotech debate intersected with a wider debate about the systemic hazards of intensive monoculture, e.g. agrichemical pollution, plant pests, pest resistance, etc. Agbiotech proponents attributed these problems to deficient inputs, which therefore must be corrected by editing their genetic information, thus making agriculture more efficient and clean. As they search for molecular knowledge for eco-efficient production,

Figure 2.1 Double-helix money tree (Credit: LGC/DTI, 1991).

proponents have often used metaphors of computer codes, combat and commodities. By attributing human qualities to natural characteristics of things, the biotechnological project naturalised a specific societal future.

Promoting biotechnology investment, the British government portrayed the DNA double-helix as a money tree, sprouting £5 banknotes (LGC/DTI, 1991: 26). Such metaphors are not merely rhetorical; rather, they represent literal investments in nature. R&D was directed towards selecting genes that could increase the market value of products.

According to European proponents, biotechnology provides 'natural methods' which avoid or even remedy agrochemical pollution. With combat metaphors, biopesticides were symbolised as a green bow-and-arrow, with the punny caption, 'Fighting for a better world, naturally', as in an advertisement from the Danish firm Novo Nordisk. In a similar vein, GM biopesticides 'attempt to do better than mother nature in designing improved, more efficacious toxins . . . ', according to the chief of a US company (Goodman, 1989: 52).

Eventually similar genes for insecticidal toxins were inserted into crops; likewise genes for herbicide tolerance. As a euphemism for GM crops, often these were called 'improved' or 'smart' seeds. According to biotechnologists, herbicide-resistant crops were genetically modified for relatively less persistent, low-dosage herbicides, and the products would reduce the need for pre-emergence spraying. Thus herbicide-resistant crops will 'reduce the dependence upon herbicides which have already given cause for concern', according to ICI Seeds (Bartle, 1991: 11). Here farmers' dependency on purchased inputs, albeit needing lower quantities, was portrayed as liberation.

Europe-wide opposition emerged in the late 1980s around the European Parliament's Rainbow Group, later called the Green Group. From an apocalyptic frame, these MEPs campaigned against biotechnology as an irrevocable, ominous, even immoral rupture: GMOs pose 'social and economic risks, as well as risks to our world view and culture'. As their brochure warned, 'This way forward can never be reversed!' 'Genetically modified nature' was depicted as pollutants, labelled as such with biohazard symbols (Rainbow Group, 1989). Soon Friends of the Earth Europe started an anti-biotechnology programme with a regular *Mailout*.

Those EU-level campaigns drew upon and coordinated efforts across several member states. In Germany the *Genethicsnetzwerk* opposed all biotechnology. In Denmark the FoEE affiliate Noah started a public debate on agbiotech, raised issues of sustainable agriculture and campaigned for strict regulatory controls. UK groups warned that GM crops would aggravate the socio-economic dependence and environmental hazards endemic to intensive monoculture (Genetics Forum, 1989). Although Greenpeace UK had no agbiotech campaign until the mid-1990s, earlier its staff characterised GMOs as 'self-reproducing pollutants' which could intensify harmful processes in agriculture.

In sum, agbiotech was variously cast as a promise or threat through political discourses which appealed to natural characteristics. Proponents invested nature with capitalist metaphors of efficiency and commmodities, thus framing agbiotech as a beneficent simulation of natural qualities. Through alternative discourses, agbiotech critics framed both nature and agriculture as vulnerable to agro-industrial methods, for which Europe needed alternatives. This polarisation signalled potential obstacles to the agbiotech project in Europe.

4 PRIVATISING LIFE AND PUBLIC RESOURCES

As a neoliberal co-production, agbiotech depended upon measures to privatise living material and public resources. Proprietary rights over genetic material were resolved in principle in 1980 in the USA—but not until 1998 in the EU, and in an unstable way. European public-sector research too was partly marketised through various incentives and pressures, especially in the agbiotech sector.

4.1 Whose Biopiracy?

As reproducible living material, seeds have been widely regarded as a common resource but have also been targeted as prospective commodities. Since the early 20th century, commoditisation has been achieved through hybrid seeds, which do not breed true and so must be bought anew each season. Since the 1980s, moreover, GM techniques have been presented as a rationale for patent rights on seeds.

In the ensuing controversy, 'biopiracy' became a common term for the theft of genetic resources—with two opposite meanings. For advocates of greater patent rights, 'biopiracy' means violating the rights of an inventor, by using patented materials without a licence agreement or without paying royalties. For opponents of such rights, 'biopiracy' means the patents themselves, on grounds that biological material should remain freely reproducible as a common resource.

The dispute entails a legitimacy problem for property rights in living material. For advocates of greater patent rights, scientists make a significant contribution to crop transgenes and improvement techniques; through GM techniques, transgenes can be invented and fully described as chemical molecules, by decoding and recombining their genetic information. For opponents, such changes are discoveries, appropriations or simulations of common resources, which have been already selected and cultivated by farmers over many generations.

These concepts have legal significance because, according to patent rules, the applicant must have contributed an 'inventive step'. Interpreting that criterion loosely, the US Patent Office initially accepted broad patent claims—e.g. on genes inserted into GM plants, on the techniques for their insertion and on substances derived from plants traditionally cultivated in developing countries, e.g. from the neem tree. The US government has sought to extend its broad patent criteria to other countries, especially by using the Trade-Related Aspects of Intellectual Property Rights (TRIPS) rules under the WTO.

Biotechnology companies claimed that plant patents are needed as an incentive and reward for improving crops. After all, they should be able to benefit from the progeny of GM crops or else restrict their use. Indeed, 'legal protection of intellectual property serves the public interest by stimulating continued investment in technological innovation', according to John Duesing (1989) of Ciba-Geigy (later Novartis).

In Europe the appropriate breadth of patent rights has been long disputed. The 1973 European Patent Convention made two major exemptions: patents were prohibited on 'plant or animal varieties or essentially biological processes' for their production, as well as inventions whose use would be contrary to public morality. Both these criteria became contentious for specific applications submitted to the European Patent Office (EPO). For example, the EPO initially granted a patent on a process for creating a herbicide-tolerant crop, but later in 1995 accepted a Greenpeace argument that GM plants were really varieties and therefore exempt. Moreover, EC member states had their own national rules, some at odds with EPO decisions.

Such disputes led to proposals for new legislation to clarify criteria for property rights in genetic material. Another rationale for a new law was the internal market: if not all EU member states accepted an EPO decision, then a patent might not be enforceable for goods traded across national boundaries. The need for European harmonisation also provided an opportunity to strengthen property rights along lines similar to US law.

In 1988 the Commission proposed a Directive granting property rights in 'biotechnological inventions', whose title incorporated a basic concept from the agbiotech industry. Its proposal gave an extra rationale:

> Failure to provide such protection for intellectual property will drive firms to protect themselves by commercial secrecy. Such secrecy will inhibit precisely the collaborative patterns of activity which are needed in this complex inter-disciplinary field. (CEC, 1988)

Thus ambiguous, diverse patent rules were represented as a threat to cooperative relations, even though stronger patent rules in the USA had stimulated more competitive relations, especially within the public sector.

In response to the Commission proposal, opponents raised the slogan, 'No Patents on Life!' Their campaign appealed to popular concepts of 'life'—perhaps not 'nature', but rather biological material as a common resource, counterposed to property. Opponents warned against several

Figure 2.2 'Criminalise biopiracy!' (Credit: *Roots* Newsletter, January–February 2003, copyleft ASEED Europe).

harmful consequences: the Directive would provide an incentive for companies to use GM techniques rather than other methods of improving seeds; plant patents would deter seed improvement, especially of non-GM seeds; and the mere prospect of litigation could deter other plant breeders from using the germplasm of GM crops (e.g. Corner House, 1997; Global 2000, 1997).

In 1995 the draft Directive was expected to prevail in the European Parliament, but it was defeated in the plenary session. The defeat resulted from strong opposition, led by the Green Group of MEPs and eventually supported by much of the Socialist Group. A broad NGO coalition generated thousands of protest messages to MEPs. The campaign appealed to public anxieties about unnatural manipulation of life, playing God and corporate greed (Jasanoff, 2005a: 220).

Seed companies, many of them already bought up by agrichemical companies, again made pleas for extending patent rights to seeds. According to a European lobby group, GM techniques enhance biodiversity and thus the public good:

> Seed companies contribute to the objectives of the Biodiversity Convention through maintaining and improving biodiversity through the maintenance of genebanks and by using genetic resources for the

> benefit of mankind. They are creating a new biodiversity by recombining genomes through traditional crossing and adding new attributes via genetic engineering. For this it is of utmost importance to be able to patent inventions in the kingdom of plants. (FEBC, 1997)

With such language, discovery of a common resource was framed as an invention, even of biodiversity; such improvements warranted proprietary rights. According to a representative of a major pharmaceutical firm, Smith Kline Beecham, 'Genes are the currency of the future' (cited in Emmott, 2001: 378). Thus new discourses naturalised the commoditisation of nature as a patentable human artifice.

After the 1995 Parliamentary defeat of the original Directive, the text was revised. The revision was presented as a 'harmonisation measure', as if it merely formalised existing practice across EU member states. In the EU's Council of Ministers the Netherlands voted against, given that its national law prohibited patents on plants or animals. Opposing the redraft, an NGO coalition used a skull-and-crossbones symbol with the slogan, 'No to biopiracy!'

Following a ten-year-long debate, and intense lobbying by proponents, the new version was finally enacted. The 1998 EC Patent Directive allows broad patents on 'biotechnological inventions', even if they comprise new varieties of plants. The wording exempted any mere 'discovery', but patent rights apply even for 'a product consisting of or containing biological material or a process by means of which biological material is produced, processed or used'. Even more explicitly, 'biological material which is isolated from its natural environment or produced by means of a technical process may be the subject of an invention even if it previously occurred in nature'. Thus invention was broadly defined: 'Inventions which are new, involve an inventive step and are susceptible of industrial application are patentable even if they concern a product consisting of biological material' (EC, 1998a).

The final version made some concessions to critics. Under a farmers' privilege, farmers could resow seed for their own use or non-commercial exchange, regardless of any patented gene. But the seed vendor could restrict resowing through a contract with farmers, as was already happening in the USA. The Directive also established an ethical review by an EU-level expert body, though this was widely understood as relevant to human biological material, not plants.

The 1998 Patent Directive has remained contentious, especially regarding human material but also GM seeds. In 1999 the Council of Europe's Parliamentary Assembly declared that material derived from plants, animals or humans cannot be considered as inventions, 'nor subject to monopolies granted by patents'. According to an NGO lobby group, 'The Directive now legalizes biopiracy' and so should be overturned. This accusation was denied by the Commission's expert group in a 2001 statement:

> Patents on genes do not apply to elements in their natural environment, but only to molecules isolated from the human body or produced by means of a technical process. As such, the phrase 'patenting of life' is misleading and should be avoided. (quoted in DG Research, 2004: 14)

NGOs continued to attack 'biopiracy and bioprospecting for plants', especially patents on substances traditionally extracted from plants in the global South (e.g. Genewatch UK, 2000: 6). Proponents had invoked a computer-code metaphor to justify patents on 'genetic information'; in response, critics characterised human DNA as 'pre-existing information that has been discovered and not invented' (cited in McGiffen, 2005: 43). While the industry claimed that patent rights encourage research, many scientists argued that broad patents were deterring research.

The 1998 Patent Directive came under a legal challenge by some member states—the Netherlands, Norway and Italy—partly on grounds that it contradicted their own national laws. The challenge was rejected by the European Court of Justice in 2001, though this did not resolve the conflict over the Patent Directive. Its legal force depends upon transposition into national law, which had been done in only six of the fifteen EU member states by early 2003 (CEC, 2003d: 15). Some member states were taken to the European Court of Justice for infraction procedures by the European Commission. Likewise such action was taken against two of the 2004 accession countries (CEC, 2005a).

In sum, biotech proponents designed the 1998 Patent Directive to harmonise proprietary rights for a European internal market, but such harmonisation remained elusive. The Directive did not entirely clarify how the European Patent Office should distinguish between a discovery and an invention, i.e. between natural and artificial forms of biological material. The Directive left interpretive ambiguity and thus legal uncertainty for patents related to GM crops. Nevertheless the Directive has provided an extra incentive to prioritise research which could obtain patent rights and gain royalties, in a policy context which blurs boundaries between the public and private sectors. At the same time, the anti-patent 'biopiracy' slogan stimulated unease about agbiotech within Left and trade-union circles which had normally supported technological innovation as progress.

4.2 Public-Private Boundaries Blurred

Until the 1980s agronomic methods were driven largely by inputs from commercial suppliers or from public-sector research conducted by agricultural universities and other institutions. Such influence formed part of a broader 'agricultural corporatism' between government and farmers, though this was already in decline in some countries (Frouws and Tatenhove, 1993). The public sector once offered farmers alternative inputs or methods, yet its role has increasingly merged with the commercial sector.

Under pressure to become more 'demand sensitive', public-sector research establishments (PSREs) have moved towards a more business-oriented organisational culture. In general, 'public agricultural scientists are being encouraged to pursue non-traditional sources of funding such as outside departments or ministers of agriculture in national governments and private corporations and producer (commodity or cooperative) groups', argue Huffman and Just (1999: 8). According to their study of agricultural research, 'The private interests of companies and commodity groups are seldom well aligned with the social good or public interest. Thus, joint public-private research ventures frequently create a major conflict with the interests of taxpayers supporting public agricultural research organisations' (ibid.: 16).

This public-private blurring became an imperative for PSREs, not just an opportunity. As some governments reduced core funding for research, public-sector scientists had little alternative but to seek greater funds from private sources. Moreover, some public funding was shifted from PSREs to universities, which can more rapidly move into new areas and can more readily utilise short-term funding, e.g. by employing contract researchers.

PSREs in Europe have undergone a general pressure to contribute to national economic performance by becoming more responsive to the needs of industry users, in turn providing a basis for private finance. Core funding was reduced or transferred to output financing, i.e. dependent upon competitive bidding for specific projects. Implicitly this simulates a 'market', with government agencies or industry partners acting as customers. In some cases, PSREs have become entrepreneurs in their own right, as potential partners or even competitors with companies. These European changes went the furthest in the UK and the Netherlands (Levidow et al., 2002).

As an extreme case, the UK system of crop research has been studied from a public-interest perspective. According to an NGO report:

> The distinction between public and private has also become increasingly blurred: public research is contracted out to private companies; public research organisations follow industry trends towards short-term staff contracts . . . This means less long-term research. It also makes it more difficult for public researchers and organisations to pursue the public interest. (FEC, 2004)

As a result of all the above changes, PSREs have faced conflicting pressures. On the one hand, PSREs are expected to contribute to national economic performance by becoming more responsive to the needs of industry users, in turn providing a basis for private finance. This has meant favouring research which can readily be turned into patents and/or commodities. For example, they may set up spin-off companies to commercialise the results of fundamental research financed by the public sector. In particular PSREs are expected to assist the major private-sector innovation trajectory involving biotechnology. Even a small proportion of industry funding can influence overall research priorities: the tail can wag the dog.

On the other hand, PSREs are expected to use their knowledge and expertise in the public interest. Such roles include the following: developing new techniques and products which have public benefits but are unlikely to be commercially attractive, e.g. Integrated Crop Management (ICM); devising risk and monitoring systems, thus performing a public 'watchdog' role and informing regulatory controls; and evaluating agri-technological developments from the perspective of potential risks and benefits.

Increased private-sector involvement undermines the public-service roles of PSREs and their public credibility as independent experts (Levidow et al., 2002).

Consequently, EU advisory expertise may have difficulties, according to a report by leading advisors in the late 1990s. When appointing scientific advisory committees, governments encounter greater difficulty in recruiting members who are independent of private interests, in at least two senses. First, even if scientists are not employed by industry, their careers and institutes are often dependent upon industry contracts. Second, their individual standing depends upon gaining such contracts through competitive tenders, conducting prestigious research, publishing high-quality papers, etc. Consequently, few are willing to serve on advisory committees. Given these pressures, 'it will prove increasingly difficult to recruit top-flight scientists', and members tend 'to consider themselves operating in a consultancy mode', therefore requiring substantial remuneration, according to leading members of EU-level scientific committees (James et al., 1999: 66). Indeed, the blurred boundary between public and private roles would eventually become a legitimacy problem in the GM debate (see Chapters 3 and 5).

5 NORMALISING AGBIOTECH THROUGH RISK REGULATION

For its success in Europe, the agbiotech project would depend upon predictable routes to regulatory approval, as a means to normalise the technology as a series of safe products. Like the US regulatory procedure, 'product-based regulation' had strong European advocates. But this policy lost out to alternative proposals, taking a more cautious approach to the possible risks and societal conflicts which may need management. In its own way, nevertheless, the EU system offered a potential basis 'to make Europe safe' for agbiotech.

5.1 Internal Market Impetus for Legislation

Early on in the European development of agbiotech, there were diverse narratives regarding its potential risks and role in sustainable development. EC regulatory proposals drew upon concepts of ecological modernisation, seeking technological solutions which anticipate and thus avoid potential damage in advance. 'Environmental protection was represented as a positive sum-game and, hence, as an administrative problem', to be solved through statutory regulation along with citizens' participation (Gottweis 1998: 232–34). This managerialist frame was elaborated mainly by DG XI for Environmental Protection.

In the late 1980s EC policy discussions on agbiotech regulation focused on how to facilitate the internal market for agbiotech products. Some Commission officials cited 1986 OECD recommendations, which saw no basis for statutory regulation of GMOs, on grounds that the GM techniques pose no novel hazards. By analogy to the USA, therefore, GMOs could be adequately regulated under 'vertical', product-based laws of the relevant sector, e.g. pesticides, pharmaceuticals, crops, etc. For example some staff in the EC's DG XII-Research advocated a general law on novel organisms, whereby regulatory judgements would be guided by an expert body under industry influence.

By contrast, the Environment Directorate-General XI advocated legislation covering all products of the GM process or technique. In its view, this 'horizontal' process-based framework was needed to evaluate any uncertainties about risk. Such legislation also would be necessary to avoid or overcome trade barriers, which could otherwise result from the restrictive legislation of some member states. In particular, strong anti-agbiotech movements in Germany and Denmark were pressing for statutory restrictions there, as cited by Commission staff (ibid.).

In the face of these problems, new legislation was proposed as an instrument of the European internal market. European Commission staff advocated this 'horizontal' framework as a basis for harmonising EU-wide regulatory criteria (Levidow et al., 1996). They even portrayed such a framework as a basis for eventually overcoming transatlantic regulatory differences (Jasanoff, 2005a: 82). Indeed, trade liberalisation and regulatory harmonisation agendas drove legislative frameworks; these more readily relegated decisions to expert bodies, thus limiting democratic accountability for regulatory standards. This tension between policy aims pervaded conflicts over statutory arrangements and their interpretation.

From the start, Commission proposals for new legislation linked scientific and political uncertainty. Draft directives treated GM techniques as a novel process generating novel products which therefore warrant special risk-assessment procedures, as well as EU-wide rules to avoid trade barriers. On this dual basis DG XI-Environment gained support from DG III-Internal Market for draft Directives regulating the contained use and deliberate release of all GMOs (Gottweis, 2005).

Most contentious was the draft Directive on the Deliberate Release of GMOs. Concerns about scientific uncertainty led the European Parliament to support relatively stringent legislation. The Commission's original version relegated GM products to separate legislation for similar ones at the commercial stage, potentially involving less scrutiny of risks. The Parliament strongly supported the DG XI proposal for horizontal process-based regulation and even voted for extending it from the experimental stage to the market approval stage of GM products (Lake, 1991). Such amendments were led by the Rainbow Group (later the Greens); their MEPs opposed any release of GMOs but pragmatically advocated stringent regulation, which gained wider support in the Parliament.

The 1990 Deliberate Release Directive 90/220 was characterised as a 'preventive' measure, which really meant 'precautionary', a term not yet part of Commission policy. According to the Directive, 'completion of the internal market would be based on a high level of protection for the environment and human health'. Every deliberate release had to undergo a prior risk assessment and authorisation. Member states had a duty to ensure that GMOs do not cause 'adverse effects' (EEC 1990). Together all those features provided a potential means to evaluate uncertain risks and to shape the technological trajectory in environmentally more beneficial ways, partly depending on how authorities interpreted its criteria, e.g. 'adverse effects' to be prevented.

However, regulatory means to influence innovation remained implicit and marginal in Commission policy. Such limitations can be seen in official policy frameworks, such as in the Fifth Environmental Action Programme. On the one hand, the document emphasised the problems of intensive agriculture and thus the need to develop more extensive methods of cultivation (CEC 1993b: 35–36). On the other hand, the same document framed biotechnological risks in a naturalistic way, separated from such agronomic issues. For example, GMOs offer environmental benefits, but 'there are concerns that this new technology might entail potential risks . . . GMOs could upset the delicate balance existing in nature or even have evolutionary impacts' (ibid.: 60).

Some policy documents emphasised the prospect that GMOs might cause an ecological imbalance, popularly known as 'running out of control'. The 'imbalance' concept provided scope for ecological risk assessment, even for precaution. Through prior risk assessments and approval, a regulatory procedure could evaluate and symbolically vindicate GMOs of this suspicion. This risk-frame did not consider agri-ecological issues, i.e. systemic effects on farming systems. Within the Commission's policy framework, the Directive could more readily normalise agbiotech products within an intensive monocultural model.

5.2 Pressures for Regulatory Approval

Agbiotech regulation was further limited by political pressures hostile to the implicit precautionary basis of the Deliberate Release Directive. The chemical industry organization, CEFIC, set up the Senior Advisory Group on Biotechnology in July 1989. SAGB represented major agrochemical and pharmaceutical multinationals which were investing heavily in biotechnology. It had unsuccessfully lobbied for the original weaker version of the Directive and soon attempted to reverse that defeat.

Shortly after the Deliberate Release Directive was enacted, SAGB criticised 'a tendency to think of "biotechnology" as fundamentally different from other technologies and therefore requiring special rules'. SAGB advocated 'product-based regulation' on the US model, as grounds to 'apply existing, non-discriminatory approaches for safety in research and industrial processes'. This would mean, for example, assessing organisms according to 'the inherent characteristics of the product'. According to SAGB, such approaches must draw upon 'professional competence of the highest standard', thus implying a deficiency in DG-Environment. SAGB argued that more favourable government policy in the USA and Japan was helping Europe's major competitors in biotechnology; indeed, over-regulation was driving its member companies out of Europe to the USA and Japan (SAGB, 1990).

In reality, the largest relevant companies were already transatlantic, so their mutual competition was being misrepresented as an economic threat to 'Europe'. In the guise of protecting Europe, SAGB was addressing quite different problems: a transatlantic company could find that its safety claims are subjected to more stringent scrutiny in Europe than in the USA. As a remedy, a shift to vertical legislation could help to minimise any extra demands upon companies.

Despite that transatlantic business agenda, the mainstream debate was framed in terms of European disadvantage. The EC's horizontal legislation was put on the defensive for impeding European innovation and investment in R&D. Such attacks were encouraged by some Commissioners and staff. Using a judicial metaphor, one later spoke about 'documented charges that over-regulation was rendering companies uncompetitive and driving investment out of the Community'; moreover, 'the emphasis on regulatory activities focused on GMOs was itself reinforcing the message of inherent riskiness, and stigmatising the technology' (Cantley, 1995: 637, 644). At the EU and national level, formal hearings solicited such accusations.

As a high-profile example, a UK Parliamentary Committee asked one-sided questions such as, 'Is there a danger that the present regulatory regime will prevent the exploitation by British industry of research conducted in the UK science base?' Perhaps predictably, its verdict found the regime guilty, along the lines of SAGB's policy frame:

> Unfortunately, in our view the UK regulations, which are in turn based on EC Directives, take an excessively precautionary line. . . . We find that the current regulatory regime is unscientific . . . any regulations which reduce competitiveness must be reviewed critically, especially when it cannot be justified on scientific or public interest grounds. (HoL, 1993: 10)

Accommodating that diagnosis, the Commission proposed a remedy: 'risk-based regulation' (CEC, 1993a). This slogan was soon elaborated as 'the need for balanced and proportionate regulatory requirements commensurate with the identified risks' (CEC, 1994), a phrase implying that all risks could be readily identified, as in the conventional meaning of risk assessment. According to this policy, any regulatory burdens on companies must be justified by prior evidence of risk, not simply by uncertainty.

Under such pressure to lighten the regulatory burdens, the Environment DG XI sought industry allies as social partners who would help to elaborate and implement the Directive rather than undermine it. Representatives of small national companies played this role as expert participants in risk-assessment discussions (Levidow et al. 1996: 142–43). The Environment DG XI also consulted environmental NGOs and funded their conferences on regulatory issues. In particular, such NGOs joined the EU-wide working group on risk assessment (ibid.). There they proposed broader criteria for evaluating GM products and opposed regulatory harmonisation as premature. Greenpeace soon withdrew from the working group, on grounds that all its arguments were ignored (FoEE, 1996). Nevertheless NGO pressure helped to limit the extent of deregulatory changes in agbiotech legislation and practice.

In the mid-1990s regulatory harmonisation was gaining extra impetus from a high-profile initiative for trade liberalisation. In 1995 the Commission's New Transatlantic Agenda (NTA) identified 'barriers to transatlantic trade' as the main problem for EU and US policymakers. Representing multinational companies, the Transatlantic Business Dialogue (TABD) helped to define the NTA's aims and the overall policy agenda. Regulatory harmonisation was expressed with the slogan, 'Approved Once, Accepted Everywhere', at least at the transatlantic level, ideally leading to a New Transatlantic Marketplace. For the agricultural biotechnology sector, TABD members identified pre-market safety assessment as the only regulatory issue (Murphy and Levidow, 2006). Within the EU political system, the TABD agenda was promoted initially by the agrichemical industry and eventually by EuropaBio, representing biotech companies which were already transatlantic in character.

As a complementary slogan of the NTA, 'mutual acceptance of data' would provide a step towards mutual acceptance of products. EU-wide and international discussions probed differences in regulatory criteria in efforts to reconcile them. Differences concerned not simply the level of scrutiny, but also framings of the relevant uncertainty which may warrant more data. For example: France emphasised genetic imprecision of the GM construct, Denmark emphasised implications for herbicide usage, the UK emphasised ecological interactions, etc. (Levidow et al., 1996). These diverse framings reflected the different composition and perspectives of expert advisory bodies in each country.

By the mid-1990s conflicts arose more starkly over how to define 'adverse effects', especially in relation to agri-environmental issues. As some cautious experts warned, insecticidal or herbicide-resistant crops could generate resistant pests. However, these scenarios were dismissed as merely normal 'agronomic problems' which also sometimes resulted from conventional pesticide usage. Thus safety claims accepted the normal hazards of intensive monoculture, whereby the 'pesticide treadmill' would be supplemented by a genetic treadmill.

As another contentious issue, herbicide-tolerant crops allow the use of broad-spectrum herbicides, designed to kill potentially all other vegetation. The EC Directive had no clear basis for evaluating those intended environmental effects. Proponents of such products again favourably compared agbiotech to the agrochemical hazards of conventional agriculture. By accepting prevalent agricultural practices as a baseline, regulatory procedures validated environmentalist critiques—namely, that GM products perpetuate the harmful effects and technological dependency of intensive monoculture (Levidow et al., 1996: 147).

Consequently, risk assessments became more contentious as an implicit standard-setting, driven forward by regulatory harmonisation. Some member states requested more evidence regarding potential agri-environmental effects prior to regulatory approval. But their concerns were marginalised in the EU-wide procedure; amid national differences in risk-assessment criteria, these were effectively levelled down. Expert safety assessments served to minimise regulatory burdens; field experiments were cited as a basis for product safety and thus commercial authorisation. Official risk assessments were conceptualising Europe as a homogeneous economic environment for an agricultural factory. The Commission went ahead and granted approval for commercial use to several GM products in 1997–98, despite objections from some member states (Levidow et al. 1996, 2000; see also Chapter 5).

New legislation for food safety likewise established a basis for minimising or levelling down regulatory standards across the EU. In 1997 Regulation 258/97 on Novel Food established a legal duty to seek approval before commercialisation of any novel food, e.g. GM food. Unlike the Deliberate Release Directive 90/220, this new law had a simplified procedure for novel foods 'substantially equivalent to existing foods or food ingredients as regards their composition, nutritional value, metabolism, intended use and the level of undesirable substances contained therein'. If a GM product was substantially equivalent to a conventional counterpart, then no risk assessment was required (EC, 1997b: 3). In the late 1990s several GM foods were approved in this way.

The simplified procedure was designed to harmonise risk-assessment criteria across the Atlantic, as well as within the EU. The US FDA did not even require safety approval for any novel food deemed 'substantially similar' to a conventional counterpart (as noted earlier). In its own European way, the

Novel Food Regulation provided a safety imprimatur for such products. Its simplified procedure helped to implement the New Transatlantic Agenda: mutual acceptance of data and safety judgements facilitated trade liberalisation for GM foods.

Beyond risk-assessment issues, also contentious was the absence of a statutory requirement for labelling of GM products. European NGOs and some member states demanded a labelling requirement so that consumers would have 'the right to choose' non-GM products. According to the Commission, however, such a label would unfairly stigmatise all products of GM techniques and thus disadvantage them in the internal market. Ultimately no label was required by EU approval decisions for the initial GM food products—a soybean, oilseed rape and maize (EC 1996a, 1996b, 1997a).

With this basis for regulatory approval, the European public was constructed in particular roles. They were to be a passive audience for safety claims, unwitting consumers of GM food, individual beneficiaries of more efficient production methods, and thus rational supporters of technological progress. US exporters could more readily mix grain from GM and conventional sources, without needing to know their origins; this basis could facilitate trade, at least initially.

6 CONCLUSIONS

At the start, this chapter posed the following questions about the period through the mid-1990s:

- How was agbiotech linked with specific forms of nature and social norms?
- How was Europe being made safe for the development and commercialisation of agbiotech?
- How did this framework model European integration—and generate political conflict?

From the 1980s onwards, a new discourse-coalition elaborated a co-production of agbiotech, nature and society. Agbiotech was promoted as a technological saviour: GM techniques would improve crops for both economic efficiency and environmental protection. A molecular-level techno-fix would overcome agricultural problems due to deficient inputs. R&D agendas invested nature with metaphors of codes, combat and commodities, as a basis for smart seeds to overcome threats from a wild, disorderly Nature. A safe biotechnologised nature would protect agriculture from such environmental threats, while also protecting Europe from competitive threats.

New policies were designed for a 'competition state', directing resources towards the domestic capacity for global competitive advantage (Cerny, 1999). This meant efforts to attract private-sector investment, to marketise public goods and to generate globally competitive products. Such a state was promoted through agbiotech, as both an instrument and symbol of societal progress. In such policy frameworks, 'the economism of globalisation discourse is combined with an authoritarian technological determinism' (Barben, 1998: 417).

The global agbiotech agenda was led by the US agri-industrial complex and its government supporters. Long beforehand, these forces had turned agriculture into a rural factory of standardised commodity production, especially animal feed for global export. Transatlantic agrichemical companies developed agbiotech for further industrialising agriculture, along with the promise of alleviating its environmental damage through eco-efficient inputs. GM crops were promoted through new policies—broader patent rights giving financial incentives to public-sector research institutes, 'product-based regulation' normalising GM crops as safe and trade liberalisation opening foreign markets to US agri-exports. Thus agbiotech was being co-produced along with neoliberal models of the natural and social order.

The US model of agri-industrial productivity was appropriated as an inevitable European future. Since the 1980s, policy documents had represented Europe as a 'Biosociety' whose inefficiencies must be overcome through technological innovation. Soon this became linked with a neoliberal agenda, simply assuming that more eco-efficient inputs would bring environmental benefits. Within this policy framework, Europe faced the risk of losing the benefits of new technology, especially through inadequate financial rewards or over-regulation. Adapting the US model, new policies sought to make Europe safe for agbiotech as normal products.

EU agbiotech policy was also linked with a trade liberalisation agenda by invoking objective imperatives of global competition. In parallel, the European Commission promoted agbiotech as essential for economic competitiveness and thus for survival of the European agri-food sector, along lines similar to the US model of industrial agriculture. By the mid-1990s EU-US discussions were identifying 'barriers to transatlantic trade', which must be removed through regulatory harmonisation, especially for biotech products.

EC policies also facilitated efforts to commoditise genetic resources. After a decade-long conflict, a 1998 EC directive extended patent rights to 'biotechnological inventions'; this broadened the scope of discoveries or techniques which could be privatised and thus accrue royalty payments. Broader patent rights raised concerns among civil society groups who were otherwise inclined to regard agbiotech as progress. The 1998 Directive extended controversy over whether 'biopiracy' meant unauthorised use of GM seeds—or rather 'Patents on Life', i.e. patent rights on mere discoveries of common resources. Several EU member states failed or refused to incorporate the Directive into national law, sometimes amidst national conflict over patent rights. The European Commission brought court actions against them; such formal trials could not harmonise national rules, nor resolve the legitimacy problem. The 'biopiracy' issue raised doubts among Left and trade-union groups which were otherwise inclined to support technological innovation as societal progress.

R&D policies created greater incentives for the use of GM techniques, partly by blurring the boundary between public and private sectors. In many EU member states, public-sector research institutes were allocated less state funds than before and were expected to substitute income from the private sector or from patents, e.g. through GM techniques. EU R&D funding priorities complemented that shift towards a marketisation policy for hitherto 'public-sector' research. NGOs raised concerns that GM crops would stimulate agricultural intensification, undermine farmer independence and jeopardise rural livelihoods. In response, biotech companies framed any socio-economic disruption as essential means for renewing democratic societies. Thus the market was idealised as a free, naturally beneficent regulator for determining societal benefits.

Amidst the early debate over potential risks to the environment and human health, the EC's 1990 Directive had established a 'preventive' framework, linked with 'completion of the internal market'. Soon after its enactment, transatlantic agbiotech companies sought to undermine the 'preventive' features. As a prime accusation, the Directive was 'over-regulation', jeopardising industry's freedom of innovation. Along similar lines, some public hearings put the EC's legislation on the defensive for impeding Europe's industrial base; the 1990 Directive was found guilty by a UK Parliamentary committee. As the EC legislative framework was symbolically put on trial for jeopardising European progress, regulatory authorities came under pressure to demonstrate its viability, especially by approving GM products.

Although the 1990 EC Directive was implicitly precautionary, its practical scope was constrained by a new policy of 'risk-based regulation'. This defined harm in narrow ways and shifted the regulatory burden of evidence towards demonstrating risks. Expert safety assessments served to minimise regulatory burdens; field experiments were cited as a basis for product safety and thus commercial authorisation. From the mid-1990s onwards, regulatory conflicts emerged over how to ensure in advance that GM crops would provide environmental improvements (cf. Weale and Williams, 1993). Some regulators sought to put these promises on trial through extra evidence of safety, but such efforts were marginalised by neoliberal policies in the mid-1990s.

'Mutual recognition' of data was promoted by transatlantic agbiotech companies, with support from some state bodies, as means to avoid trade barriers. Official EU risk assessments accepted the normal hazards of intensive monoculture, thus complementing the policy framework of higher productivity for economic competitiveness. This political agenda was depoliticised by invoking objective imperatives such as globalisation, treaty obligations and 'risk-based regulation'. Thus early EU regulatory procedures incorporated policy assumptions of agbiotech promoters.

Through a technicist harmonisation agenda, Europe was being deterritorialised as a purely economic zone, devoid of cultural identities (Barry, 2001: 70). In this way, safety assessments could symbolically normalise GM products. Despite demands for special GM labelling, this was rejected on several grounds: for lacking any scientific basis, unfairly impeding the internal market and making the EU vulnerable to a US challenge under WTO rules. These policies prevailed against significant dissent in all EU institutions.

In all those ways, by the mid-1990s EC policies were making Europe 'safe' for agbiotech to achieve commercial success, while modelling European democracy along neoliberal lines. Under 'risk-based regulation', societal decisions on agbiotech were reduced to a case-by-case approval of GM products, with a relatively narrow definition of risks. The 1990 EC Directive had resulted from the EU's co-decision procedure and was then implemented through the comitology procedure, representing EU member states, whereby dissent was marginalised.

State accountability meant regulatory procedures for authorising 'safe' GM products, which could then enter the EU internal market as extra options for farmers' free choice of seeds. As unwitting consumers of GM food, the public would effectively support a beneficial technology serving the common good of Europe; publics had little scope to act as citizens. These arrangements provoked significant dissent from some member states as well as from civil society.

In sum: Three discourse-coalitions framed the issues in contending ways (as shown in Table 2.1). In each frame, agbiotech was linked with accounts of social order and disorder. Agbiotech promoters portrayed GM crops as eco-efficient agri-inputs which would also address global competitive threats. Through an apocalyptic frame, critics promoted a different social order of common resources. State authorities elaborated a managerialist frame for mediating the societal conflicts.

Through interactions among those frames, the EU devised a distinctive neoliberal approach facilitating an agbiotech market, while adapting elements of US agbiotech policy. For its democratic legitimacy, this approach depended upon the EU co-decision procedure, where dissent over product approvals was procedurally marginalised. As the EC legislative framework was symbolically put on trial for jeopardising European progress, regulatory authorities came under pressure to defend its viability by approving GM products. Early dissent indicated fractures that would eventually crack the EU policy system: its pro-agbiotech policies and procedures too would be put on trial, as shown in the next chapter.

3 Opening Up Risks, Disputing Un/Sustainable Agriculture

INTRODUCTION

EU agbiotech policy invoked economic threats which could be overcome through the opportunity of agbiotech. As a techno-fix, GM crops would help solve problems of environmental sustainability and economic competitiveness—in turn framed to justify neoliberal policies (see Chapter 2).

By the late 1990s, however, agbiotech was facing mass opposition. Critics were accused of unfairly turning the technology into a proxy for extraneous issues, such as sustainable agriculture and globalisation. Yet agbiotech itself was being promoted in such terms—which became more explicit through public controversy.

Through those links, protest challenged agbiotech as an inevitable European future. From the mid-1990s to the end of decade, 'widespread public ambivalence about GM foods . . . gave way to widespread public hostility' (Gaskell et al., 2000: 938). This greater opposition coincided with wider framings of the issues at stake. As agbiotech was put on trial, the charge-sheet was expanded.

For the European conflicts over agbiotech in the late 1990s, this chapter explores contentious links between agbiotech and sustainable development, in particular:

- How did early dissent expand into mass opposition?
- How did agbiotech opponents put promoters and their policy agenda on the defensive?
- How was agbiotech linked with un/sustainable agriculture?

To answer those questions, the story begins with divergent accounts of sustainable agriculture, amidst a broader conflict over European futures. Subsequent sections show how national debates over agbiotech linked accounts of risk, sustainable agriculture and globalisation. The chapter briefly mentions regulatory issues which are taken up in Chapter 4, covering a similar historical period (from the mid- to late 1990s).

1 AGBIOTECH AS UN/SUSTAINABLE AGRICULTURE

In the European agbiotech controversy, contending accounts of sustainable agriculture became more explicit in the late 1990s. By analogy to the issues which arose in the early debate, 'sustainable agriculture' can be analysed through a tripartite typology of discursive frames: eco-efficient intensification; precautionary regulation; and quality agriculture (compare Tables 2.1 and 3.1). Each informed a discourse coalition. The 'quality' frame valorises agri-environmental public goods and consumer-producer links; these cultural values resonated with the apocalyptic frame opposing agbiotech.

Table 3.1 Divergent Framings of Agbiotech vis à vis Sustainable Agriculture

View *Issues*	*Sustainable Intensification*	*Multifunctional Agriculture*	*Quality Agriculture*
General problem	genetic deficiencies of crops; inefficient inputs which limit farm productivity and economic competitiveness	depletion of agri-environmental resources; trans-boundary risks of GM crops; regulatory differences across countries	intensive monoculture which generates hazards; farmer dependence on multinational companies
Environment to be protected	natural capital as a production asset; biodiversity simulated and extended by lab techniques	all resources that could support agriculture or that it could enhance	agriculture as a social-territorial commons; biodiversity of cultivars and biocontrol agents
Economic Aims	Compete for sales of 'green' commodities.	Avoid trade barriers through common regulatory standards.	Link producers and consumers through quality production. Protect those markets from agri-industrial threats.
Solution	input-output eco-efficiency to replace energy and materials with genetic information.	precautionary research and control measures to conserve natural resources, especially biodiversity.	less-intensive, skilled methods aided by green supply chains, farmer cooperation and government
Expertise	Develop GM crops which reduce agrochemical usage.	Compare biophysical effects of GM/non-GM crops.	Develop farmers' knowledge of biodiversity and local resources.

1.1 Agbiotech as Sustainable Intensification

Since the mid-1990s the biotechnology industry has appropriated 'sustainable agriculture', cast in its own image of intensive monoculture. From an eco-efficiency frame, proponents emphasised benefits of minimizing agrochemical usage, deploying resources more efficiently, increasing productivity, and so enhancing economic competitiveness. In this frame, society faces a common problem: the risk of failing to reap the benefits.

In particular, Monsanto raised the slogan, 'Creating value through sustainability'. According to its *Report on Sustainable Development,*

> Our products create value for our customers by helping them to combine profitability with environmental stewardship. For product impact, this means: more productive agriculture, more soil conservation, less insecticide use, less energy, better habitat protection. (Monsanto, 1997)

By incorporating genetic knowledge for pest control, GM crops substitute intelligence for energy and materials, argued the report. For example, 'inbuilt genetic information' helps GM crops to protect themselves from pests and disease. Herbicide-tolerant crops facilitate no-till agriculture, which 'decreases soil erosion, nutrient and pesticide runoff, as compared to conventional tillage' (Magretta, 1997).

Moreover, such products help to avoid the need for difficult trade-offs:

> The problem is often framed as a choice: either feed a rapidly growing population . . . or preserve natural habitats for biodiversity. But we can do both by continuing the progress of high-yield agriculture. (Monsanto, 1997: 16)

Promoting agbiotech as a global solution, Monsanto linked market competition, use values, environmental protection and food security.

Likewise, according to Novartis, GM insecticidal maize 'contributes to sustainable agriculture through savings on mineral fertilisers, fossil fuels and pesticides' (Novartis, 1998). Such arguments exemplify the industry's general perspective on intensifying agriculture in beneficent ways:

> Sustainable intensification of agriculture can be defined as follows: The use of practices and systems which maintain and enhance: a sufficient and affordable supply of high quality food and fibre, the economic viability and productivity of agriculture, the natural resource base of agriculture and its environment, and the ability of people and communities to provide for their well-being. (Imhof, 1998)

EU policy likewise supported agbiotech as an eco-efficient innovation, as theorised by ecological modernisation (as sketched in Chapter 2). The EU's

Economic and Social Committee celebrated such benefits, even before GM crops were being commercially grown in Europe: Biotechnological solutions are 'guaranteeing yields, helping to cut the use of plant health products in combating pests and diseases, and creating quality products'. Thanks to its precise techniques, moreover, genetic engineering 'allows more accurately targeted risk prediction', argued the committee (EcoSoc, 1998). In its view, biotechnological precision could readily clarify any uncertainties in risk assessment. Thus a high-profile EU body made explicit some optimistic (even omniscient) assumptions which had been implicit in regulatory procedures.

1.2 Public Distrust of Agri-Efficiency

In the 1990s such assumptions were becoming a greater focus of European public distrust towards regulatory authorities. Expert denial of uncertainty became a high-profile issue through various food and medical scandals. In particular, the 'mad cow' (BSE) epidemic had resulted from animal feed containing unknown infected material, still biologically active due to a deregulatory change in requirements for heat treatment. The 1996 scandal undermined official images of policy-neutral expertise at both national and EU levels. Moreover, the controversy was turned into a crisis of industrial agriculture and its regulatory controls.

As a further basis for the scandal, UK expert advice had made policy assumptions, especially that real-world practices would follow risk-management guidelines, and that explicit uncertainty would alarm the public (Jasanoff, 1997; Millstone and van Zwanenberg, 2001). The EU expert committee initially accepted the UK's optimistic assumptions. The Commission covered up the problem, for fear that public concern about the BSE crisis would endanger the European beef market, according to a report by the European Parliament (EP, 1997).

Subsequent revelations led to a legitimacy crisis, sometimes diagnosed as a 'democratic deficit'. EU regulatory procedures were illegitimately equating expert advice with science, as a basis to pre-empt or conceal political decisions. For these reasons, the BSE crisis has been partly blamed for public reaction against agbiotech.

However, the crisis per se does not explain public attitudes. According to focus-group research with ordinary citizens in five European countries, public concerns about agbiotech express resentment towards decision-making procedures, not an absolute opposition to GM products as such. People were expressing unease at the prevalent direction of the agro-food system, which remains beyond democratic control; they could see no political means to influence decisions. They regarded the decision-making procedure as relying too much upon particular expert claims, whose risk-benefit judgements lack transparency (Wynne et al., 2001).

Comments in the focus groups were paraphrased as follows:

- 'Progress' is being defined according to particular technological trajectories, thus marginalizing or pre-empting alternatives.
- Proponents offer no evidence for environmental benefits (e.g. reducing harm from agrochemicals) or social benefits (e.g. feeding the world).
- Regulatory institutions downplay uncertainty about risks, especially long-term and irreducible uncertainty, and exclude such consideration from their decision-making.
- Regulatory decisions claim to be entirely 'science-based' yet implicitly incorporate non-technical concerns, e.g. economic and ethical criteria.

Such views make people suspicious towards 'scientific' claims for safety, given the various implicit criteria considered in risk evaluations and regulatory decisions. People demand that these criteria be more transparently explained (ibid.).

According to the comments of ordinary citizens, moreover:

- Lessons of the BSE crisis have not been learned by policymakers and their scientific advisors, nor have they been applied to agricultural biotechnology.
- Risk management and communication continues to assume that all risks can be anticipated, or even that they have been.
- Decision-making includes no (or inadequate) measures to reduce and monitor risks of new products and technologies after they have been approved for commercial use (Wynne et al., 2001).

Thus the BSE crisis symbolised a deeper malaise of the agri-industrial system and regulatory claims for objective knowledge, rather than simply causing or explaining distrust towards agbiotech. Nevertheless policymakers generally diagnosed the problem as public distrust of agbiotech and 'science' in general, as an imperative to enhance public confidence. Institutional reform could more readily attempt to re-establish cognitive authority than to address the sources of malaise (Levidow and Marris, 2001).

1.3 Biotechnological Efficiency on Trial

The 1996 BSE crisis offered a new opportunity for opponents of agbiotech. Activists undermined the credibility of official safety claims for food products, while aggravating suspicion towards intensive agricultural methods. In Europe GM crops were approaching the commercial stage amidst a wider debate over the hazards of intensive agriculture and possible alternative futures. Anti-biotechnology activists drew analogies to BSE (e.g. Greenpeace, 1997); they cited unpredictable risks of GM food and feed as grounds for a moratorium on GM product approvals. Two GM products became test cases for divergent framings of agbiotech; indeed, these products were turned into high-profile symbols of dangerous, disorderly government and agbiotech.

In 1996 Monsanto's GM soybean received EU-wide commercial authorisation for food and feed import, without any requirement for GM labelling. On 16 October 1996, World Food Day, Greenpeace held a demonstration at Unilever offices around Europe, and the European consumer federation BEUC demanded mandatory labelling of GM food. When US soya shipments arrived in late 1996, these provided a high-profile target for agbiotech opponents, as well as a basis for Europe-wide attention through the mass media. A French newspaper article was headlined 'Alerte au Soja Fou'—mad soya alert (*Libération,* Paris, 01.11.96). This metaphor highlighted disorders of government responsibility and product behaviour in the BSE episode. At several ports, Greenpeace staged a symbolic blockage with rubber dinghies, temporarily delaying the shipments, thus gaining publicity for its anti-GM message. NGOs accused companies and governments of 'force-feeding us GM food'.

In January 1997 the Commission approved Ciba-Geigy's insecticidal maize, Bt 176, for import and cultivation, despite opposition from most member states. According to EU expert committees, there was no evidence of risk from the product. Many national experts dissented from that judgement and its optimistic assumptions about scientific evidence. Attention focused on the antibiotic-resistance gene, which had been used as a marker to create the GM crop; if the marker spread to pathogenic microbes, it could undermine the clinical efficacy of the antibiotic. Analogies were drawn to animal husbandry excessively using antibiotics, thus spreading resistance. NGOs and some member states also demanded a 'GM' labelling requirement; this demand led to disagreements among Commissioners and procedural delays (Rich, 1997).

When Bt 176 maize was eventually authorised in January 1997 by the Commission, its decision was criticised by a broad range of civil society organisations. These included consumer NGOs, which did not necessarily oppose agbiotech but demanded more rigorous risk assessments and GM labelling for consumer choice. Newspapers gave front-page coverage to such critics. In a Belgian newspaper, the Commission was denounced for 'recidivism' (Rich, 1997). This metaphor turned the Commission into a symbolic defendant, by analogy to criminal collusion in allowing British beef to be sold for so long, despite danger signs for public health. In April 1997 the European Parliament denounced the approval decision of the Commission, thus formalising its guilt. The regulatory arena was turned into an opportunity for political-cultural discourses, thus delegitimising safety claims as well as the Commission.

As highlighted in the controversy over Bt 176, those issues of consumer choice and health risks struck a public chord, though mass opposition came from deeper issues. A campaigner describes how NGOs appealed to widespread European attitudes through the following campaign messages: Tampering with food may not be safe. Too much technology and industry in your food may threaten your health. If eating habits support intensive agricultural systems, then these harm conservation. Organic food is better for health and the environment, yet threatened by genetic pollution. Labelling is a right for consumers to make an informed choice. American companies are force-feeding us GM food and meddling with European cultures. Better to be safe than to trust regulatory claims for 'sound science' and technological advance. For all these reasons, GM food and feed are morally wrong, they argued (Schweiger, 2001: 364).

Through the agbiotech issue, diverse European movements coalesced. They 'found a unifying topic like no other', helped by 'the fact that genetic engineering touches virtually all areas of life', according to that campaigner. NGO agbiotech campaigns crossed the usual boundaries between environmental, consumer and farmer issues. National NGOs intervened at the European level. All shared a common aim: 'stopping the technology from infiltrating the food and agricultural sectors' (Schweiger, 2001: 371).

Protest was driven mainly by activists from environmentalist and farmer groups, which catalysed broader societal debate and opposition networks. In the late 1990s an anti-GM movement emerged, led by environmentalist groups, especially Greenpeace Europe, FoEE and their national affiliates. Another key opponent was the Coordination Paysanne Européenne and its national affiliates, representing relatively less-intensive or small-scale farmers; they opposed the entire agri-industrial model, while counterposing extensification measures as an alternative. Agbiotech opponents also stigmatised GM crops as 'contamination' which jeopardised benign alternatives.

Although not opposing agbiotech, consumer NGOs took up agri-environmental issues as well as GM food safety. Some consumer groups linked GM food with potential environmental risks of cultivating GM crops. Many people boycotted GM food in shops, as a way to 'vote' against agbiotech, or at least its commercialisation, in lieu of a clear democratic procedure for a societal decision about a contentious technology. Anticipating greater boycotts, European supermarket chains decided to exclude GM grain from their own-brand products by the late 1990s (see Chapter 7).

Previously, in the mid-1990s, European agbiotech critics had focused their efforts upon mobilising counter-expertise, especially for criticising regulatory failures of responsibility. They questioned the adequacy of the available science for regulatory decisions. They also challenged the draft Patent Directive as 'biopiracy' (Purdue, 2000; see also Chapter 2).

When mass protest emerged in the late 1990s, activists gained greater opportunities to criticise regulatory gaps and weaknesses. Amidst health scares over agbiotech 'tampering with our food', European supermarket chains decided to exclude GM grain from their own-brand products (see Chapter 7); this also set back prospects of a European market for GM seeds.

Soon the 'Life Sciences' strategy went into reverse with demergers. This reversal had many sources: expected synergies between pharmaceutical and crop research had been overly optimistic (Tait et al., 2002; Wield et al., 2004: 1052). Moreover, pharmaceutical companies sought to avoid the economic liability of agri-supply companies, as well as the stigma of GM crops.

Moreover, the agbiotech controversy intensified a wider debate over technological pathways, agricultural futures and alternatives. Oppositional 'risk' discourses framed productive efficiency as a problem, while counterposing 'sustainable agriculture' as an alternative. This was given diverse meanings, e.g. organic farming, Integrated Crop Management, peasant skills, etc. Agbiotech critics diagnosed the agricultural problem as intensive monocultural practices, global standardisation and farmer dependence upon multinational companies. The controversy also stimulated alternative agricultural methods for pesticide-reduction and thus environmental quality. These feature a shift towards biological crop-protection agents, along with changes in agronomic practices and farm structure, in order to avoid pests and disease. These changes drew upon and stimulated research into agro-ecology (e.g. Greens/EFA 2001).

2 SUSTAINING WHICH AGRICULTURE FOR EUROPE?

Agbiotech opponents have been criticised for unfairly targeting the technology as a proxy for issues of sustainable agriculture, or even for attempting to undo the history of industrial agriculture. As a US official lamented, for example, the European biotechnology debate 'is a retroactive referendum on a century of industrialization of agriculture' (Gifford, 2000). Perhaps, yet the debate also served as a prospective referendum on future agriculture and globalisation—which were put on trial as potential future which could be different, at least in Europe.

As in the US and European debates over agbiotech, 'sustainable agriculture' has become more prominent but also more contentious. The term has been used to sustain a future high-yield agriculture driven by agribusiness. For many proponents of sustainable agriculture, however, it means a critique of past practices, especially of intensive cultivation methods and farmer dependence on the agro-food industry (Peterson, 1997).

In Europe 'sustainable agriculture' has been increasingly defined by distinct cultural values, linking the quality of food products, rural space and livelihoods. Although chemical-intensive methods still prevail, the countryside has been increasingly regarded as an environmental issue, variously understood—e.g. as an aesthetic landscape, a wildlife habitat, local heritage, a stewardship role for farmers and their economic independence. These cultural-economic meanings conflict with neoliberal accounts of agriculture as an arena for economic competitiveness through higher productivity.

This conflict can be analysed as three divergent frames of sustainable agriculture, related to those which explicitly take up agbiotech. From a sustainable intensification frame, the problem is genetic deficiencies of crops which therefore must be corrected and improved for greater eco-efficiency. From a quality agriculture frame, the problem is intensive monoculture, control by multinational companies and globalisation of bulk commodity markets. From a precautionary regulation frame, mediating conflicts between the other frames, the problem is potential harm to environmental resources. (Compare Tables 3.1 and 3.2; also elaborated in Chapter 1, section 5.)

Table 3.2 Divergent Framings of Sustainable Development

View	*Neoliberal Eco-efficiency*	*Managerialist*	*Community*
Led by	multinational companies	government agencies	small-scale producers and environmental NGOs
Problem-definition	inefficient use (and depletion) of environmental capital	environment/development falsely separated; global interactions	unaccountable institutions; profit-driven innovation
Concept of nature	capital to be invested; assets providing environmental services	ecological support system and human habitat	a commons to be protected and shared; a harmonious balance to be maintained
Sustain what?	natural capital, substitutable by human capital, as a means to productive efficiency	optimum resource usage, natural capital as a public good; legitimacy of the state	communities as guardians and beneficiaries of commons
Economic aims	compete better in market for green commodities	economic growth through socio-technical re-organisation to increase carrying capacity	enhance livelihoods of small-scale producers
Solution	eco-efficiency to reduce pollution and reap cornucopia	negotiated rules and standards; international cooperation	link producers with consumers
Expertise	R&D for clean products	interdisciplinary expertise to model environmental effects	use local resources; know & work with nature

Note: This table draws upon Woodhouse (2000) and Dobson (1996).

Such divergent frames have arisen in conflicts over how to reform the Common Agricultural Policy (CAP), the EU system of agricultural subsidy. Such reform has attracted efforts to reinforce intensive methods or else to challenge and supersede them. This policy arena also sets a broader context of analogous issues and stakeholders for conflicts over GM crops. For these reasons, this section will sketch conflicts over CAP reform.

Historically, agricultural subsidies have been justified as sustaining common benefits of European agriculture—e.g. food self-sufficiency, cultural heritage and rural livelihoods, especially from small-scale and family farms. On that basis, the CAP has subsidised otherwise 'uncompetitive' farmers, whose products are exported at low prices. Consequently, 'the accompanying regulations have placed a considerable additional burden on subsistence farmers in the poor countries', e.g. by restricting markets for their high-value products (Redclift, 1987: 91).

Since the 1980s intensive methods have come under greater criticism for other reasons: environmental damage, over-production and increasing subsidies. These reasons intersected with another problem: US pressures upon Europe to eliminate production-based subsidies under the 1994 WTO agreement. For both problems, a possible European solution was extensification, i.e. reducing agrochemical inputs, thus reducing surplus production too.

For that aim, among others, CAP reform was linked to environmental policy. Extensification was promoted to reduce environmental harm and to protect ecological systems. Previously, environmental protection had been regarded as constraining agriculture; now it was regarded as re-legitimising subsidies, by assigning farmers the role of environmental stewards (Clark et al., 1997). Such measures were formalised in a 1992 EC Regulation on agri-environmental schemes. These schemes provide financial incentives for farmers to reduce pesticide usage, to make environmental improvements, to remove land from cultivation, etc. Subsidy would be shifted away from a production basis, towards a land-use basis.

A similar diagnosis was elaborated in the EC environmental policy. In the Fifth Environmental Action Programme, 'Towards Sustainable Development', environmental problems were attributed to agricultural intensification,

> . . . leading to over-exploitation and degradation of the natural resources on which agriculture itself ultimately depends: soil, water and air. In crop protection, systematic use of plant protection products has led to a relative resistance in parasites [pests], increasing the frequency and cost of subsequent treatments and causing additional soil and water pollution problems. (CEC, 1993a: 35–36)

As the corresponding solution, the EU undertook to promote extensification, e.g. cultivation methods which decrease inputs of agrochemicals, as well as agri-environmental schemes. In practice, however, earlier subsidies have been simply re-labelled as 'agri-environmental', while effectively perpetuating intensive cultivation practices. Extra payments are used and needed, paradoxically, to mitigate harmful effects of the overall subsidy system. Moreover, environmentalist groups have supported the continuation of subsidies in the name of environmental protection and sustainable development (Toke and McGough, 2005). However, conflicts over CAP reform eventually provided a new basis for environmentalist groups to support alternative agricultures.

In a 1997 proposal for reforming the CAP, *Agenda 2000,* 'sustainability' broadly defined the relevant environment to encompass the overall socio-economic effects of agriculture, not simply agrochemical pollution. The document emphasised the priority to enhance rural livelihoods, the quality of food production and its 'environmental friendliness'. In passing, it suggested, 'The development of genetic engineering, if well controlled, could enhance production but may raise questions of acceptability to consumers' (CEC, 1997: 27, 29).

The new policy aimed to establish 'a multifunctional, sustainable and competitive agriculture' throughout Europe. According to the Commission, agriculture harms the natural environment, but 'abandonment of farming activities can also endanger the EU's environmental heritage through loss of semi-natural habitats and the biodiversity and landscape associated with them'. Consequently

> The CAP's objectives include helping agriculture to fulfil its multifunctional role in society: producing safe and healthy food, contributing to sustainable development of rural areas, and protecting and enhancing the status of the farmed environment and its biodiversity. (CEC, 2003c: 2; also CEC, 2004)

From that perspective, subsidies would become conditional upon measures to conserve environmental resources, e.g. by using more eco-efficient methods which minimise agrichemical inputs and pollution. With lower support prices, farmers would need to produce more competitively at world prices. Through a neoliberal eco-efficiency framework, agriculture was being conceptually separated from the wider environment.

This future vision was made more explicit by a UK government advisor, Lord Haskins: Through CAP reform, subsidy would be reduced. More fundamentally, it would be transferred to 'funding to sustain and enhance the environment', while 'ensuring that environmental regulations do not stifle global competitiveness'—understood as dependent upon greater productive efficiency. Land use would be designated for either industrial agriculture or other uses such as environmental enhancement:

> . . . where European agriculture can be competitive, this competitiveness should, within environmental limits, be maximised. Where it cannot be competitive, farming *per se* should be downgraded behind good environmental husbandry as the linchpin of a subsidy/welfare system. (Haskins, 2002: 7–9)

Opposing that potential future, small-scale producers denounced the new policy, especially the plan to reduce the support prices. Such reductions would make farmers more dependent upon direct payments for their overall income. This 'will benefit agri-industry and the distribution sector rather than consumers', argued the Coordination Paysanne Européenne, representing relatively less-intensive farmers. Under the new reform, moreover,

European taxpayers will continue to:

- pay huge subsidies to huge farms, while driving the small ones out of business;
- support increasingly industrial farming methods, to the detriment of employment and the environment (CPE, 1999).

In the *paysan* view, such reform would encourage large-scale farms to continue intensive methods, while paying them to steward an 'environment' outside farming. This would 'accelerate the disappearance of multi-functional family farms'. They counterposed de-intensification measures, based on 'remunerative agricultural prices and sustainable family farming, with multiple benefits for society' (CPE, 2001). In other words, multifunctional skills should be sustained within agriculture through farming communities and agri-environmental biodiversity.

Thus proposals for CAP reform have generated distinctive policy stances. These have parallels and institutional links with discourse coalitions in the agbiotech debate. Not coincidentally, for example, EC policy on both CAP reform and agbiotech has promoted agricultural intensification, while the Coordination Paysanne has attacked both those policies in similar terms.

These issues found a central place in the 'anti-globalisation' movement, later renamed the 'global justice' movement. The European Social Forum adopted the slogan, 'Another World is Possible', defying neoliberal claims that objective imperatives constrain possible futures. That slogan was adapted as 'Another Agriculture Is Possible' for sessions on agri-food futures, led by the Coordination Paysanne. Agbiotech was put on trial as a symbol of unsustainable agriculture and globalisation.

3 'SUSTAINABLE AGRICULTURE' IN NATIONAL AGBIOTECH CONTROVERSIES: BRIEF EXAMPLES

Agbiotech controversy has focused on different issues across EU member states, each conflicting with the EC's technicist harmonisation agenda. By the late 1990s, greater opposition led some national authorities to shift their agbiotech policy or to reinforce changes already under way. They evaluated GM crops on a relatively broader basis, encompassing various public goods or common resources—e.g. safe drinking water, organic agriculture, local specialty products, etc. These goods were seen as under threat from industrial agriculture in general and GM crops in particular. Implicitly or explicitly, those national regulatory frameworks linked biotechnological risk with unsustainable agriculture.

In such ways, each national authority elaborated a distinctive managerialist frame, diverging from the EC policy frame of eco-efficiency models and optimistic safety assumptions. At the same time, EC member states diverged from each other. This section illustrates some regulatory changes and differences through four brief examples: Germany, Denmark, Austria and Italy. Subsequent sections take up two longer examples, the UK and France. Those two have special significance because their governments had led the EU-wide pressure to approve GM crops for cultivation uses—until the late 1990s, when their support weakened.

3.1 Germany: *Agrarwende* versus *Agrarfabriken*

German society has been polarised by biotechnology since the 1980s. Along with the UK, Germany was a leading driver of the early EC policy promoting biotechnology. Unlike the UK, however, Germany had a well-organised opposition campaign against 'gene technology', the more common term there. Led by social movements and the Green Party in particular, this campaign warned against environmental, health and ethical dangers, while demanding debate on alternative solutions. Although Germany's 1990 Gene Technology Act imposed stringent administrative requirements on biotech activities, the law combined the roles of promotion and protection in the same authority; this dual role came under criticism. NGOs denounced the Act for maintaining the government's bureaucratic-technocratic structures of control, while excluding demands for greater public participation (Jasanoff, 2005a: 59–61).

In the early 1990s Germany's policy framed biotechnology as a *Hoffnungsträger* or hope carrier: biotech would provide an essential tool for R&D investment, innovation, a stable job market and international competitiveness. Protest emphasised that GM crops threaten 'nature', popularly associated with forests in Germany. Accepting agbiotech as eco-efficient and safe, the Germany regulatory framework cast critics as irrational. Such polarisation continued through the 1990s, largely around a debate on uncertain risks to nature (Dreyer and Gill, 2000).

Since the late 1990s agbiotech has been turned into an issue of agricultural sustainability, especially through a change in government policy. In 2000 the BSE crisis in Germany provided a new opportunity for critics of the *Agrarfabriken,* i.e. factory farming, a phrase pejoratively linking intensive agriculture with animal diseases. Facing these concerns and criticisms, the government transferred agricultural policy to a new Federal Ministry for Consumer Protection, Food and Agriculture. Its new policy undertook to evaluate GM crops in relation to sustainable agriculture, beyond simply product safety. Agbiotech proponents claimed that GM crops encourage more sustainable agricultural practices, while critics counterposed alternatives as more sustainable. This debate went beyond the earlier 'risk' focus, towards choices for future agriculture, as a basis for new alliances between environmentalist and some farmer groups.

In 2001 a new policy, the *Agrarwende* ('agricultural turn') shifted agricultural policy towards consumer interests, informed choices and sustainable methods. Led by a Green Party politician, the new Ministry promoted organic agriculture as a model for more sustainable forms of farming, and aimed to increase its share to 20% within ten years. Any new technology had to demonstrate its contribution towards the goal of sustainable agricultural change (Boschert and Gill, 2005).

For the *Agrarwende* the Ministry also funded research to develop and promote less-intensive agricultural methods. The programme counterposed public-interest research to the mainstream agenda of 'public-private partnerships' and agbiotech (Öko-Institut, 2002b). In particular, this research emphasised agrobiodiversity: 'Besides its direct value in supplying food, as well as for livelihoods, habitats and ecosystems, diversity in agriculture lowers certain production risks and is still today a form of insurance against poor harvests and susceptibility to pests and illnesses' (Öko-Institut, 2004: 14).

3.2 Denmark: Herbicide Effects Sustainable?

Denmark's environmental legislation has affirmed the general aim of 'sustainable development' since the 1980s. It also had a policy to reduce agrochemical usage, especially so that groundwater could be used safely as drinking water. The Danish approach valued groundwater as a common resource, thus favouring more extensive cultivation methods which would use fewer pesticides.

Citing that policy aim, NGOs criticised the long-term implications of GM herbicide-tolerant crops for herbicide usage and residues, especially in groundwater. In the mid-1990s they successfully pressed the Danish Parliament to raise such questions about herbicide-tolerant crops within Danish regulatory procedures. Under this pressure, the Environment Ministry adopted broad risk-assessment criteria along those lines, thus providing a basis for a broad national consensus (Toft, 1996). This regulatory scope went beyond the risk evaluation in most other EU member states.

Later the herbicide issue became more complex for GM herbicide-tolerant sugarbeet, when risk research investigated effects on farmland biodiversity as well as groundwater. The results identified a dilemma for farm-management practices. Early-season spraying meant killing all weeds and thus harming biodiversity, but it also meant less herbicide usage and thus less impact on groundwater; by contrast, late spraying meant more weeds and thus benefited biodiversity but also meant more herbicide usage and thus more impact on groundwater (Toft, 2005).

Meanwhile the agbiotech debate became more and more polarised. NGOs and many farmers denounced GM crops as unsustainable, while industrial and agricultural decision-makers advocated these products as useful tools for sustainable agriculture. Facing all these conflicts, Danish authorities deferred any clear stance on supporting or opposing EU-wide approval of GM herbicide-tolerant sugarbeet.

3.3 Austria: Organic Versus GM Agriculture

In Austria agbiotech was turned into a symbolic threat to organic agriculture. Even before GM crops became a high-profile issue there in the mid-1990s, the Austrian government was promoting organic farming—as ecologically sound, as quality products and as an economically feasible market-niche alternative for an endangered national agriculture. This 'competitiveness' strategy conflicted with the pro-biotechnology imperative to increase agricultural productivity.

Some government officials regarded agricultural biotechnology as a threat to the environment and an obstacle to sustainability. Austrian regulators unfavourably compared potential environmental effects of GM crops with methods which use no agrochemicals, as grounds to oppose commercial approval. When NGOs campaigned against agbiotech, they effectively reinforced the government's stance (Torgerson and Seifert, 2000).

As a GM-free Austria nearly became a national consensus, the government sought stronger means to justify this policy, especially given its conflict with EU legislation. Austria banned several GM crops after they obtained EU approval, while referring to detailed criticisms of the official risk assessment. In Austria's own risk-benefit analysis, risks were always uncertain, while benefit was understood as promoting the political aim of a society oriented towards sustainability.

To justify restrictions on GM products, civil servants linked the precautionary principle with sustainable development—a link already in the 1992 Rio Declaration. In addition, Austria's law on biotechnology had a 'social sustainability' clause, which prohibits 'inappropriate disadvantages' for societal groups through biotechnology. Civil servants anticipated using this clause to justify strict rules for segregating GM crops, thus deterring their cultivation (Torgerson and Bogner, 2005).

3.4 Italy: *Prodotti Tipici* Versus Agbiotech

Italian agbiotech opponents sought to protect the agro-food chain as an environment for craft methods and local specialty products, known as *prodotti tipici*. In the late 1990s the Italian Parliament had already allocated subsidies to promote such products and foresaw these being displaced by GM crops. According to a Parliamentary report, the government must 'prevent Italian agriculture from becoming dependent on multinational companies due to the introduction of genetically manipulated seeds'. Moreover, argued the report, when local administrations apply EU legislation on sustainable agriculture, they should link these criteria with a requirement to use only non-GM materials. Parliament endorsed such proposals (Terragni and Recchia, 1999).

Such anti-agbiotech demands gained widespread support, especially from the Coltivatori Diretti, a million-strong union of mainly small-scale farmers. Environmental NGOs, farmers and food retailers built a national network seeking to exclude GM products from Italian agriculture. This network successfully maintained Italy's political and commercial opposition through government changes. When Romano Prodi's *L'Ulivo* (Olive Tree) coalition was replaced by Berlusconi's *Casa delle Libertà* coalition in 1996, its policy generally shifted along neoliberal lines; and the new government included strong advocates of agbiotech. Yet Italian regulatory officials continued to deter or block field trials and to oppose commercial authorisation of GM products at the EU level.

That policy was often translated into risk arguments in EU-level regulatory procedures. When a company requested authorisation to import GM rapeseed in 2003, for example, Italy argued that any escaped seed could contaminate related plants and thus undermine centres of diversity for *Brassica* crops. This risk argument effectively served to exclude GM crops and grain—framed as a threat to Italian food products, their wholesome image and small-scale producers.

4 FRANCE: GLOBALISATION ON TRIAL

Since the 1980s the French government had provided strong support for the EC promotion of agbiotech. The government's advisory body was dominated by molecular biologists, who focused on the genetic precision of GMOs as a prime basis for risk assessment. Agronomic and ecological experts were marginalised, so they sought allies to promote more cautious assessments and ecologically-oriented risk research.

In the mid-1990s, however, critics found common cause with some cautious scientists over environmental risk issues. In 1996 the NGO Ecoropa initiated a petition emphasising unknown risks of GM crops, as a basis to advocate a moratorium. It was signed by several hundred scientists, many seeking more stringent regulation rather than a ban. Soon critics were putting the government onto the defensive for failing to protect France from risks of GM crops.

A particular GM crop was turned into a symbolic and literal trial of responsibility for risks. In 1997 greater controversy emerged over GM glufosinate-tolerant oilseed rape, a crop which has great capacity to spread its pollen to related plants. Expert advisors anticipated that weeds would eventually acquire resistance to glufosinate herbicides, thus jeopardising and complicating methods for weed control. Since the mid-1990s France had led the EU-wide procedure for authorisation to cultivate the crop, while regarding the weed-control implications as merely agronomic issues, irrelevant to environmental risk assessment. This stance was represented as simply the law rather than an interpretation.

After the Commission decision to authorise the product (EC, 1997d), its validity depended upon final approval by France. But the government now declined, thus reversing its earlier stance in response to expert disagreements and public controversy. Invoking the precautionary principle, moreover, in November 1998 the government announced that this product would not be approved for commercial use. Courts were asked to interpret the discretionary scope for an EU member state to act in this way (Marris, 2000; Marris et al., 2004).

In the wider public debate, moreover, the Institut National de la Recherche Agronomique (INRA) was put onto the defensive for pro-biotech bias, given its joint innovation research with seed companies on GM herbicide-tolerant oilseed rape. In early 1998 INRA abandoned this research programme, partly in order to protect the neutral reputation of its research on environmental risks of GM crops. In March 1998 Agrevo decided to destroy its own field trials of GM glufosinate-tolerant oilseed rape in France, in order to avoid further unfavourable publicity (Roy and Joly, 2000).

Such crops were kept on trial as a sustainability problem for weed control. INRA expanded ecological field research on GM herbicide-tolerant oilseed rape, especially gene flow and hybridisation with weedy relatives. When this turned out to be more extensive than anticipated, the results were further cited to justify the French rejection and to raise doubts elsewhere.

Agbiotech critics put the French government on the defensive over another GM crop, Novartis' Bt 176. As already mentioned, this had gained EU approval (EC, 1997a), amid controversy about several risks including its antibiotic-resistance marker gene. The French government was accused of favouring commercial interests over scientific criteria. According to Ecoropa, 'Obviously, the French government surrendered to interests of multinational agrochemical companies and its decision is entirely commercially motivated' (quoted in FoEE, 1997). As their specific criticism, the antibiotic-resistant marker gene had been ignored in the risk assessment:

> This is in spite of the fact that the most recent scientific evidence (by Professor Courvalin, May 1998) shows that the return of antibiotic-resistant marker genes by horizontal transfer to soil bacteria and to the gut of mammals is possible, with all the negative consequences that this implies for public health and the environment. (Ecoropa, quoted in FoEE, 1998)

When France proposed to approve maize varieties derived from Bt 176 in 1998, the government was more literally put on trial, as critics entered the judicial arena. Ecoropa and Greenpeace filed a challenge at the Conseil d'Etat, the administrative high court, on several grounds—that the risks had not been properly assessed, that the correct administrative procedures had not been followed, and that the Precautionary Principle had not been properly applied. Their arguments gained some support in the court's interim ruling in September 1998, though not in the final one (Roy and Joly, 2000; Marris, 2000). This court case stimulated public debate over the appropriate role of scientific uncertainty, precaution and societal considerations. By highlighting policy aspects of expert risk judgements, this debate challenged the government's efforts to represent its policy in scientific terms (Marris et al., 2005).

In the late 1990s the French agbiotech debate expanded from 'risk' to sustainability issues, featuring divisions among farmers. The Fédération

Nationale des Syndicats d'Exploitants Agricoles (FNSEA) represented industrial-type farmers, who sought access to GM crops as a means to enhance their economic competitiveness. In the name of 'sustainable production', they also anticipated environmental benefits such as reductions in the use of pesticides and water.

That sustainability perspective was opposed by the Left-wing farmers' trade union, the *Confédération Paysanne* (henceforth *Conf*). Its activists denounced agbiotech as a threat to their skills and livelihoods. According to their spokespersons, GM crops pose risks to their economic independence, to high-quality French products, to consumer choice and even to democracy. Those values were expressed in the *Conf* slogan, 'Pour une autre agriculture: produire, employer, preserver.' This slogan resonated with *produits de terroir,* a marketing label which denotes products' origins from specific localities and producers.

A high-profile sabotage action landed defendants in court, where they attempted to put agbiotech and the government on trial. As a defiance of the law, approximately a hundred *Conf* activists forced entry into a Novartis warehouse and destroyed a stock of Bt 176 seed in January 1998. Three activists were arrested and charged—including José Bové and René Riesel, who thereby gained a higher profile in the anti-agbiotech movement. When their trial began in Agen the following month, they turned it into a trial of GM crops. As a defence argument, they denounced Bt 176 GM maize as 'the very symbol of a type of agriculture and society that we reject'. As an alternative future, they argued, 'Today, more and more farmers lay claim to a farmer's agriculture, which is more autonomous, economic, and which integrates problems associated with the environment, employment, and regional planning' (Bové, 1998).

Through the Agen trial, the defendants broadened the public debate beyond biophysical risk. Although they called upon scientists as witnesses in their defence, they also invoked their own expertise as *paysans* and as trade-union workers uniquely situated to speak about the issues: food quality, farmers' duties to protect and develop French seeds, and the implications of industrialised agriculture for rural peoples and cultures. In this trial and subsequent ones, such arguments were rejected by the court but gained credence and legitimacy outside that arena, notably in the mass media.

By bringing the debate into new public arenas, they went beyond the risk framework of official procedures. According to one analysis:

> As the French debate illustrates, the struggle over agricultural biotechnology is not just about genetic technoscience or risk assessment. In this struggle, there is a collision between two framings and their forms of expertise. When initially framed as a 'risk' issue, the GMO debate invoked scientific expertise for assessing the environmental and health hazards associated with the technology. When later defined as a 'food quality' issue, the debate shifted to *paysan* expertise. (Heller, 2002: 5)

Figure 3.1 'Mac Do out—Protect Roquefort'. (Credit: Confédération Paysanne).

Conf activists found extra opportunities to draw links with globalisation through an earlier trade dispute. Since the 1990s the US government had been pursuing its complaint that the EU had banned US beef exports, without adequately justifying a scientific basis in risks of hormone-treated beef. In 1998 the WTO Appellate Body ruled in favour of the US government and then authorised higher tariff barriers on some European food exports, as compensation for the USA's economic loss. The tariff list included specialty products such as Roquefort cheese, being produced by French *paysans*. As a symbolic retaliation, in 1999 *Conf* activists symbolically dismantled a McDonald's restaurant under construction in Millau. Their statements associated GM crops with *malbouffe,* i.e. junk food such as hamburgers, hormone-treated beef and GM food. With their slogan 'Gardarem Roquefort', they adapted the 1980s slogans—'Gardarem Larzac and Plogoff'—opposing government plans for a military base and nuclear plant in those towns, respectively.

When prosecuted in court, the defendants put 'globalisation' and government policy symbolically on trial, attracting large protests in the town and nationwide attention in the mass media. Democracy too was put on trial, as noted by an academic analysis:

> In this way, activists have used the courts as a public arena to confront their arguments with other actors, and to promote their cognitive and normative framework for the GM debate away from simply 'risk'

Figure 3.2 'Gardarem Plogoff': farm animal appeals for solidarity in 1980s campaign poster.

> issues, and towards globalisation, *malbouffe,* a challenge to the legitimacy of public sector research, and a debate about what constitutes a 'democratic debate'. (Marris et al., 2004: 20)

Through such protests, the French public controversy was extended to agri-innovation choices, far beyond environmental risk issues. Agbiotech opponents framed the ideal agriculture as a common good linking producers with consumers, through an agrarian-based rural development. They targeted agbiotech as a threat which would reduce agriculture to another branch of industry. Against the commoditised inputs of multinational companies, they counterposed a *paysan savoir-faire,* as a basis for a different societal future. In those ways, they also 'set in motion a discourse and an activist strategy that would later counter the risk hegemony of the French GMO debate' (Heller, 2002: 16).

5 UK: 'CONTAMINATION' ON TRIAL

Since the early 1990s, small NGOs in the UK had criticised the government's agbiotech policy, especially for lax regulation. An opposition movement first appeared in 1996: activists carried out secret night-time attacks on field trials of GM crops, which made them vulnerable to being seen as vandals. They soon shifted to overt, playful, even theatrical activities. Activists adopted an 'X' symbol, evoking alien imagery from the *X-Files* TV series. In April 1997 they dressed up as cartoon characters, 'Superheroes against GenetiX', for a brief takeover of Monsanto's headquarters.

In the run-up to protests against the G8 Summit in Birmingham in May 1998, an activists' meeting set up 'GenetiX Snowball: a campaign of civil responsibility'. Their name and tactics were inspired by peace activists who had each made small cuts in fences at nuclear missile bases. By analogy, the Snowballers would collectively, openly 'decontaminate' GM maize fields, thus encouraging others to follow their example. GenetiX Snowball leaflets and collection bags portrayed GM crops with a biohazard symbol. Activists produced a handbook 'for safely removing genetically modified plants from release sites in Britain', along with the slogan: 'Freedom from American corporations!' (Thomas, 2001: 338).

To claim legitimacy, GenetiX Snowball quoted the UK Deputy Minister of Agriculture: 'The government is not in the driving seat'. Here the Minister meant that commercialisation was driven by companies and by EU decisions to approve their GM products, thus allowing little choice for member states. According to the activists, 'Our democratic system has failed us; government has waived its responsibility . . . Meanwhile transnational corporations hold the reins and pull the strings of power' (GenetiX Snowball leaflet, 1998). Thus the technology and its authorisation were framed as an undemocratic, sinister control.

Monsanto sought to deter attacks on its field trials but did not press for criminal charges, which would have entitled the activists to a jury trial. Instead Monsanto sought a civil injunction banning anyone from interfering with its crops. GenetiX Snowball protested against this attempt to 'privatise justice'. They 'attempted to turn the civil proceedings into a full trial on the issue of genetic engineering, turning their media-savvy attack from the crops to the behaviour of the bullying corporation itself', as an activist later put it (Thomas, 2001: 342).

Indeed, the defendants successfully argued that the case warranted a full trial, on grounds that they were protecting non-GM crops from damage. According to the judge, they had made sufficient arguments about a 'justification in the public interest'. As a headline writer put it, 'Crops case puts GM food on trial' (Brown, 1999). Likewise, according to a trade journal, the case was likely to be billed as 'GM foods on trial' (Anon, 1999). Monsanto obtained a temporary injunction against GenetiX Snowball, pending the full trial. Meanwhile activists had already used the opportunity to gain campaign advantages, even to reverse the original charges, at least in the jury of public opinion.

The opposition movement was joined by large NGOs, especially Greenpeace and Friends of the Earth, which metaphorically put agbiotech and its inadequate regulation on trial. In July 1999 Greenpeace volunteers were

Figure 3.3 Snowballing genetic decontamination. (Credit: Paul Fitzgerald).

arrested for removing a crop of GM maize. At the subsequent trial for criminal damage, the legal defence argued that those involved had a 'lawful excuse', as the defendants had reasonable beliefs about the risks. The defence submitted statements by scientific experts about inadequate science and regulatory scrutiny; these were not challenged by the prosecution, so the scientific arguments were never aired in court. In lieu of such a hearing, the expert statements were published as a booklet, *GM on Trial,* scrutinising scientific uncertainties and weaknesses of safety claims (Greenpeace, 2000).

Among other theatrical stunts, Greenpeace dumped a lorry-load of soybeans at 10 Downing Street in early 1999. As a message to the Prime Minister, the lorry bore the slogan, 'Tony, don't swallow Bill's seed', alluding to the Monica Lewinsky hearings. The slogan humorously linked US commercial aggression with sexual contamination and political subordination to a foreign power.

Through various pollution metaphors, more generally, activists stigmatised all institutions which might promote, authorise or sell GM products. 'GM contamination' had diverse meanings, for example: unnatural genetic combinations posing unknown ecological risks, money interests perverting science, multinational companies controlling seeds, etc.; globalisation corrupting national democratic procedures; intensive methods further industrialising agriculture and perpetuating technological dependence; and pollen flow contaminating non-GM crops, thus denying consumer choice (Levidow, 2000).

Figure 3.4 Dumping soya in Downing Street (Credit: © Greenpeace UK/ Nick Cobbing).

Monsanto carried out an advertising campaign in mainstream newspapers in order to address criticisms. However, this generated negative comment in the mass media, depicting the putative biotechnological cornucopia as a threat. Some newspapers even ran campaigns against agbiotech. In the *Daily Mail,* a headline warned against 'Frankenstein foods', later shortened as 'Frankenfoods'. In the satirical magazine *Private Eye,* a cartoon showed an enormous Father Christmas rampaging through the streets, with the caption, 'Run for your lives, it's Monsanta!' The Monsanto slogan—'Food, Health, Hope'—was parodied by *The Ecologist* magazine: 'At Nonsanto, opinion is something that we buy . . . Fraud, Stealth, Hype'.

Moreover, political cartoons regularly portrayed Prime Minister Tony Blair as an agent of the biotechnology industry. His tomato-shaped head symbolised Zeneca's GM tomato, a product which had been sold and clearly labelled without any controversy. A metal bolt through Blair''s head reinforced the 'Frankenfood' slogan. In one cartoon, the Environment Minister keeps a 'safe distance' from him, by analogy to the standard isolation distance for testing GM crops (*The Independent,* 18.06.99). These themes were again linked during the Gulf War, when a cartoon portrayed President George W Bush as a maize-headed master giving his maize-headed poodle Blair an order to 'Heel' (*The Independent,* 10.03.04). In such images, the Prime Minister is contaminated—and contaminates democracy.

In opposing agbiotech, NGOs promoted organic food as a wholesome alternative. With the warning, 'Protect your food', Greenpeace launched a True Food Campaign. Friends of the Earth launched a Real Food Campaign. Another key opponent were organic farmers, whose Soil Association campaigned especially against 'GM contamination', which would disqualify crops from organic certification.

A loose network of activists, the Genetic Engineering Alliance, proposed a 'Five Year Freeze' on the commercial use, import or patenting of GM products. Its February 1999 manifesto criticised shortcomings of the regulatory system and demanded public involvement in such decisions (GEA, 1999). Soon the coalition had attracted more than forty members, including consumer, environmental, development and quasi-governmental organizations.

Facing intense public suspicion towards agbiotech, farmers became divided or ambivalent. Like their European counterparts, UK organic farmers opposed GM crops, especially by setting a 'zero tolerance' policy for organic certification in 1998. This policy raised the stakes for the spread and presence of any GM material. The National Farmers Union initially supported GM crops as an important tool for economic competitiveness, but later its members became more cautious. Early dissent came from a split-off group called the Small & Family Farm Association (later FARM), which advocated government support for extensive cultivation methods.

Agbiotech was framed in ominous ways even by mainstream bodies. Some drew an analogy between GM crops, industrialised agriculture and the BSE crisis. The Director of the Consumers Association ridiculed the agro-food industry for its 'unshakeable belief in whizz-bang techniques to conjure up the impossible—food that is safe and nutritious but also cheap enough to beat the global competition' (McKechnie, 1999). According to a report of the UK Environment Agency, agri-biotech products became controversial because they are designed for an 'increasingly intensive monoculture'. Therefore GM crops should be evaluated in a wider debate about sustainable agriculture, 'not just relative to today's substantially less-than-sustainable norm' (Everard and Ray, 1999: 6).

Environmental issues were framed by different accounts of sustainable agriculture, especially regarding the potential effects of GM herbicide-tolerant crops, designed to replace specific herbicides with broad-spectrum herbicides which kill all vegetation. According to proponents, these crops would help farmers to minimise herbicide sprays and so protect wildlife habitats in or near agricultural fields. According to critics, broad-spectrum herbicides could increase such harm.

Given the intense public controversy over agbiotech, the herbicide implications could not be contained within the existing regulatory framework and expertise. In 1997 the government's own nature conservation advisors advocated a delay in commercial use of herbicide-tolerant crops, pending additional research (see Chapters 5 and 6). Expert advisors noted a conflict between the two government objectives: on-farm biodiversity protection and 'building a competitive industry', i.e. between environmental protection and more efficient production. Consequently, the advisors argued, the

Figure 3.5 Bush telling Blair to 'heel' (Credit: © Dave Brown, *The Independent* newspaper, London, 10 March 2004).

government should establish a stakeholder forum 'to discuss and balance the inevitable conflicts' (ACRE/ACP, 2001). Such proposals led the government to establish the Agriculture and Environment Biotechnology Commission, which in turn supported demands that the government sponsor a public debate on GM crops (see Chapter 4).

6 UK PUSZTAI AFFAIR: SCIENTIFIC INTEGRITY ON TRIAL

As public unease about GM foods intensified in the late 1990s, the UK government funded additional research on risk-assessment methods, including efforts to improve and standardise whole-food tests on animals. This large project was based in the Rowett Research Institute (RRI) and led by Arpad Pusztai, an internationally renowned expert on lectins—naturally occurring toxins that protect plants from insects. Intended to enhance the public credibility of regulatory science and oversight, the project took a course which eventually undermined that aim.

The RRI project used GM potatoes containing a transgene for a lectin that was understood to be harmless to mammals. After ingesting the GM potato, however, rats apparently suffered damage to their immune systems and organ development. With encouragement from the RRI, Pusztai announced these unexpected experimental results on a UK television programme in 1998, before his report had undergone peer review. The transgene itself was not a plausible cause of damage, but he raised the possibility that the genetic modification process had led to an unknown change in the potato. This hypothesis raised doubts about the safety of GM foods already on the market, thus intensifying the overall controversy

Following the television programme, the institutional response became as important as the research results. The RRI ended its support for the research group's work, terminated Pusztai's employment and denied him access to his research data. He was also told to make no more public statements about his research. Such restriction can be imposed under the normal terms of public-sector research, though it was not imposed on scientists who make safety claims. Such responses led to speculation about the RRI's motives. According to some critics, the RRI were more interested in research contracts than in independent science (quotes cited in Levidow, 2002).

When an RRI Audit Committee reported that Pusztai's claims lacked a solid basis in his lab results, he wrote an Alternative Report. International networks of scientists made statements for and against the validity of Pusztai's work. An open letter supporting him was signed by twenty scientists from nine countries, coordinated by Friends of the Earth. Pro-agbiotech scientists subjected Pusztai to character assassination and maligned his motives, as if his questions about GM food safety were a desperate attempt to extend his research funding beyond retirement age.

Expert claims became overtly politicised. The Royal Society convened a special committee, which concluded, 'there is no convincing evidence of adverse effects from the GM potatoes in question' (Royal Society, 1999). In response, *The Lancet* criticised the Royal Society for a 'breathtakingly arrogant' approach to risk research on GM safety (Editorial, 1999). *The Lancet* also published a paper based on the GM lectin research, along with various commentaries (Ewen and Pusztai, 1999). Soon afterwards, according to the *Lancet* editor, he had been threatened by a Royal Society member that his job would be at risk if he published the paper (Flynn and Gillard, 1999).

Some scientists indicated the need to repeat the RRI experiments with more rigorous methods, but this was not done. The original team was not permitted to continue its research. There was no funding for further research—e.g. in order to investigate the uncertainties raised by the RRI experiments, or to overcome any methodological limitations.

An expert controversy quickly turned into a broader debate about the politics of regulatory science, vested interests and 'Frankenfears'. Journalists highlighted those issues, as in the following examples:

> Hostility to genetically modified food has exploded in Britain amid claims that it is being rammed down the public's throat without proper safety testing. (Coghlan et al., 1999)

> Consumers find neutral advice elusive . . . In the face of an impassable morass of scientific dispute, consumers are forced to search for impartial voices to guide them on to solid ground. (Fenton and Irwin, 1999)

> If a renowned expert working in a publicly-funded institute can have his reputation destroyed and his research suppressed, what message does this give to other scientists working on research that may be controversial? (Anderson, 1999)

Through the GM food controversy, then, agbiotech critics put UK institutions on trial by questioning their financial independence and scientific integrity. Official experts were put onto the defensive for their safety claims about GM food. Their scientific basis was put under further suspicion—not only as inadequate knowledge, but also as driven by pro-biotech bias.

7 CONCLUSIONS

At the start, this chapter posed the following questions:

- How did early dissent expand into mass opposition?
- How did agbiotech opponents put promoters and their policy agenda on the defensive?
- How was agbiotech linked with un/sustainable agriculture?

Through the EU policy framework, in the mid-1990s agbiotech was being co-produced along with a marketisation of nature and society, in the name of eco-efficiency improvements for agriculture. Here eco-efficiency meant an input-output efficiency for producing standard global commodities. These links to neoliberal policies provided an extra target for a political-cultural conflict over a new technology. Agbiotech and globalisation became each other's Achilles' heel (cf. Buttel, 2000). Opponents could link many policy issues and institutions, as a basis for a broad European opposition.

Since the mid-1990s agbiotech critics had been developing a counter-expertise for questioning official safety claims. They challenged regulatory systems for narrow criteria and inadequate science for risk assessment. Critics readily highlighted regulatory gaps. But these were not unique to agbiotech; they cannot fully explain the intense European controversy that emerged in the late 1990s, nor the policy changes that followed. Rather, activists mobilised new social actors who demanded a voice in agbiotech policy (cf. Gottweis, 1998: 318–19).

Controversy was extended to more arenas—from the cultural, to the regulatory, economic and judicial—where agbiotech was put on trial. In regulatory arenas, normally favouring scientific expertise, official expertise was attacked for optimistic assumptions and inadequate scientific methods.

In judicial arenas, legal expertise took up precautionary perspectives on regulatory issues. Each arena has its own implicit rules for admissible arguments or favoured expertise (Renn, 1992), yet those rules were circumvented by agbiotech opponents through cultural discourses. By entering and linking all these arenas, agbiotech opponents gained a mass audience.

Agbiotech opponents challenged the EU's models of agri-innovation and political decision-making. In their apocalyptic frame, agbiotech was turned into a symbol of multiple anxieties: the food chain, agri-industrial methods, their inherent hazards, global market competition, state irresponsibility and political unaccountability through globalisation. These issues were linked by drawing analogies between agbiotech and the 'mad cow' scandal BSE crisis. Protest stigmatised GM food as dangerous, as unsustainable agriculture and even as pollutants (FoEE, 2000). These attacks gained mass support through a collective action frame resonating with intuitive public meanings (cf. Benford and Snow, 2000).

In 1996–97 the first US shipments of GM soya and maize were turned into symbols of such threats, thus gaining Europe-wide mass-media attention and putting institutions on trial. After authorising Bt 176 maize in early 1997, the European Commission was widely accused of favouring commercial interests over safety—even 'recidivism', i.e. repeating its criminal collusion in the BSE scandal. That symbolic guilt was soon formalised when the European Parliament denounced the Commission. When France proposed to approve maize varieties derived from the same Bt 176 in 1998, NGOs filed a challenge at the Conseil d'Etat. The French government underwent a judicial trial for several crimes, especially failure to apply the precautionary principle. These accusations gained a hearing in the court, while providing a basis for a wider political trial in the mass media and civil society.

Through informal and formal trials, Governments were put onto the defensive for regulatory approval decisions. When activists were prosecuted for sabotaging field trials, they symbolically reversed the charges, by accusing the state of failure to protect society from agbiotech. In parallel with UK court proceedings, for example, a Greenpeace booklet explicitly put *GM on Trial* by highlighting scientific weaknesses of safety claims.

Lab and field experiments were intended to provide evidence for product safety, and thus to justify commercial authorisation. Field trials were designed to demonstrate that GM crops were being kept safe, while demonstrating the diligent responsibility of the authorities. Yet safety science came under criticism for ignorance and dangerously optimistic assumptions. Some experimental results were discursively linked with death images. The authorities were put symbolically on trial for failure of responsibility to develop adequate knowledge and precautionary methods.

Two experiments found evidence of risk, in turn generating high-profile disputes over the scientific basis for safety claims. When a Swiss lab experiment found harm to beneficial insects from an insecticidal GM maize, agbiotech proponents questioned its methodological rigour. In response, agbiotech critics raised similar questions about the rigour of other experiments which had supposedly demonstrated safety. As another high-profile case, UK lab experiments found harm to rats from GM potatoes, leading the project leader Arpad Pusztai to question the safety of GM foods. When he was marginalised and subjected to character assassination by agbiotech proponents, his supporters put UK expert bodies symbolically on trial by questioning their financial independence and scientific integrity, with sympathetic reportage in the mass media. The Pusztai affair was turned into a symbol of precautionary science being suppressed for commercial or political reasons. In both cases, surprising experimental results were turned into symbols of government and expert irresponsibility.

An anti-agbiotech discourse-coalition gained widespread support in civil society. Similar issues and discourses circulated across conventional boundaries of NGOs—environmentalist, consumer, farmer, etc.—as well as across national boundaries. Beyond simply 'activists', a wider societal participation took various forms such as public meetings, protest actions, consumer boycotts, sabotage of GM field trials, etc. In France and the UK, saboteurs portrayed themselves as defenders of democracy and the environment against globalisation.

The defendant on trial was expanded—from product safety, to biotech companies, their innovation trajectory, regulatory decision-making, expert advisors, government policy and its democratic legitimacy. As activists-cum-prosecutors expanded their accusations, these were taken up by the broader civil society. Acting as a pervasive mobile 'jury', civil society groups further accused and judged the various defendants. Applying a cultural rationality, they raised questions about whose control and interests were driving the technological development (cf. Krimsky and Plough, 1988; Fischer, 2005). Through links across diverse issues, arenas and constituencies, opponents stimulated a civil society networks to which governments could be held accountable.

Risks were opened up beyond the terms of official safety assessments. Regulatory decisions were attacked for too narrowly defining 'harm', for ignoring or accepting undesirable environmental effects and for failing to investigate plausible risks. Official expertise was criticised for optimistic assumptions and inadequate scientific methods. For example, when France led the proposal for the EU to authorise glufosinate-tolerant oilseed rape, the French biosafety committee treated the weed-control implications as irrelevant for this decision. Amidst expert and public debate over that issue, the French government reversed its stance in 1997, so authorisation was withheld, leading to judicial disputes over the basis for this impasse. Moreover, the French agricultural research institute (INRA) abandoned its innovation research on similar crops, partly to protect the neutral image of its risk research. GM glufosinate-tolerant oilseed rape was kept on trial as a sustainability problem for weed control, especially through ecological field research on gene flow and hybridisation with weedy relatives.

Agbiotech intersected with a prior conflict over the meaning of sustainable agriculture. The EU subsidy system conceptually distinguished agriculture from the wider environment; greater agri-industrial efficiency was promoted as a solution to competitive economic pressures. Small-scale, less-intensive farmers opposed these policies for intensifying dependence upon agribusiness and globalised commodity-production. Thus divergent accounts of sustainable agriculture favoured different future scenarios for what should be sustained—what kind of economy, environment and society (see Table 3.2).

In sum: Risk issues were linked with sustainable agriculture in contending ways. From an eco-efficiency frame, agbiotech was promoted for a sustainable, high-yield intensification of agriculture, as a means to avoid unsustainable agrichemical practices; this policy gained support from organisations of agri-industrial farmers and many European authorities. Eco-efficiency claims really meant an input-output efficiency of resource usage; this framework remained at odds with agri-ecological perspectives on sustainable agriculture.

From the apocalyptic frame of opponents, GM crops threatened the European countryside—e.g. as an aesthetic landscape, a wildlife habitat, local heritage, involving a stewardship role for farmers and contributing to farmers' economic independence. Agbiotech opponents initially counterposed organic agriculture as truly sustainable; later they supported 'quality' alternatives, as a basis for alliances with organisations of less-intensive and small-scale farmers. Such alliances emerged especially in Austria, Italy and France.

As GM crops were put on trial for contributing to an unsustainable agriculture, the societal criteria were elaborated in various arenas. Institutions faced greater pressure to test claims that GM crops would provide agri-environmental improvements. Thus informal-symbolic trials were shaping and pervading formal ones.

Agbiotech opponents were criticised for unfairly targeting GM crops as a proxy for extraneous issues such as globalisation and sustainable agriculture. Yet agbiotech was already being promoted along such lines—as an essential means for responding to market forces, as an environmentally friendly techno-fix, and as an objective imperative to marketise public goods. Thus the EU policy framework facilitated a neoliberal co-production of agbiotech, nature and society. By the late 1990s mass opposition was undermining that project. Governments were put on the defensive for irresponsibly failing to exercise adequate control over agbiotech. Citizen action was making Europe relatively less 'safe' for agbiotech to be commercialised—and more open to alternative futures. In parallel, citizens' capacities and potential roles were being tested in participatory procedures, as shown in the next chapter.

4 Channelling Participation, Testing Public Representations

INTRODUCTION

Agbiotech was challenged early on as a choice for future agriculture—first in Germany and Denmark, and later in other countries such as France and the UK (as analysed in Chapter 3). Through autonomous initiatives such as open meetings, protest, boycotts, mass-media stunts and sabotage, citizens' collective action challenged government promotion of agbiotech. Societal conflict potentially jeopardised the policy framework on which agbiotech depended.

In responding to or anticipating such conflict, state bodies established extra procedures for risk regulation and public consultation. They also sponsored formal participatory events, which can be seen as technology assessment exercises. Some were called citizens' or consensus conferences, where small pre-selected groups deliberated together on expert claims about agbiotech, resulting in a group report. Other events have been called simply 'public debates', which occurred alongside autonomous, sporadic public debates that were happening anyway. These formal exercises had a potential scope to consider all aspects of agbiotech, yet the outcomes generally focused on risk issues.

To analyse those participatory state-sponsored exercises as a specific arena, this chapter discusses the following questions:

- How and why did state bodies sponsor participatory technology assessment (TA) of agbiotech?
- What aims arose in designing, managing and using those exercises?
- How did each exercise test models of public representation and citizens' roles?
- How did the process bear upon the accountability of representative democracy?

Spanning over sixteen years, this chapter overlaps with the period of the previous one on protest movements, as well as the next one on regulatory conflicts. After explaining some analytical concepts and EU-wide contexts, the chapter surveys participatory TA exercises in Denmark, Germany, the UK and France, as well as EU-level consultation events where similar issues arose.

1 CONTESTING BOUNDARIES IN PARTICIPATORY TA

A meaningful role for public participation depends on which issues are open for government decisions, as well as which voices which must be heard. For agbiotech policy in Europe, both were increasingly disputed in the 1990s, especially around the following boundaries:

- Imperatives/alternative options: agbiotech innovation as a presumed imperative for a common European future, versus alternative understandings of agri-environmental problems (as mentioned in Chapter 2);
- Scientific/policy issues: issues deemed 'scientific' and thus reserved for expert judgements, as distinct from policy issues open to public involvement;
- Expert/lay status: voices which were deemed adequately expert for involvement in decision-making, as distinct from lay voices which could judge (at most) which experts to trust.

These tensions arose in related national arenas: regulatory decision-making, consultation procedures and special participatory exercises. Risk regulation was the only official basis for government decisions on agbiotech products and thus the main formal opportunity for public involvement, e.g. through comments on draft rules and regulatory decisions. As a fundamental constraint on meaningful participation, however, dominant policy frameworks had a strong commitment to agbiotech as societal progress, to be realised through a series of useful eco-efficient products. Delay or refusal of a product could be justified only by evidence of risks (as sketched in Chapter 2).

The 1990 EC Deliberate Release Directive mandated an environmental assessment of risks only. The benefits of GM products were taken for granted in general terms, as solutions or improvements in the context of intensive monoculture, and were left for later judgements by individual consumers in the EU internal market. At the same time, expert advisors implicitly anticipated environmental benefits, e.g. through pesticide reduction, as grounds to accept potential undesirable effects or to disregard uncertain risks. This implicit risk-benefit framework was not open to public scrutiny, much less change (Levidow et al., 1996).

Within that framework, the EU regulatory procedure was designed for flexibly accommodating public participation, new issues and potential controversies. The 1990 EC Deliberate Release Directive set requirements for information disclosure by national Competent Authorities (CAs). Article 19 specified minimum details that must be disclosed regarding each notification for GMO field trials and each proposal to grant commercial authorisation. Under Article 7, each CA could decide whether or how to consult the public.

Regardless of formal consultation, information disclosure potentially gave citizens a basis for public comment or protest. Even by the mid-1990s, however, few CAs were disclosing adequate information as a basis for meaningful public comments. So NGOs demanded more—or sometimes obtained information through other member states which had greater disclosure. In the Netherlands, for example, all documents were placed on open-access shelves of the Environment Ministry library.

That full disclosure complemented the Dutch political culture of *Overleg,* an open deliberative approach which gained general prominence in the 1990s. The Dutch government subsidised new actors, such as environmental groups, which were promoting more critical policy frames (Halffman and Hoppe, 2005). For the agbiotech sector, the Dutch government's expert advisory body Cogem included a vocal critic of agbiotech from an environmental NGO. This arrangement was meant to deliberate conflicting views and to seek consensus—even if resulting in overt disagreements, as often happened (von Schomberg, 1996).

As an opposite extreme, Spain disclosed almost no basic information on GMO releases, even when the first GM products were proposed for commercial authorisation, aimed especially at the Spanish market. This virtual secrecy complemented the priorities of the official expert body: namely, to avoid domestic dissent or disputes, rather than take a leading European role in issues of scientific uncertainty (Luján et al., 1996). France likewise disclosed little information, consistent with its technocratic political culture, restricting involvement to official expertise (see later section).

In such ways, each country implemented EU legislation according to its own strategy for anticipating, avoiding or managing societal conflict. Each strategy involved a choice of government department to act as CA for the EC Directive, along with a structure and composition of an expert advisory body. Information disclosure was designed to restrict or structure public involvement, especially through boundaries between expert and lay roles.

Overall these measures provided a national strategy for representing agbiotech, the public and the relevant expertise, as means to legitimise decisions on agbiotech:

> Democratic engagement with biotechnology was shaped and constrained by national approaches to representation, participation, and deliberation that selectively delimited who spoke for people and issues, how those issues were framed, and how far they were actively reflected upon in official processes of policymaking. (Jasanoff, 2005a: 287)

Such representations of the public were devised and contested in many arenas—the mass media, regulatory procedures and public consultation. In some countries, state bodies sponsored participatory technology assessment (TA) exercises, sometimes called consensus or citizens' conferences. Individuals were pre-selected to deliberate together on expert claims about agbiotech, in order to reach conclusions that would be publicised and potentially influence a wider debate.

The rest of this chapter surveys such exercises. The events differed greatly in several respects—their historical periods, their policy contexts, their basis for selection of participants, their remit and their problem-definitions. In the Danish and French cases, for example, a Parliamentary body hosted the participatory initiative in a crisis period, when government decisions were imminent and Parliament sought a more authoritative role in agbiotech policy. In the German and UK cases, the exercises remained relatively more distant from any government decisions. In most cases the participants were chosen as ordinary citizens, ideally with no prior involvement in agbiotech issues. In the German case, by contrast, participants were quasi-expert representatives of stakeholder groups. In all cases the thematic focus was agbiotech—GM crops in the UK and French cases, and more specifically herbicide-tolerant crops in the German case. In all cases, some participants wanted the procedure to consider alternative options, leading to overt conflict in the German case.

Alongside those differences in context and structure, the exercises had some common features. Conflicts arose over their remit and evaluation criteria; such conflicts were generally channelled into regulatory issues. In some cases, especially Denmark and France, public participation anticipated, stimulated or reinforced policy changes which enhanced government accountability for regulatory decisions. Societal conflict across several countries was briefly sketched in the previous chapter on protest activity. Here each consultation exercise is placed in its national context of political culture, in order to analyse representations of agbiotech, the public and lay/expert boundaries.

2 DENMARK 1987: SUSTAINABLE AGRICULTURE?

In the late 20th century, the Danish government was seeking to make policy through consensus-seeking, especially on environmental issues. Relevant stakeholders were consulted on legislative proposals, in order to find a solution that would be acceptable to most of them. Given that no political party had a Parliamentary majority, negotiation among parties was necessary to build a majority. During the 1980s–1990s environmental protection laws were achieved by a 'green majority', linking parties inside and outside the government.

Not simply a pragmatic necessity, this consensus-seeking expressed democratic aspects of Danish political culture. Its origins lay in an adult education movement, which has a long history of public enlightenment activities. Initiated by farming communities, this was extended more widely through government-funded Day Folk High Schools, run by the Dansk Folkeoplysnings Samråd (DFS), the Danish Adult Education Association. Beyond simply the competences needed for employment, the DFS promotes education for citizen involvement:

> It improves the ability and motivation to assume self-administration and responsibility and it also provides people with the necessary competences that enable each individual to take active part in society. . . . Non-formal adult education is based on community, equality, participatory democracy, responsibility and voluntary activity. (DFS , 2006)

From this cultural basis, political negotiation could be held accountable to an informed citizenry, not simply political parties or established stakeholder groups.

That political tradition also provided a cultural basis for public participation in technology issues. After Parliament established the Danish Board of Technology, it held a series of consensus conferences, starting with the topic of agbiotech. In general the DBT has timed such events to coincide with Parliamentary debates on the same issue; this linkage has helped to stimulate wider public involvement, to broaden the issues, and thus to influence the overall policy debate, according to a long-time promoter there (Klüver, 1995; also Joss, 1998).

The consensus conference centres on a lay (non-expert) panel of 'interested citizens'; they are selected to represent diverse views, though not necessarily to represent the overall population. The panel question and evaluate expert views, including scientists, opinion-formers and anyone whose knowledge goes beyond general knowledge (Grundahl, 1995). Through such scrutiny, the panel seeks to reach a consensus on practical recommendations. Minority views are reported only in rare cases where important differences remain unresolved (ibid.; Klüver, 1995: 47).

For all those reasons, the Danish consensus conference has been advocated as a 'counter-technocracy'—a means to challenge expert claims through a deliberative process. The lay panel has no vested interest different than the general public, and its report helps to promote TA as a broad societal process. It extends a Danish tradition of *folkeoplysnig*—people's enlightenment through an adult education network which builds a reflective, informed citizenry (Joss, 1998: 20).

As its guiding principle, 'a well-functioning democracy requires a well-educated and engaged population'. Successful participation is understood in those terms: as a participant commented, for example, 'We initiated a really good assessment process among the public' (cited in Klüver, 1995: 41, 43). In the Danish consensus conference, then, 'interested citizens' personify a political culture in which technological decisions are held accountable to public debate, mediated by Parliament.

Denmark's debate on agricultural biotechnology was initiated in the mid-1980s by environmental NGOs. Several 'debate booklets' were issued by NOAH, the Danish affiliate of FoE, proposing new legislation to regulate GMO releases. In response to public concerns, a Parliamentary 'green' majority imposed a statutory ban in the 1986 Gene Technology Act; releases would not be permitted unless there was sufficient knowledge about the ecological consequences (Toft, 1996). With this wording, the government could be held accountable to demonstrate such knowledge for risk assessment; this burden of evidence meant a *de facto* ban for several years.

Parliament also mandated funds for an information campaign on biotechnology. Some funds were specially earmarked for NGOs, especially NOAH and some trade unions, in order to stimulate further debate on advantages and disadvantages of biotechnology. In these ways, environmental NGOs gained extra resources and political opportunities to frame the issues for further public debate. NOAH organised ten public conferences on the wider environmental consequences, on sustainable agriculture including organic agriculture, on food labelling, on animal welfare and ethics, on the Third World, on seed diversity (including patents) and on biological warfare. These debates were reported through a series of publications and statements from NOAH.[1]

- 1987 Consensus Conference and beyond

In that context the Danish Board of Technology held its first consensus conference in 1987 on 'Gene Technology in Industry and Agriculture', timed to coincide with Parliamentary debate on the issue (Hansen et al., 1992; Klüver, 1995: 44). In its report the lay panel took up risk issues as well as ethical ones (Teknologinævnet, 1987). Accepting a key recommendation, Parliament voted to exclude animals from the 1987–90 national R&D programme for gene technology. The conference eventually had more profound effects on the Danish regulatory regime through wider public debate.

A further information campaign was coordinated by the Board of Technology and Danish Adult Education Association. During the period 1987–90 they supported more than 500 local meetings all over the country in order to stimulate debate on human and non-human uses of biotechnology, including concerns about risk and ethics. Environmental NGOs were often invited to speak, as the most visible critical actors on the scene.

The government also funded a subsequent programme, organised by trade unions, to stimulate further debate on advantages and disadvantages of agbiotech. Their educational materials posed questions about sustainable agriculture: For example, would genetically modified crops alleviate or aggravate the existing problems of crop monocultures? (Elert et al., 1991: 12). Through that wider debate, the consensus conference indirectly influenced Parliament and thus regulatory policy.

In the EU-wide regulatory procedure, dominant member states implicitly took for granted eco-efficiency benefits of herbicide-tolerant crops, while disregarding the herbicide implications or assuming them to be benign (Levidow et al, 1996, 2000). By contrast to those EU-level assumptions,

---

1 Much information here, supplied by Jesper Toft, is not available in English-language documents.

Danish regulators were held accountable for assessing the broad implications of GM crops for agricultural strategy, herbicide usage and the environment. Such judgements were scrutinised by the Parliament's Environment Committee, often by drawing upon specific questions from NGOs. Under such domestic pressures, Danish representatives in turn proposed that risk assessments evaluate those implications at the EU level (Toft, 1996, 2000).

Thus citizen participation enhanced government accountability for regulatory criteria, going beyond optimistic assumptions about environmental benefits. GM crops were subjected to criteria of sustainable agriculture, which in turn were opened up to the lay expertise of agbiotech critics. Environmental NGOs found greater scope to influence regulatory procedures and expertise.

Agri-innovation priorities became more contentious in the late 1990s, however; NGOs demanded alternatives to agbiotech and to intensive agricultural methods. In a 1999 consensus conference, the lay panel asserted the need for extra measures—not only for product safety, but also to prevent GM products 'becoming controlled by monopolistic companies', as well as measures to evaluate ethical aspects (Einsiedel, 2001). As the conference organisers emphasised, those proposals were expressing citizens' viewpoints, thus providing a basis for dialogue with decision-makers (Teknologinævnet, 1999). The panel's proposals challenged some assumptions and limitations of the EU legislative framework. Yet such demands for greater accountability were being channelled into more stringent measures to regulate biophysical risks. This pervasive tension between societal demands and incorporation has analogies to later TA exercises.

3 GERMANY 1991–93: PARTICIPATION TRAP

German political culture has been generally theorised as a *Rechtstaat*—a 'state of law'. For technology policy, regulatory decisions follow logically from the current scientific knowledge, as determined by official expertise. This procedure can be used strategically to demarcate between expert and lay status, even between rational and irrational views. In Germany,

> expert committees are usually constituted as microcosms of the potentially interested segment of society; judgments produced in such settings are seen as unbiased not only by virtue of the participants' individual qualifications, but even more so by the incorporation of all relevant viewpoints into a collective output . . . They become perfectly enclosed systems, places for a rational micro-politics of pure reason, with no further need for external accountability to a wider, potentially excluded, and potentially irrational, public. (Jasanoff, 2005b: 220)

The *Rechtstaat* model was extended to the new sector of agbiotech, which faced significant opposition from the start.

Since the time that the German government promoted agbiotech in the 1980s, this policy provoked widespread protest—e.g. from the Green Party, environmentalist groups and local campaigns. Although critics gained high-profile attention in the mass media and civil society, their views remained marginal to official procedures, unlike German corporatist arrangements for labour issues. Opposition to agbiotech split civil society and the major political parties (Gill, 1996).

Under the 1990 Genetic Engineering Act, risk regulation had to be based on 'the state of the art in science and technology'. Although this criterion could be interpreted as a requirement for better ecological knowledge and technical detection methods, in practice it meant optimistic analogies between GMOs and conventional crops. Applicants' safety claims were routinely endorsed by the government's expert advisory committee, in which molecular biologists had a dominant voice. Questions about risks and scientific uncertainty were officially dismissed as superfluous or even irrational (ibid.).

Some opportunities for public participation were available initially but were soon closed down. The 1990 Act required some information disclosure about field trials, as a basis for comment at public hearings, mainly on grounds that local people could be affected. These hearings attracted political and ethics arguments over agbiotech, in lieu of any other formal opportunity to raise issues beyond biophysical risk (Gill, 1996). When the 1990 Act was amended in 1993, it abolished the mandatory hearings, partly on grounds that they had been abused for irrelevant purposes. The 1993 amendment also accommodated industry claims that 'over-regulation' of GMOs was impeding industrial investment and job-creation in Germany (ibid.).

Official procedures offered no meaningful basis for public comment on GM crops proposed for commercial authorisation, despite the greater potential consequences at that stage. Thus the official procedures defined the risk problem, the scientific issues and relevant expertise in ways which cast critics as irrelevant outsiders. The public could either accept safety claims or else become 'irrational' dissenters.

- TA exercise

German public controversy focused on herbicide-tolerant crops, given their potential for spreading that trait and for changing patterns of herbicide usage. To address such conflicts, the government sponsored a TA exercise on GM herbicide-resistant crops in the early 1990s. Funding came from the Ministry of Industry and Research, which was strongly promoting biotechnology. It was initiated and coordinated by the Berlin Wissenschaftszentrum (Science Centre) as an experiment in environmental conflict management. The fifty-odd participants had quasi-expert roles; they included overt proponents and opponents of HR crops, as well as representatives of regulatory authorities, agricultural associations, consumer organisations, etc.

From the start, conflict erupted over how to define the relevant scientific issues and the expertise needed to evaluate them.

A broad participation was needed to deliberate the arguments arising in the polarised public debate on agbiotech, according to the organisers. The TA was designed to evaluate those arguments for and against herbicide-resistant GM technology, especially its possible consequences—but not alternative options for weed control in agriculture. Thus the procedure was 'a technology-induced TA, not a problem-induced TA' (van den Daele, 1995: 74).

Environmental NGOs counterposed the latter approach. They wanted the TA to compare biotechnology products with other potential weed-control methods, as alternative solutions to agricultural problems. However, the NGOs' proposal was rejected by the organisers (B. Gill, 1993). Consequently, the narrow remit set difficult terms for participation by the broadly representative individuals from NGOs—indeed, terms for their expert status.

As the organisers acknowledged, 'The TA implicitly accepted the matter-of-course development of technology as the starting point', as well as possible risks as the main grounds for state restrictions: 'If critics fail to provide evidence of relevant risks, the technology cannot be banned'. So critics held the burden of evidence for any risks. Advocates held the burden to demonstrate benefits, though failure to do so would have no bearing upon regulatory decisions (van den Daele, 1995: 75). This framework marginalised alternative agronomic solutions, while reinforcing the dominant system: 'intensive farming as the reference system'. Within that framework, participants themselves defined their controversies as debates about empirical evidence, e.g. regarding the possibility of environmental damage—not about values and goals (ibid.: 76, 77).

The organisers aimed to include and deliberate all viewpoints on the risk-benefit issues. By subjecting expert views to scrutiny, the TA could reach conclusions about empirical claims, rather than political or ethical ones. 'This procedure placed participants under massive pressure either to admit consensus or justify dissent', especially through detailed empirical evidence (ibid.: 80).

From NGOs' standpoint, the technology-induced TA framework effectively favoured experts in specialised technical areas, e.g. gene flow and herbicide effects. In practice, the TA exercise set a lower burden of evidence for demonstrating benefits than for demonstrating risks, in a period before much empirical research had been done on risk scenarios. Consequently, the discussion emphasised environmental benefits, especially the prospects for farmers to use less harmful herbicides and/or lower quantities of them (B. Gill, 1993).

On the basis of the expert reports, the TA symbolically normalised any risks. According to agbiotech proponents, echoing the government's advisory body, any risks from GM herbicide-tolerant crops were similar to those from conventional crop plants and herbicide usage. 'In many areas it was argued that there was no need for political action because the identifiable problems could be dealt with in the established registration procedures. . . . if one agreed to the "normalisation" of the risks' (van den Daele, 1995: 82). In this way, the exercise undermined NGO claims about novel or unknown risks; once normalised, any risks would be manageable through regulatory procedures, even contemporary ones.

- Science court or parliament?

The technology-induced TA framework posed a dilemma for participation by agbiotech critics. Once inside such an exercise, 'They have to criticize a technology which promises to satisfy some needs which may even be produced by the technology itself . . . ' (B. Gill, 1993: 74). That is, putative benefits satisfy 'needs' which are predefined by biotechnological solutions for intensive monoculture. Thus a technology-induced TA tends to accept and reproduce the social vision built into the technology.

Environmental NGOs and their associated research institutes faced a difficult choice: either play a quasi-expert role within that framework and thus help legitimise it, or else abandon that role and be treated as merely lay voices. After much conflict, they withdrew before the TA exercise could report its conclusions. They gave several reasons for withdrawal, e.g. that their voluntary participation was occupying too much time, especially the task of commenting on long expert reports (van den Daele, 1995: 81). According to an NGO expert, 'I had not imagined that you could destroy participation by throwing paper on top of people' (cited in Charles, 2001: 107). By withdrawing from the TA, they could devote greater resources to public protest and preserve their credibility with NGO members and activists (B. Gill, 1993: 81–82).

After this withdrawal decision, they were criticised by the WZB coordinator:

> One cannot present one's position in public as scientifically substantiated and then cast fundamental doubt on science as neutral . . . Participation in the procedure implies the readiness to submit oneself on the empirical issues to the judgement of science. (van den Daele, 1995: 84; also 1994)

As the WZB coordinator told the story many years later, he had been sceptical of claims that herbicide-tolerant crops had special risks or special benefits, so he saw NGO arguments about risks as a proxy for political ones:

> . . . the idea of special risks is not a good argument. We should turn to the issues of democracy and who's going to decide how society develops . . . Apparently it would have been difficult for them [NGOs] to declare explicitly that the conflict was not about risks, but about social goals and political reforms. . . . (van den Daele, cited in Charles, 2001: 107)

However, that distinction was not so clearly drawn by the organisers beforehand; it became more explicit in later retelling the story. According to a social scientist who attended the TA exercise, some NGO participants saw it as analogous to a parliament which could evaluate agbiotech in terms of societal goals. However, van den Daele retrospectively portrayed it as a science court, whose remit the NGOs did not understand or accept; this portrayal offers a *post hoc* legitimation for the failure to integrate them (personal communication, B. Gill, 2006).

Moreover, the distinction between a science court and parliament is not so straightforward; neither is the distinction between risk assessment and socio-political goals. At issue was the range of questions to be answered by science, their normative assumptions and the alternative technological options to be considered as comparators for agri-environmental assessments. Some questions from participants were pre-empted or marginalised by the TA exercise, especially by constructing particular boundaries between expert and lay voices.

Societal futures were reduced to scientific issues, readily assessable by experts in 'the state of the art'. Civil society representatives found themselves in a 'participation trap'; they could either participate within the government's risk-benefit framework for GM crops per se, or else be marginalised. Overall the exercise reinforced the government's policy framework and its public unaccountability. In a similar way, societal conflict over agri-innovation issues was channelled into risk assessment through regulatory procedures. Together these practices extended and reinforced the *Rechtstaat,* at least until government policy began to change in 2002.

4 UK 1994: RISK-BENEFIT FRAMEWORK

In the 1980s UK environmental regulation was theorised as a 'consultative' style. Backed up by broad discretionary powers, the state guided industry in regulating itself; civil servants negotiated specific control measures according to flexible criteria. Through 'responsible co-optation', erstwhile outsiders were incorporated into official deliberations—and thus into the consensual atmosphere, political constraints and confidentiality of Whitehall (Vogel, 1986: 51–52). Although policymakers sometimes claimed to base their decisions upon 'risk assessment', this denoted a qualitative exercise, with 'no fixed methodological connotation' (Jasanoff, 1986: 29). Through the consultative style, regulators generally avoided public disputes over the scientific basis of policy; expert roles were limited to those individuals being officially consulted.

By the late 1980s pressures for institutional change were coming from EC legislation and domestic civil society. In response, the UK regulatory style shifted towards a licensing system based on formal risk assessments, along with greater public transparency. This change was made for both integrated pollution control and GMOs in the 1990s Environmental Pollution Act.

Under this new regime, run by the Environment Ministry, regulatory procedures routinely disclosed proposals for GM field trials. To justify this change, regulators cited the 1993 EC Directive on Access to Environmental Information, as well as the 1990 Directive on GMOs. Such disclosure had been advocated by the Royal Commission on Environmental Pollution, on grounds that 'Secrecy breeds distrust'.

Indeed, the new UK system had a strategy to achieve the converse: that transparency would gain public trust, while incorporating potential conflicts within the regulatory procedure. This implicit role of official expertise was complemented by the UK's information disclosure and its Advisory Committee on Releases to the Environment (ACRE). Although all members were formally appointed as individual experts, ACRE included an officer of a mainstream environmental NGO, the Green Alliance, a Parliamentary lobbying and consultancy organisation. With this representation, as well as a scientific ecologist, regulators portrayed the committee as a 'broad-based consensus which includes green interests'.

This apparent consensus was devised by marginalising difficult issues. ACRE gave advice only on the risks of each proposed release, generally under some confinement measures. The advice deferred issues that might arise at later stages with fewer control measures, especially in commercial use. Even at the commercial stage, the environmental issues were defined narrowly. UK regulatory policy was officially based on a precautionary approach for protecting 'the natural environment'. This meant ignoring or accepting so-called 'agronomic effects' and herbicide implications in the agricultural environment—issues which drew criticism from NGOs (Levidow and Carr, 1996). The environmentalist member accepted that narrow remit as a basis for consensual advice from ACRE, while raising wider issues in parallel. In such ways, UK 'decision making was soon channelled into a framework of carefully structured expert committees that provided assurance by internalizing dissent' (Jasanoff, 1995: 328).

Although NGOs questioned safety claims, their statements emphasised issues which were being evaded or denied by government policy—GM product labelling, herbicide implications of herbicide-tolerant crops and corporate control over the agri-food chain. They raised these issues in consultation meetings with regulatory officials, but with no result. They generally saw no advantage in submitting comments on official risk assessments (Levidow and Carr, 1996).

- National Consensus Conference

In that national policy context, before any significant public debate on agbiotech, the UK's National Consensus Conference on Plant Biotechnology was held in 1994. Proposed by staff at London's Science Museum, it was funded by the Biotechnology and Biological Science Research Council (BBSRC). Initially reluctant to sponsor the event, the BBSRC was persuaded by the focus on GM crops as 'the least contentious' area of biotechnology, especially as compared to animal biotech. Yet civil servants criticised that focus because agbiotech was not being considered in policy debate at that time (Joss, 2005a: 211).

The exercise was coordinated by the Science Museum, whose staff implicitly diagnosed the problem as public misunderstanding or anxiety. The coordinators had previously obtained funds in the name of diagnosing and overcoming public unease about biotechnology. At the beginning and end of the Consensus Conference, the funders made clear their aim to enhance 'public understanding' of biotechnology and thus support for it. Underlying the exercise was a presumed cognitive deficit of the public.

The Consensus Conference centred upon a lay panel of relative newcomers to the biotechnology debate; they would question and learn from designated experts—whose selection was contested within the Steering Committee. Two members attempted to exclude representatives of 'extreme' anti-biotech groups from expert status—and thus from a list prepared by the organisers—though this effort did not prevail (Joss, 2005b: 211). The organisers portrayed themselves as neutrally mediating between experts and the public. However, the exercise demarcated a boundary between 'expert' and 'public-interest' views, thus demoting the latter (Purdue, 1995, 1996).

A particular lay/expert boundary was performed by expert witnesses, in the process of being questioned by the lay panel. The panel expressed views about economic, political, legal and ethical issues of agbiotech.

> Yet the key questions—and the experts' responses—were largely framed within the technocratic discourses of specialist expert knowledge . . . It was largely taken for granted that the task of technology assessment depended primarily upon the technical and professional skills of research scientists. (Barns, 1995: 203)

The structure implied that experts are needed to help overcome the deficient understanding of the public, though the lay panel often challenged the supposed neutrality of official expertise (ibid.).

> [This] set up a functional division of labour: 'lay' people ask questions, while 'experts' provide the answers. Indeed to play out their 'lay' role properly, the 'lay' panel was obliged . . . to show appropriate deference to the 'experts and the organisers. The 'lay' panel was thus encouraged to take on the challenge of investigating biotechnology, but from an exaggerated position of innocence and ignorance. (Purdue, 1995)

> The whole construction of their layness induced an undue deference to the experts, irrespective of the expert's actual level and area of competence. (Purdue, 1996: 533)

> The lay/expert boundary was reinforced in the final, public stage of the process. There the chairman tended to give pro-biotechnology speakers the status of 'mobile experts', knowledgeable on diverse aspects. By contrast, NGO activists were put on the defensive to demonstrate their expertise. (ibid.)

The process raised wide-ranging questions and disagreements, even within the Panel. Nevertheless, the organisers instructed the panellists to present a single report, permitting no minority views (Purdue, 1996: 537). Consequently, some critical views were marginalised in the panel's report, as if there were consensus on how to define risks and benefits.

Particularly marginalised were concerns about who would legitimately direct biotechnological innovation. Among themselves, panel members raised issues about who was 'in control'—e.g. concerns about R&D priorities, environmental monitoring and accountability (Joss and Durant, 1995: 82). In the panel's report, these issues were largely reduced to safety controls and patent issues.

Having listed potential benefits and risks, the report concluded: 'Biotechnology could change the world, but in order for it to be used effectively—maximising benefits and minimising risks—we also need to adapt economic and social structures to take account of the changes it might produce'. By contrast to government policy, the panel opposed any extension of patent rights; it also advocated mandatory labelling of GM food for the public right to choose. In particular: 'Regulatory control in the UK is among the most stringent; however, there is still room for improvement' (Science Museum/BBSRC, 1994: 7, 14). Although questioning some probiotech arguments, the report reinforced a common societal problem of product safety, while adding the principle of consumer choice.

- Whose consensus?

After the panel presented its final report, the document was interpreted in divergent ways. According to the organisers, 'the lay panel has given the field of plant biotechnology its qualified support' (Science Museum/BBSRC, 1994: 2). However, the report could just as well be read as sceptical; it emphasised not only risks, but also predictable disadvantages of agbiotech. It also criticised inadequacies of government regulation, along lines similar to criticisms by NGOs. One excerpted the report as campaign material, entitled 'Whose consensus?', emphasising differences between the panel's report and government policy (Genetics Forum, 1994).

The UK exercise sought mainly to explore 'the public understanding of science' in Britain (Joss and Durant, 1995: 76, 96, 104- note 14), according to the conference organisers. They claimed 'to adopt the Danish model of the consensus conference', yet this aims to generate a wider societal debate that could influence the Parliament and government. The UK exercise anyway had little potential for such influence: Parliament had no relevant policy decision at that time (ibid.: 99), and there was little public debate on agbiotech.

In any case, the lay panel had little means to challenge the UK risk-benefit framework, even if it had presented minority views. A more significant policy challenge was coming from the opposite direction. UK regulatory procedures then were facing deregulatory pressure from the agbiotech industry and other Ministries, amidst a Europe-wide campaign against 'over-regulation' (see again Chapter 2). Environment Ministry officials saw the lay panel's report as helpful for protecting their regulatory procedures and expertise from such pressure.

In all those ways, the UK Consensus Conference reinforced an expert/lay boundary within the UK's risk-benefit policy framework. The Panel recommended regulatory adaptations to ensure that agbiotech would be kept beneficial and safe. Although individual panel members raised issues about corporate-biotechnological control over the agri-food chain, these were reduced to regulatory control measures, e.g. safety regulation and product labelling. This framework implied little scope for public participation in definitions of risk or benefit, much less in innovation priorities. Policy issues could be implicitly delegated to expert bodies through normative assumptions in their advice.

5 FRANCE 1998: A BENIGN TECHNOCRATIC STATE

French political culture has been theorised as an elite centralised technocracy. Claiming a Parliamentary mandate to represent the general good, regulatory procedures readily exclude or marginalise dissent. This pattern was exemplified by the agbiotech sector. Molecular biologists dominated the expert advisory body, the Commission du Génie Biomoléculaire or CGB, hosted by the Ministry of Agriculture. This arrangement was complemented by confidentiality. Regulatory procedures disclosed little information prior to decisions, despite EC requirements to do so.

By 1997 French regulatory policy faced a legitimacy crisis. France had led efforts to gain EU-wide approval for GM crops, yet these were now opposed by a broad range of organizations. The Confederation Paysanne, representing farmers who elaborated a peasant identity, opposed agbiotech while counterposing 'quality' alternatives to industrialised agriculture (Heller, 2002). An oppositional petition was signed by many prominent scientists, not necessarily anti-agbiotech, but all of them concerned about regulatory failures to develop appropriate ecological expertise and risk research (Marris, 2000).

In February 1997 the Prime Minister decided not to authorise commercial cultivation of Ciba-Geigy's Bt 176 GM maize in France, even though French regulators had led EU authorisation of the same product (EC, 1997d). This unstable policy indicated a crisis of official expertise within an elite-technocratic political culture. According to some critics, an official 'objectivity' too narrowly defined the relevant expertise. As an alternative approach, expert procedures would open up a scientific critique of possible options; this space would provide the expertise necessary for decisions (Roqueplo, 1996: 67, my paraphrase). By incorporating counter-expertise, regulatory procedures would develop an *expertise contradictoire* (contradictory expertise), which would enhance democratic debate and state accountability for decisions.

Perhaps illustrating those concepts, in November 1997 the government announced a set of new measures. It would finally approve cultivation of the blocked Bt maize, while establishing a Biovigilance Committee including overt opponents of GM crops. This Committee would oversee efforts to monitor Bt insecticidal maize, including pollen flow and effects of the Bt toxin (Roy and Joly, 2000). This plan was not operationalised, partly because few commercial fields were ever planted with Bt maize.

- 1998 Consensus Conference

The November 1997 measures also included a plan to sponsor a consensus conference on GMOs, by reference to the Danish Model. This event was later officially called a Citizens' Conference. As an official rationale, this event would provide 'a new way of elaborating decision 'and a means to implement 'participatory democracy', according to the Ministry of Agriculture. Yet the government never clarified the relation between the citizens' conference and its own decision-making procedure (Marris and Joly, 1999). This relation was subtly played out within the conference process, especially by defining expert roles.

From the start, the conference was designed to re-assert the benign expertise of the state, especially the Parliament, which saw itself as the only legitimate representative of the Nation. Organisation of the citizens' conference was delegated to a Parliamentary unit, Office Parlementaire d'Évaluation des Choix Scientifiques et Technologiques (OPECST), which symbolised a political neutrality separate from the government. OPECST appointed the steering committee, which in turn decided that the panel membership should represent diverse views of ordinary citizens—rather than stakeholders in the debate. It also decided which 'experts'—all of them scientists—would give briefings or testimony to the panel, thus framing the issues in advance (Marris and Joly, 1999). The organisers saw those arrangements as necessary 'to prepare a public debate which is not taken over by one side or the other', i.e. to correct or avoid biases in the existing public debate (OPECST, 1998a). Implicitly, such biases included anti-agbiotech NGOs on one side and Monsanto on the other side, especially from the perspective of the Left-Green Parliamentary majority.

Held in 1998, the conference included different framings of the policy problem. At the public hearings, the citizens' panel often challenged claims by experts about risks and benefits of GM crops. According to the panel's report, control by multinational companies could threaten farmers' independence. Genetically altered species pose a risk of standardisation. And GM rapeseed poses known risks of uncontrolled proliferation, through both pollen and seeds. Nevertheless GM crops could bring economic benefits to European agriculture (OPECST, 1998b; Boy et al., 1998). Together these arguments implied the need for national public-sector expertise in agbiotech innovation.

The panel's recommendations focused on institutional arrangements for better managing agricultural biotechnology. Such measures included the following: greater social participation in scientific advice; public-sector research on ecological risks and agbiotech innovation; a system to ensure traceability of food derived from GM crops; and adequate labelling to inform consumer choice. 'Until these conditions are satisfied, part of the panel believes that a moratorium would be advisable' (ibid.). By advocating state funds for agbiotech innovation, the panel accepted the government's problem-definition of a national technological gap whose solution requires public-funded science, presumed to be benign. The panel's concerns about rapeseed complemented the French government's decision to oppose approval of GM herbicide-tolerant rape, on grounds that gene flow could complicate weed control (Marris and Joly, 1999).

The panel's conclusions were translated into policy advice by the Parliamentary organisers, as if they were neutral experts in the public good. Moreover, having attended the proceedings, the OPECST President presumed to speak for the panel:

> Taking all these views into account he then himself adopted a position on a number of topics . . . He has identified the issues and looked into peoples' fears and concerns. (OPECST, 1998b)

This translation can be illustrated by the strategic issue of how to structure expert advice. The panel had proposed that a citizens' commission should be part of the scientific advisory committee. Yet OPECST recommended instead that it be kept separate; this proposal could better perpetuate a neutral image of scientific advice, thus reinforcing a boundary between expert/lay roles.

The panel's advice anticipated the general direction of government policy: more stringent regulatory criteria, risk assessment by a broader scientific expertise and 'independent' risk research, which was equated with public-sector institutes. It helped to legitimise and reinforce such initiatives, which had not been universally accepted within the government beforehand. In June 1998 the government announced measures along those lines (Marris and Joly, 1999). Institutional reforms emphasised expert procedures to minimise the risks and enhance the benefits of a controversial technology.

Despite its limitations, the citizens' conference initiated a new form of active public representation and knowledge-production. Panel members explored techno-scientific and social aspects together from the perspective of ordinary citizens. They sought to inform decision-makers about the views of those who do not normally speak out—and who do not feel represented by political parties, trade unions or environmental and consumer NGOs. This potential for participatory evaluation, especially for considering alternative options, was limited by the overall structure, especially the small opportunity to interact with designated experts (Joly et al., 2003).

Overall the citizens' conference was used to legitimise state claims to represent the public good, especially through expert roles. OPECST selectively promoted some accounts of agbiotech and its regulation as the expert ones, while explicitly speaking on behalf of citizens. The Agriculture Ministry had claimed to implement 'participatory democracy', yet the exercise extended the French tradition of technocratic governance (Marris and Joly, 1999).

Within this framework, expert roles remained the exclusive realm of the state authorities and their officially designated advisors. Ordinary people could question experts and recommend institutional reforms, but Parliamentary experts would officially speak for them. Thus the process reinforced lay/expert boundaries, in the face of public challenges to the official expertise for agbiotech.

6 EU-LEVEL NGO PARTICIPATION 2001–02

As the European Commission sponsored discussions over 'science and governance' from the late 1990s onwards, greater efforts were made to consult EU-level NGOs on technology policy. They participated in several consultations on agbiotech regulatory issues during the transition period of the de facto moratorium of 1999–2003. Tensions arose over the appropriate scope of public involvement in expert issues. These difficulties would intensify later when the new Directive was being implemented (see Chapter 5).

In particular NGOs were invited to take part in expert discussions of risk research and assessment. As a high-profile example, DG-Research organised Round Tables on Biosafety in 2001 and 2002, discussing evidence on risk issues of Bt maize and herbicide-tolerant crops, respectively. Opening the first Round Table, DG-Research Commissioner Philippe Busquin appealed for a reasonable, scientifically measured compromise between GMO crusaders and radicals, as he called them.

Yet NGOs were ambivalent about such participation, partly because they were dependent upon their scientific advisors for expertise. Consumer and environmental NGOs had this problem in common. At the same time, they used expertise in quite different ways, according to their respective policy agendas, which had little scope for the 'compromise' that was requested by Busquin.

Since the late 1990s environmental NGOs had emphasised scientific risks or uncertainties which could justify a ban, especially on cultivation of GM crops, while demanding non-GM alternatives. At the EC Round Tables on Biosafety, national environmental groups across Europe were represented by the FoEE speaker. He emphasised unpredictable and unknown risks (e.g. Ritsema, 2002).

Afterwards the FoEE speaker reflected on NGOs' roles in expert arenas:

> We enter the GM issue because of scientific doubts about safety, but we intervene mainly on the policy level. Politicians will eventually take the decisions, not simply on the basis of science. Our role is somewhere between science and politics. We commission scientific reports and cite them. We base our campaign work on science. We monitor scientific work, as does DG-Environment. We promote 'non-scientific' aims, e.g. a clean environment and sustainability. We should be open to the possibility that some GMOs won't be dangerous, but the opposite is also possible. (interview, FoEE, 2002)

Exemplifying this counter-expert role, FoEE commissioned plant scientists from a mainstream institute to identify uncertainties about effects of GM crops (de Visser, 2000). Later this report was cited in campaign literature by FoEE for its proposal to maintain the *de facto* EU moratorium on approval of any new GM products.

Unlike environmental NGOs, large consumer NGOs did not challenge the safety of GM food already authorised by the EU, though they did request an explicit account of scientific uncertainties. At the EC Round Tables on Biosafety, national consumer groups across Europe were represented by the Bureau Européen des Unions de Consommateurs. The BEUC speaker emphasised the need for a better relationship between such organisations and the risk-assessment process, especially through greater transparency about uncertainty and expert judgements. This proposal expressed an aim to enhance public confidence in the regulatory process, thus defining public distrust as a common societal problem to be jointly overcome.

Later the BEUC speaker reflected on NGOs' difficult roles in expert arenas:

> For the Round Tables hosted by the Commission, the organisers wanted a more targeted discussion [on risk research and assessment of GM products]. This role is more difficult for an ordinary NGO because we cannot be expected to know the dossiers in detail. The quality of our presentation may not meet their expectations . . . If the organisers want a focused discussion on specific risk-assessment issues, then we must ask our own technical experts on GMOs. We must be careful about what we can contribute, and the organisers must be careful about any conclusions. (interview, BEUC, 2002)

In sum, environmental and consumer NGOs participated in expert arenas with quite different agendas—to undermine or to enhance public confidence in safety claims, respectively.

At around the same time, a stakeholder dialogue was convened by the European Federation of Biotechnology (EFB), with funding from the European Commission. Convenors aimed to focus on 'environmental risks and safety of GM plants', but the discussion could not be contained within those terms. A key point from the plenary sessions was that participatory experiments need to involve the public in defining scientific research priorities; public concerns must be taken into account 'so as to democratise decision-making processes' (SBC, 2001).

Participants disagreed over how GM crops relate to sustainable agriculture. According to some industry and government representatives, agbiotech products could facilitate Integrated Crop Management systems, but NGOs regarded them as incompatible. NGOs sought a greater knowledge-base for adequate regulation and for a different innovation trajectory:

> . . . generation of data on the impacts of different agricultural systems would provide a context for evaluation of the impacts of GM crops, and would make it easier to judge their significance.
>
> What are the relevant agricultural systems to compare? . . . Options are organic, extensive/integrated, or intensive/conventional agriculture. . . . (SBC 2001, summary of Working Group III)

In a later EFB workshop, on 'Public Information and Public Participation' in agbiotech regulatory issues, sharp disagreements again arose on the expert basis for regulatory procedures. Participants disagreed about whether scientific risk assessments are value laden; likewise whether environmental and health issues are also ethical issues (SBC 2002: 9). NGO views converged on these issues, while together disagreeing with industry representatives, whose arguments relegated risk issues to objective expertise. NGO arguments implied greater potential scope for public participation in regulatory procedures.

Beyond those procedures, moreover, NGOs proposed research on less-intensive crop-protection methods—as a more stringent comparator, and as an alternative societal choice. 'Public participation might lead to better identification of research needs, e.g. comparison of agro-ecological consequences of conventional agriculture, IPM with/without GM crops, and organic agriculture', according to the workshop report (ibid.). Environmental and consumer NGOs had quite different regulatory agendas, though they all sought ways to open up innovation to broader expertise and alternative futures. Thus stakeholder consultation challenged the state's commitment to agbiotech as a European imperative, as well as the expert/lay boundaries which underlay regulatory procedures.

As a longer-term response to the wider controversy over agricultural futures, since 2000 the European Commission has sponsored a series of conferences on agricultural research priorities. NGO delegates proposed alternatives to the dominant agri-industrial model, along with democratic accountability of research agendas. However, these proposals were sharply rejected:

> [Some participants were] objecting to the notion that agricultural research should be 'under democratic control', preferring terms such as 'governance' or 'participation' . . . A tentative consensus was reached about what is needed—i.e., not democratic control, but transparency, democratic dialogue and involvement of the public in issues of scientific importance. (DG Research, 2002: 10)

This meant governing the societal conflicts over the EU's deep commitment to agbiotech within a wider neoliberal agenda.

7 UK 2003 PUBLIC DIALOGUE: POLICING BOUNDARIES

From the late 1990s onwards, the UK had a widespread public controversy over agbiotech. Protest actions and attacks on field trials gained public support by linking GM crops with various issues—BSE, other food scares, globalisation, 'pollution', etc. (see Chapter 3). The government faced an impasse over regulatory decisions, especially the criteria for permitting a GM herbicide-tolerant maize which the EU had approved in 1998. As a key issue, conservation agencies had warned that changes in herbicide usage could harm farmland biodiversity, so the government funded farm-scale trials to monitor such effects (see Chapter 6).

To address wider issues beyond risk regulation, the government had created the Agricultural and Environment Biotechnology Commission in 2000. Its report, *Crops on Trial,* advised the government to initiate an 'open and inclusive process of decision-making' within a framework that extends to broader questions than herbicide effects. It proposed a 'wider public debate involving a series of regional discussion meetings' (AEBC, 2001: 19, 25). The government was persuaded to sponsor this—alongside the intense, sporadic debate which was occurring anyway.

Called 'GM Nation?', the official public debate was carried out in summer 2003. Beforehand the government vaguely promised 'to take public opinion into account as far as possible'. The exercise was intended for the organisers to gauge public opinion, rather than for participants to deliberate a collective view on expert matters (Horlick-Jones et al., 2006). 'GM Nation?' also aimed to elicit views of the ordinary public, rather than organisational representatives—an artificial distinction, given that most civil society organisations and wider social networks had discussed agbiotech in previous years.

An overall Public Dialogue had a tripartite structure which explicitly distinguished between lay and expert issues. 'GM Nation?' was designed mainly for the lay public. An expert panel carried out a Science Review of literature relevant to risk assessment. And a government department carried out a Costs and Benefits Review of GM crop cultivation in the UK.

The Public Dialogue was designed in those three separate parts, with an explicit aim that they would work closely together. The three procedures were kept formally separate, yet the supposedly lay and expert issues became intermingled in practice. The official boundaries were both challenged and policed, thus constructing the participants in contradictory ways.

7.1 Representing Public Views?

'GM Nation?' featured several hundred public meetings open to anyone interested, drawing over 20,000 participants (DTI, 2003). When participants in 'GM Nation?' largely expressed critical or sceptical views towards agbiotech, arguments ensued over whether they were representative of the public. According to a pro-agbiotech coalition, the Agriculture and Biotechnology Council, the exercise was hijacked by anti-biotech activists, so the format was not conducive to a balanced deliberation of the issues.

According to academic analyses, however, that criticism frames the public as atomised individuals who have no prior opinion. The exercise predictably drew a specialised public which was largely suspicious or hostile to agbiotech. Participants represented both themselves as individuals and wider epistemic networks. The debates were filling an institutional void, in the absence of any other formal opportunity to deliberate the wider issues (Reynolds and Szerszynski, 2006).

The government sponsors had asked the contractors to involve 'people at the grass-roots level whose voice has not been heard'. As the official evaluators noted afterwards, however, it was problematic to distinguish clearly between 'an activist minority' and a 'disengaged, grass-roots minority'. Many participants in 'GM Nation?' were politically engaged in the sense that their beliefs on GM issues formed part of their wider worldview. Yet policymakers tend to construct 'the public' as an even-handed majority—and therefore legitimately entitled to participate in engagement exercises (Horlick-Jones et al., 2004: 135; Horlick-Jones et al., 2006). Indeed, 'grass-roots' conventionally means local organised activists, yet this term was strangely inverted to mean a passive, uninformed public.

As envisaged by the sponsors, separate focus groups would allow the public to frame the issues according to their own concerns, yet special measures were needed to realise the policymakers' model of the public. They saw the open meetings as dominated by anti-biotech activists, unrepresentative of the general public. Politically inactive citizens were seen as truly representative and thus as valid sources of public opinion, by contrast to 'activists'. To exclude the latter individuals from focus groups, candidates underwent surveillance and screening. 'Perhaps paradoxically, the desire to allow the public to frame the discussion in their own terms led the organisers to rely on private and closely monitored forms of social interaction'. According to this ideal model of the focus groups, the organisers would be listening to the *idiotis,* by analogy to ancient Greek citizens too ignorant to fulfil their responsibilities (Lezaun and Soneryd, 2006: 22–23). In this way, the more informed, expert citizens would be excluded from representing the public.

'GM Nation?' was intended to canvass all views and concerns about agbiotech, yet there were boundary disputes over issue-framings, admissible arguments and participants' roles. Some used the opportunity as politically engaged actors in their own right, not just as indicators of public opinion. Attending shortly after the US-UK attack on Iraq, some participants drew analogies between government claims about agbiotech and about Weapons of Mass Destruction. They suspected that the government was concealing or distorting information in both cases; they wondered whether it would ignore public opinion towards agbiotech, as in the attack on Iraq. Initially the chair tried to steer the discussion back to agbiotech, on grounds that 'GM Nation?' was not about the Iraq war, though participants still elaborated the analogy. Thus the public consultation had a disjuncture between public politics and government policy as understood by the sponsors of the exercise (Joss, 2005b: 181).

7.2 Expert/Lay Roles

For the carefully selected focus groups, the organisers commissioned 'stimulus material', so that participants would have a common knowledge-basis for discussion. The Steering Group asked the contractors to supply 'objective' information. Yet there were grounds to include 'opposing views' because this is often how people encounter information in real life', according to the official evaluators of 'GM Nation?' The ultimate material did include divergent views, but their sources were removed from the workbook for focus groups. Afterwards the official evaluators questioned 'the extent to which information is meaningful if it is decontextualised by stripping it from its source' (Horlick-Jones et al., 2004: 93–94; Walls et al., 2005).

Indeed, people often make judgements on the institutional source of expert views, but they had little basis to do so in the 'GM Nation?' focus groups. Omission of the sources was not simply a design deficiency in the exercise. By default, the issue of expert credibility was diverted and reduced to scientific information about biophysical risk. Participants had little basis to evaluate such information, so the exercise constructed a lay/expert boundary, constraining public roles even more narrowly than in the wider public debate.

Separate from 'GM Nation?', the GM Science Review was officially limited to a panel of experts evaluating scientific information. At the same time, relevant NGOs were consulted about experts who could represent their views on the panel. In this way, panel members were selected along relatively inclusive lines, encompassing a wide range of views about GM crops. As these selection criteria recognised, the public did not regard scientific expertise as a neutral resource (Hansen, 2006: 580), so the Panel's public credibility would depend upon a diverse composition. Although the Panel's report identified no specific risks, it emphasised uncertainties and knowledge-gaps important for future risk assessment of GM products (GM Science Review, 2003). These uncertainties implied scope for a wider public role in expert judgements.

As a high-profile part of the GM Science Review, the Royal Society announced a meeting to 'examine the scientific basis' of various positions. Opening the event, the chair announced the laudable aim 'to clarify what we know and do not know' about potential effects of GM crops. In the morning, agro-ecological issues were analysed in a rigorous way, especially for their relevance to the prospect that broad-spectrum herbicides may be widely used in the future. But those complexities were ignored when considering GM herbicide-tolerant crops in the afternoon (Levidow, 2003). By downplaying expert ignorance, the overall structure did not facilitate a debate about knowledge versus ignorance, nor provide much basis for public involvement.

Moreover, the boundaries of 'science' were policed along pro-biotech lines. Inconvenient issues, findings or views were deemed non-scientific. For example, speakers freely advocated the need for agbiotech to solve global problems, e.g. environmental degradation, the food supply, etc, but the chair cut off anyone who questioned these claims—for going beyond science (ibid.). Thus biotechnological framing assumptions were reinforced as 'science', along with the expert status of their proponents—while sceptics were marginalised as merely expressing lay views on extra-scientific issues.

In sum, the UK Public Dialogue involved a struggle over how to construct the public, especially in relation to expertise. The structure and management imposed boundaries between apolitical grassroots versus activist, as well as between lay versus expert status. Nevertheless participants challenged those boundaries, performed different models of the public and questioned dominant expert assumptions.

8 CONCLUSIONS

At the start, this chapter posed the following questions:

- How and why did state bodies sponsor participatory technology assessment of agbiotech?
- What aims arose in designing, managing and using those exercises?

- How did each one test models of public representation and citizens' roles?
- How did the process bear upon the public accountability of representative democracy?

Since the 1980s various state bodies in Europe have sponsored participatory technology assessment (TA) of agbiotech. In most national contexts, agbiotech was being officially promoted as safe eco-efficient products, essential for economic prosperity and environmental improvement. This policy framework was increasingly challenged by autonomous citizen initiatives.

In responding to or anticipating such conflict, participatory TA exercises were promoted for diverse aims, even contradictory ones. Sponsors and other advocates variously sought to democratise technology, to educate the public, to counter 'extreme' views, to gauge public attitudes, to guide institutional reforms and/or to manage societal conflicts. Such aims informed and shaped each exercise. The process tested diverse accounts of technology, the public, expertise and democracy (cf. Joss, 2005a). As a distinct arena, these participatory cases have some common features across different national contexts, especially where EU member states have had strong commitments to agbiotech.

In these TA exercises, individuals were selected for a process seeking to reach a group view. Participants deliberated the normative, value-laden basis of expert claims, thus developing a lay expertise; they went beyond simply questioning experts. By contrast to a negotiation among interest-groups, participants addressed the public good by appealing to common societal interests and problems (cf. Hamlett, 2003).

However, some problems were treated as common ones for group deliberation, while others were ignored or marginalised as uncommon ones, inconvenient for a group consensus or for a thinkable government policy. Issues were narrowed by the design and management of the exercise, as well as by self-censorship of participants. Some questioned whether agbiotech would provide a means for sustainable agriculture and a benign control over the agri-food chain; some suggested the need for alternatives, thus implying broader citizenship roles. However, these questions were generally channelled into biophysical risk issues and were reduced to regulatory control measures. Thus agbiotech was largely spared the deeper trial being undergone in the public debate.

Despite some aspirations to democratise agbiotech, then, participatory TA exercises biotechnologised democracy. Discussion generally focused on appropriate regulatory arrangements for agbiotech, represented as a series of potentially beneficial products; at issue was how to minimise their risks and maximise benefits. Conflicts over societal futures could be absorbed into regulatory issues (cf. Gottweis, 1998). In analogous ways, regulatory procedures attracted broad comments questioning the innovation; these likewise conflicted with official regulatory criteria (Bora and Hauseldorf, 2006; Ferretti, 2007).

In the TA exercises, any wider deliberation was constrained—by a search for consensus, by the design of each exercise, by the government policy framework and sometimes by overt policing. This context limited what could be said with influence on the process, and thus what roles could be credibly performed by participants, regardless of their views on the issues (cf. Hajer, 2005). The process internalised and reinforced policy assumptions about agbiotech as essential progress—albeit perhaps warranting more rigorous, publicly accountable regulation. Through a discursive depoliticisation, contentious issues were displaced onto the management problems of an inevitable common future (cf. Goven, 2006; Pestre, 2008). Consequently, tensions arose between a 'common' problem—how to make agbiotech safe or acceptable—versus deeper issues of political-economic control, innovation priorities and societal futures.

Those tensions took the form of various boundary conflicts, which erupted more starkly in some national cases. In the German TA exercise, NGO representatives demanded a comparative assessment with alternative options, but they could maintain their official expert status only by accepting the risk-benefit framework imposed by the organisers; their risk claims were being put on trial. Instead they withdrew from the exercise and so were relegated to the status of lay public or irrational objectors. In the 2003 UK Public Dialogue, the official structure nominally separated all relevant issues into three components—public concerns, scientific risk assessment and economic benefits; accordingly, expert matters were formally separated from the issues for discussion by lay participants. Despite that official tripartite structure, all the issues became mixed in practice; their boundaries were both contested and policed.

Models of public representation and citizen roles were also being tested in these participatory TA exercises. Their design and management structured relations between expert and lay roles. Participants could question, evaluate, anticipate or even simulate expertise. However, such discussion generally focused on appropriate regulatory and expert advisory arrangements for GM products, perhaps as a means to influence official agendas by accepting their limits. By variously accepting or challenging expert/lay boundaries, participants were performing different models of the public and its representation. Sometimes innovation issues were opened up, thus implying broader citizenship roles. These performative interactions generated different understandings of the policy problem (cf. Hajer, 2005).

State-sponsored participatory TA generally complemented a wider policy process channelling public dissent into regulatory procedures. In each TA exercise, discussion went beyond the government policy framework but marginalised wider issues from the public debate. Ultimately the process reinforced official boundaries between expert and lay roles. These boundaries took forms specific to each national case; agbiotech was being co-produced as progress along with particular models of expertise, publics, citizenship and democracy.

In some cases, state-sponsored participatory TA exercises anticipated, stimulated or reinforced policy changes which enhanced the state's accountability for regulatory frameworks. Such outcomes depended upon a longer-term socio-political agency beyond the TA exercise and its panel, such as in the Danish case. However, the exercises did not help publics to hold the state accountable for its commitment to agbiotech, being represented largely as an imperative rather than as a choice. Meanwhile this choice was being challenged and put on trial by autonomous forms of public participation—neither welcomed nor sponsored by state bodies.

These tensions matter for the accountability of representative democracy. According to an advocate of participatory TA: 'In practice, the relationship between representative democracy and participatory methods becomes most clear and complementary, when engagement is approached as a means to open up the range of possible decisions, rather than as a way to close this down' (Stirling, 2006: 5; likewise Stirling, 2005: 229). In the cases analysed here, however, participatory methods and representative democracy were not complementary. Or perhaps they were perversely so: the process reinforced the public unaccountability of representative democracy for innovation choices.

Thus state-sponsored participatory TA complemented neoliberal forms of democracy, especially the dominant form of European integration (Levidow, 2007). Agbiotech supporters opposed democratic control over research priorities, instead preferring participation and 'governance'. By default, if not by design, conflict over innovation trajectories was channelled into regulatory issues. These carried the burden of societal conflict, as shown in the next two chapters.

5 Regulating Risk, Testing EU Reforms

INTRODUCTION

In early EU policy, agbiotech innovation was nearly removed from political debate. Decisions were left to 'risk-based regulation', a policy conflating risk assessment, the available science and official expert advice. In this scenario, EC expert committees would provide a scientific basis for the Commission to approve GM products, which could then freely circulate on the internal market.

From the mid-1990s onwards, these arrangements were destabilised, for several related reasons. The BSE (mad cow) scandal undermined the public credibility of EU risk regulation and its expert basis, at least in the agri-food sector. These difficulties stimulated reforms aiming to create a truly independent expert basis for EU regulatory decisions. Yet official expertise again faced criticism, especially for its advice on GM products.

The EU Council's 1999 *de facto* moratorium blocked approval of any new GM products, eventually leading the Commission to revise agbiotech legislation along more stringent lines than it had originally proposed. Meanwhile some member states were imposing bans on GM products already approved by the EU, often by invoking the precautionary principle. This EU-wide conflict linked several arenas: risk regulation, risk research and EU expert advice (see Table 5.1).

This chapter discusses the following questions:

- What aims drove EU reforms for regulating agri-food products?
- How did the agbiotech controversy test those reforms?
- How were state bodies put onto the defensive?
- What roles were played by expert advice in various trials?

This chapter tells a decade-long story from the late 1990s onwards in roughly chronological order; sections alternate between general EU reforms and the agbiotech sector. The earlier story overlaps with public controversy in Chapter 3; the later story overlaps with other regulatory conflicts over commercial scale-up in Chapter 6.

Table 5.1 EU Institutional Reforms for Agri-Food Regulation

	Agbiotech risk legislation	*GM food regulation*	*Scientific uncertainty: policy role*	*EU advisory expertise: general arrangements*	*EU expert advice on GM products*
Mid-1990s	1990 DRD bases the internal market on 'a high level of protection' for environmental and health risks; aims to harmonise criteria.	Substantial equivalence can justify GM food safety.	Safety is/should be based on 'sound science'. Claims for 'no evidence of risk' ignore uncertainty.	Hosted by the Directorate-General responsible for legislation and product approvals	[No EU-level expert advice on GM products]
Post-BSE crisis (1996–97)	Proposals to streamline DRD procedures and to lighten regulatory burdens (CEC, 1996)	NFR bases the simplified procedure on substantial equivalence.	Debate over how expert procedures should address uncertainty about risks.	Scientific advice must be based on the principles of excellence, independence and transparency.	'No evidence of risk' from each GM product to be approved. 'Adverse effects' are defined narrowly.
Late 1990s	2001 DRD broadens risk assessment and requires market-stage monitoring.	Some member states apply more stringent criteria for 'substantial equivalence'.	Guidelines for triggering the precautionary principle, i.e. measures to manage uncertain risks (CEC, 2000)	Functional separation of risk assessment from risk management, thus protecting the scientific integrity of expert advice	No reason to indicate that (each) GM product will cause adverse effects.
Since 2001	2003 GM F&F Regulation centralises expert advice to facilitate harmonisation.	2003 GM F&F Regulation abandons the simplified procedure for GM food.	Any precautionary restriction must be justified by a risk assessment indicating uncertain risks.	2002 Food Law: EFSA to clarify different views of expert bodies and to help harmonise precaution	Safety claims undergo pressures to acknowledge uncertainties and normative judgements.
Explicit policy aims	Exploit the potential of biotechnology while taking account of the precautionary principle and social concerns.	Substantial equivalence remains a 'dynamic concept' for assessing GM food safety.	PP must be used in ways compatible with EU treaty obligations.	Independent, objective, transparent advice should inform decisions which can gain public confidence.	Obtain adequate data to clarify uncertainties and overcome expert disagreements in risk assessment.

Abbreviations
DRD = Deliberate Release Directive 90/220 (EEC, 1990), 2001/18 (EC, 2001)
EFSA = European Food Safety Authority (EC, 2002a)
GM F&F = GM Food & Feed Regulation 1829/2003 (EC, 2003a)
NFR = Novel Food Regulation 258/97 (EC, 1997a)
PP = Precautionary Principle (CEC, 2000)

1 BROADENING RISK ASSESSMENT, SHIFTING SCIENCE/POLICY BOUNDARIES

After the BSE crisis, EU-level reforms were meant to demarcate more clearly between science-based risk assessment and policy-based risk management. This distinction became contentious in the agbiotech sector, as more 'risks' than before were being debated as policy issues. In the late 1990s those disagreements were eventually translated into regulatory changes—initially at the national level, eventually leading to EU-level change. The EU legislative framework was opened up to broader environmental assessments and public comment on regulatory issues. This section analyses how these changes were designed to help legitimise EU regulatory decisions.

1.1 Reforming Advisory Expertise, Separating Science/Policy?

Amidst a series of food and health scandals, the 1996 BSE scandal aggravated credibility problems of EU regulatory arrangements, especially in the agri-food sector. Citing regulatory failures and deceptions, critics undermined official images of policy-neutral expertise and its discursive association with science. In the run-up to the BSE scandal, expert advice on animal feed downplayed scientific uncertainties and made optimistic assumptions that real-world practices would follow risk-management guidelines. Moreover, experts aimed to give advice that would be politically acceptable to regulators, while avoiding public alarm about any risks (Millstone and van Zwanenberg, 2001). The Commission likewise covered up the BSE problem, for fear that public concern would endanger the European beef market, according to a report by the European Parliament (EP, 1997).

Politicians came under criticism for avoiding full responsibility for decisions, instead citing expert advice whose assumptions may not be clear. Expert advisory bodies were hosted by the same Directorate-General responsible for relevant legislation and product approvals. Consequently, EU safety claims came under greater suspicion.

To address the legitimacy problem, institutional reform aimed to restructure EU advisory expertise in ways which could gain acceptance by the corresponding expert advisory and regulatory bodies in member states. As an official rationale for reform, the Commission needed 'to obtain timely and sound advice'. For matters relating to consumer health, 'scientific advice must, in the interests of consumers and industry, be based on the principles of excellence, independence and transparency'. In the new procedure, committee members would be selected through a call for expressions of interest (EC, 1997c), rather than from government nominees as previously.

The new EU scientific committees were meant to be more independent in three respects: from member states, from Directorates-General which propose and implement legislation, and from material interests. To pursue those aims, in 1997 all the expert advisory committees were transferred to

DG 24 for Consumer Affairs, later restructured and renamed DG-SANCO. Nominees were asked to declare any material interests, e.g. sources of research funding, in an effort to enhance expert independence. Overall the EU attempted to separate risk-assessment advice from risk-management bodies and policy influence. The new arrangement was later formalised as a policy: 'that those responsible for scientific assessment of risk must be functionally separate from those responsible for risk management, albeit with ongoing exchange between them' (EU Council, 2000a: 3).

1.2 Disputing 'Adverse Effects' of GM Products

As a centrepiece of the new Commission strategy, that official distinction ran up against disagreements over how to define environmental harm and where to assign responsibility. According to the 1990 Deliberate Release Directive, 'completion of the internal market would be based on a high level of protection for the environment and human health'; member states must ensure that GM products do not cause 'adverse effects', which were left undefined (EEC, 1990). When the first GM crops were proposed for commercial cultivation in the mid-1990s, such effects were being narrowly understood by proponents and expert advisors. Conflicts emerged over what standards would shape the internal market and over whether these were policy issues.

Official risk assessments accepted or ignored foreseeable agri-environmental effects as 'agricultural problems'—as normal, albeit undesirable effects of intensive monoculture. For example, if weeds acquired tolerance to herbicides, or if insects acquired resistance to Bt insecticidal toxins, then such effects were regarded as acceptable or irrelevant to agbiotech legislation. According to safety claims by the UK and France, other pest-control methods would still be available, so current pest-control options were regarded as dispensable (Levidow et al., 1996). 'Risk' was thus being framed by a neoliberal eco-efficiency account of sustainable agriculture, accepting a genetic treadmill as normal.

A similar policy framework underlay advice from EU-level expert bodies whose remit included GM crops. Operating before the 1997 EU reforms, the Scientific Committee on Pesticides had not regarded Bt resistance as an 'adverse effect'. Likewise its successor, the Scientific Committee for Crops, accepted pest resistance to Bt or to herbicides as irrelevant to EU risk legislation, though the Committee recommended measures to limit volunteer herbicide-tolerant plants which could cause weed problems (SCP, 1998a, 1998b). Citing this favourable advice from EU experts, the Commission granted approval for commercial cultivation of Bt 176 maize and seed multiplication of herbicide-tolerant oilseed rape (EC, 1996a, 1997a). Thus the 'genetic treadmill' was symbolically normalised along with GM crops; designed to sustain intensive monocultural practices, agbiotech was being co-produced with official expertise.

Another contentious issue was the harm to non-target insects which would count as 'adverse' or unacceptable. In the late 1990s lab experiments indicated that Bt toxins could cause such harm to lacewings, a pest predator (Hilbeck et al., 1998a, 1998b). Experts disagreed over the methodological basis for predicting harm, as well as over the normative baseline for its acceptability. According to EU experts, Bt maize would anyway cause less harm to non-target insects than 'that from the use of conventional insecticides' (e.g. SCP, 1998b). Moreover, the choice of comparator for the effects of GM crops was a purely scientific matter, argued a spokesperson:

> We have to evaluate potential effects on the basis of existing agricultural practices. A comparison with chemical insecticides makes the potential harm acceptable . . . This is a scientific issue . . . We are asked only scientific questions. (interview, Chairman, SCP Environmental Sub-Committee, June 1998)

Thus the official risk assessment implicitly assumed that Bt maize would always replace conventional maize sprayed with chemical insecticides, which thereby could provide a normative baseline for GM crops. In most European maize fields, however, such sprays were little used, partly because they cannot reach the corn borer inside the stalk.

As another contentious boundary, herbicide-tolerant crops could stimulate changes in how farmers use herbicides, but EU-level criteria and procedures took no responsibility for the potential agri-environmental effects, partly on grounds that such effects would not be caused by the crop. Since the mid-1990s several member states demanded that the risk assessment should encompass the overall implications of growing GM crops, including any consequent changes in pesticide use. This demand originally came from Denmark, which had a policy aim of reducing herbicide usage, partly in order to protect groundwater for drinking purposes. A similar demand came from Austria, which regarded any increase in herbicide usage as unacceptable, relative to its baseline of organic agriculture. By the late 1990s those countries were joined by more powerful ones in demanding a broader risk assessment (see next section).

1.3 Legislating Broader Risk Assessments

Public protest changed the original direction of EU legislative changes. Since the mid-1990s the Commission had been discussing ways to 'streamline' legislation on GMOs, especially to facilitate commercial approval, along lines proposed by the agbiotech industry. Eventually it issued such a proposal (CEC, 1996). However, this was soon abandoned in the face of greater public protest and contrary demands from some member states.

Through the risk issues described above, GM crops were being put on trial with a more severe charge-sheet of potential harms. Disagreements concerned how broadly to define 'adverse effects' and whether this was more than a scientific issue for official experts. In 1998 the French government reversed its previous support for herbicide-tolerant oilseed rape, on grounds that herbicide-tolerant weeds could complicate methods of weed control. Meanwhile the UK was facing intense domestic conflict about whether broad-spectrum herbicide sprays could increase harm to farmland biodiversity (see Chapter 3). In 1998 the UK persuaded the EU Environment Council that GM crop assessment under the Deliberate Release Directive 90/220 should encompass the effects of agricultural practices, e.g. any changes in herbicide usage resulting from GM crop management.

EU policymakers were facing similar pressures from more and more member states. Eventually the regulatory criteria were broadened in a new proposal for revising the Deliberative Release Directive (CEC, 1998b). Now the definition of 'harm' was acknowledged as a policy issue:

> All three types of judgment—scientific, legal, and political—are involved in any judgement on defining 'adverse effects'. It involves considerations broader than science, e.g. by interpreting the law, and taking on board public concerns. (interview, DG XI/Environment, 20.01.98)

Meanwhile greater conflicts among member states were impeding the EU decision procedure. In June 1999 several Environment Ministers signed statements opposing the approval of any more GM products until regulatory criteria were strengthened, including a requirement for traceability and labelling of all GM material, as well as precaution as the basis of risk assessment. Such changes were necessary 'to restore public and market confidence', according to their statements (FoEE, 1999: 3). Widely known as the *de facto* moratorium, this regulatory blockage stimulated changes in EU law.

With strong support from the European Parliament, the Deliberate Release Directive was revised along more stringent, precautionary lines. Risk assessments now had to evaluate any long-term effects and changes in herbicide usage. They also had to acknowledge uncertainties about risk: 'It is important not to discount any potential adverse effect on the basis that it is unlikely to occur' (EC, 2001: 20).

Moreover, under the new Directive, GM crops could be kept on trial indefinitely during the commercial stage. Commercial approval would be granted for only a ten-year period. Each marketing notification must include a plan for case-specific monitoring, with this aim: 'the monitoring should confirm that any assumptions regarding the occurrence and impact of potential adverse effects of the GMO or its use in the environmental risk assessment (e.r.a.) are correct' (EC, 2001: Annex VII; see also our Chapter 6).

Means for formal public involvement were also expanded. Under the new Directive, member states must consult the public on each proposed field trial. There was also opportunity for public comment on commercial authorisation of products:

> Member States shall lay down arrangements for this consultation, including a reasonable time-period, in order to give the public or groups the opportunity to express an opinion.

> Member States shall also establish registers for recording the location of GMOs [commercially] grown under Part C, *inter alia* so that the possible effects of such GMOs on the environment may be monitored. (EC, 2001: Article 9)

Thus new requirements for risk assessment, information disclosure and precaution were linked with public accountability for regulatory decisions. Before these new arrangements were implemented for new GM products in 2003, however, further EU reforms sought to centralise expertise and thus enhance the Commission's authority over regulatory procedures.

2 HARMONISING PRECAUTION?

In the late 1990s, EU agbiotech regulation underwent pressures for greater precaution, as in the revised Deliberate Release Directive 2001. Further institutional changes were designed to avoid or overcome EU-wide regulatory conflicts. The mid-1990s slogan, 'risk-based regulation' (CEC, 1993a), was now recast as 'science-based regulation', thus placing a greater burden and reliance upon advisory expertise in the name of science. Changes in advisory expertise and regulatory procedures were meant to harmonise risk assessment, and even precaution, as analysed in this section. Parallel legislative changes required labelling and traceability of GM agri-food products (analysed in Chapter 7).

2.1 Enhancing Expert Authority?

After the 1997 reforms of EU advisory expertise, more difficulties were identified. According to leading members of EU-level expert committees, their role was hindered by the lack of in-house scientific expertise at DG-SANCO, and often their own advice conflicted with national expert views (James et al., 1999: 8). Functionally separating risk assessment from risk management created new problems: 'the current risk assessment process . . . has negligible input from those dealing with issues of risk management, on practical options for change or on the validity or effectiveness of control measures'. Therefore the overall procedure needed 'to ensure articulation between these two components of the risk analysis process'. Moreover, public-interest groups had little access to the process and judgements which formed expert advice (ibid.: 43). This diagnosis indicated the need for greater transparency, with systematic links between advisory expertise, risk managers and stakeholders.

As another problem, greater political dependence upon expert advice meant greater difficulties for regulatory harmonisation. Various European and national expert committees were conducting risk assessments, often with different outcomes, so that expert advice readily became a political tool within EU-national conflicts. Some member states created independent agencies, whose risk-assessment advice often questioned applicants' safety assumptions, even the expert advice from EU-level scientific committees. 'This is the source of much confusion and tends to undermine the credibility of the risk assessment process', so the EU should take firm steps 'to harmonise the process', argued the EU's Scientific Steering Committee (SSC, 2003).

All those diagnoses informed proposals to create an independent agency for EU expert advice. In its 2001 White Paper on Food Safety, proposing what became a 2002 food law, the Commission outlined its plan for a European Food Safety Authority (EFSA). Equipped with its own in-house expertise, EFSA was designed to achieve a greater cognitive authority for accommodating and/or challenging national expert bodies.

In the 2002 food law establishing EFSA, the new body had this official rationale: 'In order for there to be confidence in the scientific basis for food law, risk assessments should be undertaken in an independent, objective and transparent manner, on the basis of the available scientific information and data'. According to its diagnosis, the precautionary principle 'has been invoked to ensure health protection in the Community, thereby giving rise to barriers to the free movement of food or feed' among EU member states; therefore precaution should be harmonised: 'it is necessary to adopt a uniform basis throughout the Community for the use of this [precautionary] principle' (EC, 2002a: 2). According to the new law, wherever 'the possibility of harmful effects on health is identified but scientific uncertainty persists', authorities may adopt 'provisional risk management measures necessary to ensure the high level of health protection chosen in the Community' (ibid.: 9); this echoed EC policy on the precautionary principle (CEC, 2000: 11). For those policy aims, EFSA would 'provide a strong science base, efficient organisational arrangements and procedures to underpin decision-making in matters of food and feed safety' (EC, 2002a: 6).

This strategy sought public confidence for regulatory decisions through independent expert advice. According to the Commission, establishment of EFSA was 'generally regarded as the most effective way to address the growing need for a solidly science-based policy and to increase consumer confidence' (EU Food Law News, 2000: 5). In formally separating science from policy through EFSA, the EU had an extra incentive in WTO rules. Although internal difficulties drove these policy innovations, they were also being 'designed to fit with the WTO model and thereby ward off trade retaliation', argues an academic (Skogstad, 2001: 496, 498, 501).

The EFSA structure was designed to help expert advice to gain an objective status by standing above policy. When EFSA communicates its results, 'the information will be objective, reliable and easily understandable for the general public' (CEC, 2002c). In a similar vein, the Commission invested expert authority in 'independent' agencies on grounds 'that their decisions are based on purely technical evaluations of very high quality and are not influenced by political or contingent considerations' (CEC, 2002d: 5). According to the relevant Commissioner, the independence of EFSA 'will ensure that scientific risk assessment work is not swayed by policy or other external considerations'. Moreover, he stated, 'the Authority's reputation for independence and excellence in scientific matters appertaining to food will put an end to competition in such matters among national authorities in the Member States' (Byrne, 2002: 4–5). In all these ways, EFSA was meant to protect the scientific integrity of EU-level expert advice, while implying that national differences and dissent had an extra-scientific basis.

As an extra means to enhance its authority, EFSA could incorporate or accommodate potential critics. Its Management Board included representatives of industry and consumer NGOs. According to an academic analysis, the new arrangements involved a wider range of EU-level stakeholders with shared understandings of policy problems, especially the need to gain public confidence. Their participatory role pluralises the policy process, while weakening corporatist relations with business and farmers (Smith et al., 2004).

EU policymakers recognised that legitimate decisions could not simply rely upon expert authority. The new arrangements had potentially democratising procedures—e.g. co-decision powers between the Commission and Council, transparency rules for expert advice on GM products and stakeholder involvement in EFSA itself. These procedures could potentially internalise or overcome regulatory conflicts.

For implementing those reforms in the agbiotech sector, expert advice came from a new expert committee. Before 2003 such advice came from the EU's two Scientific Committees on Plants and on Food, respectively. In early 2003 EFSA replaced all the former EU-level scientific committees with new ones, including a single integrated Scientific Panel on GMOs (henceforth 'the GMO Panel'). Centralisation through EFSA was designed to overcome regulatory conflicts, but critics of centralisation anticipated difficulties for that strategy:

> EFSA could become politicized, i.e. leading to battles over control and expertise. The distinction between risk assessment/management is already shaky and could become more blurred. (interview, GMO advisor of EP Greens, 2002)

Indeed, further moves towards centralisation would provoke greater conflict over the official distinction between science and policy.

2.2 Centralising Procedures, Shifting Responsibility

A 'GM contamination' scandal became another turning point in Commission policy. Starlink GM maize had been legally cultivated by US farmers prior to US government approval for food and feed purposes. Such approval was ultimately denied because of concerns about possible allergenicity. In 2000 Starlink maize was found to be illegally present in the food chain, especially in North America but also in Europe. Citing this problem, opposition groups in Europe demanded a ban on GM products.

The Commission instead used the problem to promote a long-standing agenda: transferring GM products from the horizontal, process-based Deliberate Release Directive to 'vertical', product-based legislation. Since the mid-1990s, it had been attempting to integrate approval for all product uses (food, feed, cultivation) under the slogan, 'One door, one key', partly as a means to avoid regulatory blockages. Now the Commission proposed vertical integration for an extra reason—to avoid another StarLink crisis. 'Learning from the US experience with StarLink, the proposal provides that GMOs likely to be used as food and feed can only be authorised for both uses or not at all' (CEC, 2001d: 4).

On that new rationale, implying an advance in sovereignty, the Commission proposed 'a centralised, clear and transparent Community procedure' for GM products (CEC, 2001d: 4). The 2001 draft GM Food and Feed Regulation would replace previous laws—the Deliberate Release Directive and Novel Food Regulation—for all GM agri-food products. EFSA would become responsible for circulating and evaluating product files. Moreover, EFSA was asked to standardise the evaluation criteria across member states: 'In order to ensure a harmonised scientific assessment of genetically modified foods and feed, such [risk] assessments should be carried out by the Authority' (CEC, 2001e: 3; EC, 2003a: 4). EFSA's advice was meant to inform decisions by the Commission, coordinated and prepared by DG-Agriculture as *chef de file* for the new Regulation. In parallel a draft Regulation on Traceability and Labelling of GM products was prepared by DG-Environment (see Chapter 7).

Among the contentious aspects was the centralised procedure. Approval decisions would be removed from the Deliberate Release Directive, thus bypassing the national authorities responsible for environmental protection, even for cultivation uses of a GM crop. Early on, environmental NGOs alerted MEPs and national authorities, which dissented from the proposal. For example:

> Denmark questions whether EFSA would have a sufficient competence to cover the natural and environmental differences in the EU regions. (Økonomi-og Erhvervsministeriet, 2001)

> Risk assessment cannot be rationalized as a factor of economic competitiveness. By analogy to investment in train systems, there is a choice between profits versus safety. Under the [new EFSA] umbrella of scientific committees, the Commission has aimed to create its own legal tool for approving products . . . EFSA experts must be geniuses if they can evaluate potential environmental effects in all EU member states. (interview, Belgian CA, 2002)

Similar conflicts arose between the Commission and Parliament, which sought to ensure a role for national authorities in the environmental risk assessment. This issue became contentious when the draft Regulation was first debated by MEPs (EP Plenary, 2002). Eventually a compromise was reached: for cultivation of a GM crop under the new Regulation, EFSA shall ask a national CA to carry out an environmental risk assessment (EC, 2003a: 14). Parliament had demanded 'a greater degree of decentralisation', but 'we have had only limited success', according to the Austrian rapporteur Karin Scheele (EP Plenary, 2003a, 2003b).

Like other key features of the Commission proposal, centralisation provoked conflicts among Parliamentary groups. The large Right-wing European People's Party, including Christian-Democrat parties, led the arguments for agbiotech, centralisation and less restrictive rules. Overt opposition came from the Greens and the United Left (GUE-NGL Group). As a way forward, many MEPs made proposals to strengthen the Commission's draft Regulation; they sought greater public access to the regulatory procedure, disclosure of all sites where GM crops were commercially cultivated and a role for national experts in EU-level risk assessment. Such efforts were led especially by Karin Scheele from the Socialist Group and Jonas Sjöstedt from the Swedish Left Party (affiliated to the United Left). Eventually anti-agbiotech MEPs were persuaded to support the resulting Regulation—as a means for anticipating and preventing any harm, as well as a means for civil society to gain accountability for the evaluation and commercialisation of GM products. Both they and the regulatory procedures could potentially be kept on trial.

Although the GM Food and Feed Regulation had overwhelming support in Parliament, the final debate in July 2003 illustrates its three contending frames, especially regarding the prospects to authorise GM products. According to an MEP of the European People's Party, the new rules attempt 'to ensure that the market will function smoothly and to achieve a high level of consumer protection'. Such changes 'will result in the lifting of the *de facto* moratorium on the approval of new GMOs and prevent a trade war . . . ' (Antonio Trakatellis, EP Plenary, 2003a, 2003b).

From an ambivalent managerialist frame, a Liberal MEP supported the legislative package as follows:

> A common framework of policies is needed if we are to avoid the risk of a trade war with the United States, with appeals to the WTO, and to curb the risk of divisions amongst ourselves across the European Union. . . . If this [legislation] ultimately slows the development of this technology while more research is undertaken, then that may be no bad thing. (Chris Davis at EP Plenary, 2003a, 2003b)

Expressing an apocalyptic frame, a United Left MEP argued that it was premature to lift the moratorium, and that 'we can never satisfy the US government in this area'. Fellow MEPs denounced multinational companies for seeking profit and spreading 'GM contamination' (Jonas Sjöstedt and others at the EP Plenary, 2003a, 2003b).

Amidst that polarisation, the Commission represented the draft Regulations as progress for European sovereignty and integration. Commissioners attempted to set a consensual tone for a common way forward. According to David Byrne of DG-SANCO, we enact this legislation 'because we believe it is right'—not because other people believe we should speed up the approval procedure, under threat of WTO proceedings (ibid.). According to Margot Wallström of DG-Environment, the final legislation left the regulatory procedure mainly to the member states: 'But if they are not willing to do it, then the Commission will certainly take responsibility' (ibid.). Indeed, EFSA's advice could justify Commission decisions to approve GM products, especially in cases where member states were deeply divided.

2.3 Tightening GM Food Assessment

Although the 2003 GM Food and Feed Regulation somewhat centralised risk assessment, it abandoned a specific harmonisation strategy for GM food products. The law being replaced, the 1997 Novel Food Regulation, had a simplified procedure for approving novel food products, including GM foods. A national authority need not carry out a risk assessment for a GM food which had 'substantial equivalence' with a non-GM counterpart regarded as safe (EC, 1997b). This option was not carried over by the 2003 GM Food and Feed Regulation, amidst great conflict over the concept of substantial equivalence.

Since the late 1990s, substantial equivalence had come under widespread criticism—even as an 'unscientific' means to avoid risk assessment and safety tests (e.g. Millstone et al., 1999). In practice the concept meant that safety was presumed by subjecting the novel GM protein to toxicological tests, while also testing the whole food for its physico-chemical composition. This framework was widely represented as a 'principle' which could be left to technical experts in risk assessment. But some national expert advisory bodies were interpreting the concept according to more stringent criteria; for example, they requested more rigorous evidence of physico-chemical composition, as well as more toxicological tests. Scientists' efforts along those lines converged with demands of consumer organisations criticising substantial equivalence (Levidow et al., 2007).

A breaking point came in August 2000, when the Italian government suspended the sale of products derived from four varieties of GM maize. In its advice to the Commission, the EU-level Scientific Committee on Food stated that the Italian authorities had not provided evidence that the GM maize posed a risk to human health. Citing this advice, the European Commission then demanded that Italy remove the ban; and it requested support from the Standing Committee on Foodstuffs, which represents EU member states. That body instead sided with Italy, stating: 'it was unacceptable that GMO-derived products were placed on the EU market under the simplified procedure, without undergoing a full safety assessment' (StCF, 2000: 2).

Such conflicts led the Commission to omit any simplified procedure from its draft Regulation on GM Food and Feed:

> The use of this regulatory short-cut for so-called 'substantially equivalent' GM foods has been very controversial in the Community in recent years

> ... and there is consensus at the international level ... that whilst substantial equivalence is a key step in the safety assessment process of genetically modified foods, it is not a safety assessment in itself. (CEC, 2001e)

In the GM Food and Feed Regulation, substantial equivalence lost its statutory role, while remaining available as a risk-assessment tool. EU-wide harmonisation might be difficult to achieve for this 'dynamic concept', whose interpretation was still under development, according to a Commission official (Pettauer, 2002: 23). Without a regulatory short-cut via substantial equivalence, test methods for GM foods would become subject to more stringent demands and argumentation in the EU regulatory procedure.

3 EXTENDING REGULATORY DISSONANCES

Having established new institutional arrangements, the Commission sought to resume the approvals procedure for GM products. To manage the conflict, the Commission invoked 'science-based regulation', implicitly equating science with advice from EU-level expert bodies. It also devised and interpreted guidelines on precaution in ways which downplayed uncertainties in risk assessment, thus placing a greater burden upon the available science. As this section shows, the agbiotech regulatory procedure extended the earlier regulatory dissonances from the late 1990s rather than harmonising risk assessment, much less precaution.

3.1 Commonly Understanding Precaution?

In 2000 the Commission issued a Communication on interpreting and implementing the precautionary principle. It aimed to guide member states so as to deflect international criticisms that the EC's precautionary approach posed an unfair trade barrier, e.g. regarding animal hormones or agbiotech. According to that policy document, the precautionary principle can be triggered only by reasonable grounds to expect 'potentially dangerous effects' that would jeopardise the 'chosen level of protection'. In risk assessment 'scientists apply caution, not to be confused with precaution', which applies only to the risk-management stage. Recourse to precautionary measures is justified only until extra scientific information is obtained to allow a more complete risk assessment (CEC, 2000). The Commission represented its own account as a common understanding (ibid.: 8). Yet other EU institutions expressed broader understandings of precaution, including its role within risk assessment (Levidow et al., 2005).

After several GM products gained EU-wide approval in the late 1990s, more and more member states were imposing national bans. In general these governments cited uncertain risks or inadequate scientific information. Going further, Austria submitted an extensive critique of the official safety assessments as grounds for imposing a ban on three GM maize products. For GM herbicide-tolerant crops, it regarded the herbicide implications as a threat to 'ecologically sensitive zones', as argued in an expert report (Hopplicher, 1999). Thus it was invoking environmental differences which could warrant a more cautious risk assessment. Austria's arguments, including new evidence of risk, were dismissed by the EU's Scientific Committee on Plants—and again later by the EFSA GMO Panel (2004a).

After a five-year blockage of the EU approvals procedure, member states were again evaluating GM products in 2003, when EFSA's scientific panels also began operating. For each GM product proposed for commercial approval, EFSA's Scientific Panel on GMOs advised that it would be as safe as its non-GM counterpart. Member states often disagreed with EFSA's safety claims. Some questioned whether the available information was adequate for a risk assessment, whether the environmental assessment adequately covered their specific conditions and whether specific undesirable effects should be regarded as acceptable. Such criticisms were often taken up by other member states. In these ways, precaution was being elaborated in practice through expert disagreements (Levidow et al., 2005).

In the overall EU-wide debate, precaution was being framed according to three distinct policy agendas:

i. *Agbiotech proponents (eco-efficiency frame):* Europe will gain environmental, agronomic and economic benefits by commercialising GM crops. Risk assessments for our GM products show that any risks are negligible, e.g. because the evidence demonstrates safety, or because any uncertainties can be managed at the commercial stage. If regulations impose precautionary measures, they should be proportionate to the risk. Consumers should have the right to choose GM products as well as non-GM ones. Regulatory delays deprive Europe of the benefits.
ii. *European Commission (managerialist frame):* EU regulatory procedures must be implemented in a scientific, rational manner. According to objective, independent advice from EU experts, there is no evidence of risk or uncertainty for GM products awaiting a decision. Precaution may be applied as management measures in cases of established scientific uncertainty about potentially dangerous risks. This policy will enable Community business to exploit the potential of biotechnology, while taking account of the precautionary principle (CEC, 2003d). Such regulatory procedures are also necessary for complying with EU and international treaty obligations.
iii. *Agbiotech opponents (apocalyptic frame):* GM products are 'genetic pollution', impose unpredictable or unacceptable hazards, deny consumer choice of non-GM products, and increase farmer dependence on multinational companies. Less-intensive agricultural methods provide benign alternatives and more appropriate comparators for GM products in risk assessment, rather than conventional agriculture. On precautionary grounds, the EU Council moratorium must be maintained until safety is shown.

Table 5.2 Precaution in Contending Policy Frames

Table 5.2a EU

Policy frame	*Eco-efficiency*	*Managerialist*	*Apocalyptic*
Major actors	agbiotech developers and promoters, e.g. Europabio + EPP in EP	legislative and regulatory decision-makers, e.g. CEC, Environment Council	agbiotech opponents, e.g. EEB, CPE, Eurocoop, Greens and United Left in EP
Trigger for PP	Evidence of risk	Uncertainty about serious, irreversible effects	Hazards of GM technology
Scientific uncertainty	Data gaps which can be readily filled	Inadequate data or test methods	Vulnerable environment, ignorance about GM technology
Non-GM comparator	Conventional cultivation methods	Conventional methods, though contingent on specific contexts.	Organic cultivation

Note: Adapted from Levidow et al. (2005).

Table 5.2b Germany

Policy frame *Issue*	*Innovation: Making agbiotech possible*	*Risks: Making agbiotech safe*	*Alternatives: Preventing agbiotech*
Trigger for PP	Positive scientific evidence of risk	Uncertainty and knowledge gaps	Systemic uncertainty
Uncertainty	Speculative uncertainty provides no basis for preventative action. Any risks will be handled by risk management practices.	Uncertainty changes risk research, risk assessment (RA) and risk management (RM) practices. Uncertainty can be reduced through research ("need to know more") and risk-reducing measures.	Irreducible uncertainty can justify preventative action, including bans.
State of knowledge	There is considerable knowledge about GM crops.	We do not know enough about the effects.	We don't know what we don't know.

Note: From Boschert and Gill (2005).

(continued)

Table 5.2c Denmark

	Narrower account of precaution	*Intermediate account*	*Broader account of precaution*
Advocated by	Industry	Regulatory agencies	Environmental NGOs and others
GM crops as promise or threat	Represent a positive future for agriculture because they could solve problems of industrialised agriculture by offering opportunities for more sustainable practices.	Can be evaluated in the same way as any other crop. Conventional non-GM crops have been chosen as the normative comparator for judging the effects of a GM crop.	Represent a threat to sustainable agriculture, including organic agriculture, as they impose unpredictable hazards.
Precaution and evidence	A general requirement for GM products to undergo a prior authorisation procedure (case-by-case, step-by-step) based on value-free science	Any evidence of risk or uncertainty warrants measures to investigate and clarify uncertain risks before final regulatory decisions.	Acknowledgement of, and much more focus on, questions of scientific uncertainty and lack of knowledge.
Market-stage criteria and control measures	No monitoring is needed except in special cases where the risk assessment has documented evidence of risks with adverse effects.	Even without clear evidence of potential risks, regulators could still impose extra market-stage controls, e.g. to monitor or prevent a potential risk.	Less-intensive agricultural farming systems, including organic farming methods, should be evaluated alongside the GM strategy.

Note: Based on Toft (2005). The three tables indicate similar lines of debate on the precautionary principle at the EU and national levels. Details are drawn from analyses in the publications cited.

In similar ways, three contending frames were promoted in some national debates, e.g. Germany and Denmark, sometimes dividing state bodies in the same country (see Table 5.2). In each national case, an anti-agbiotech stance challenged the optimistic assumptions of agbiotech promoters, thus leading regulators to broaden their responsibility for uncertain risks. Unsurprisingly, agbiotech proponents espoused relatively narrow accounts of precaution, while opponents espoused broader accounts and emphasised unknown risks. Some EU member states elaborated the latter view, as a means to justify bans on GM products or regulatory delays. Most member states developed more diverse, flexible accounts of precaution. That flexibility provided a means to accommodate consumer groups, which did not oppose agbiotech per se, while potentially supporting approval of specific GM products.

3.2 Ending the Moratorium?

By revising the Deliberate Release Directive along more precautionary lines, the Commission sought to accommodate member states in ways which could overcome national bans and resume the EU approvals procedure. Before its enactment in 2001, five countries—Austria, Italy, Greece, Luxembourg and Germany—had banned at least one product already approved by the EU. After the new Directive in 2001, no national bans were lifted, and even more were imposed—by Greece, Poland and Hungary. Some member states were seeking to exclude all GM products from their national territory, though not yet by declaring a 'GM-free zone'.

When proposing approval of new GM products, the Commission sought to counter public suspicion that its policy was being driven by external pressures, especially the WTO dispute that had been initiated by the US government in May 2003 (see next section). Since the late 1990s some politicians had cited the threat of such a dispute as an extra reason to approve GM products, but this argument had backfired politically. Eventually the Commission promoted the new EU procedures as a better regulatory model, whose global credibility would depend upon timely implementation: 'We have to start because we want to demonstrate to the rest of the world that our way of taking decisions about GMOs works. Otherwise they will not believe us', stated Margot Wallström, the DG-Environment Commissioner (Geitner, 2004).

As an early test case for EU regulatory approval, Syngenta's Bt 11 sweet maize was a hybrid of conventional sweet maize and GM field maize; only the latter had already undergone safety tests. In 1998 a notification for Bt 11 sweet maize had been submitted to the Netherlands under the Novel Food Regulation. The national advisory committee requested and obtained more data on chemical composition, e.g. secondary metabolites—but not additional data requested by other member states. The EU's Scientific Committee on Food, then advising on products under the Novel Food Regulation, acknowledged that the company data 'provide only limited evidence of safety' but accepted its adequacy for a favourable risk assessment (SCF, 2002). Some member states continued to request more rigorous data.

In 2003 DG-SANCO pressed ahead with the Syngenta Bt 11 dossier, though with no new data—on grounds that the earlier SCF advice had identified no risks. Member states raised numerous doubts about the quality and relevance of the available data, e.g. the molecular characterisation, chemical composition and the use of field maize rather than sweet maize for toxicity tests. Moreover, some asked whether the novel cross could generate unknown novel proteins. French advisors demanded that toxicity tests be redone using the sweet maize, on grounds that differences from field maize could lead to unexpected effects (AFSSA, 2003). This view was published shortly before the EU regulatory committee was to decide on the Commission proposal for approval. Syngenta regarded extra toxicity tests as unnecessary, though without explicitly claiming that the hybrid was equivalent to the GM field maize. Thus this product generated expert disagreements around the practical meaning of precaution in risk assessment, as well as criteria for substantial equivalence.

This Bt 11 sweet maize became the first test case for lifting the *de facto* moratorium. In November 2003 the product went to a vote in the EU Standing Committee on the Food Chain and Animal Health, representing EU member states under the Novel Food Regulation. There was no qualified majority to support market authorisation; member states reiterated earlier objections (SCFCAH, 2003). At the next EU Council meeting, Ministers had no qualified majority for approval, so a decision fell to the Commission under the comitology rules. A paradoxical headline read, 'EU ministers sound death knell for GM moratorium', on the basis that EU farm ministers 'failed to agree on the fate of Syngenta's Bt 11 sweet corn', thus allowing the Commission to grant approval (Anon, 2004a). It did so in May 2004, despite dissent from several member states.

Environmental NGOs and many politicians denounced the decision as an anti-democratic surrender to the USA. According to a journalistic account:

> The decision triggered a backlash from the environmental community. Green MEPs and the Commission had committed a 'grave political error' and had 'sold out democracy' by saying yes to the approval where governments had not. (ENDS, 2004a)

In response the DG-SANCO Commissioner asserted the democratic legitimacy of their decision:

> . . . the fact that ministers were unable to make a logical response is a matter for them. The decision is fully in conformity with democratic systems we've put in place in the EU . . . I don't believe there is any democratic deficit. (David Byrne, quoted in ENDS, 2004a)

Soon another GM product was turned into a symbol of political surrender versus democratic sovereignty. Monsanto's GT73 oilseed rape, proposed for grain import only, became controversial for several reasons. The company had designated part of the dossier as 'confidential business information'. The risk assessment assumed that all the grain would go directly into processing, yet spilled grain at ports could plausibly escape into the environment and create weeds, thus needing additional herbicide sprays to control them (see also Chapter 6). Greenpeace attacked EFSA's favourable opinion, 'which did not consider the environmental implications'. For all these reasons, 'The Commission should withdraw the application rather than pushing it forward to satisfy the US in the WTO case', stated Greenpeace (Agence Press Europe, 2004). Similar criticisms were soon extended to the entire regulatory procedure.

The decision rules had been changed by the 2001 revised Directive, potentially shifting responsibility to member states. The Council could previously block a product only by unanimity, while the new rules allowed a qualified majority to do so (EC, 2001). If the Council had no qualified majority for or against a Commission proposal, however, then the Commission could enact it into EU law. According to the Green/EFA Group in the Parliament, the procedure was therefore 'incompatible with the democratic ideals promoted by the Union'. The Group requested a change in the comitology procedures in order to make them more democratic (Anon, 2004b). In cases where the Council has no qualified majority, decisions should be made by ministers and parliamentarians, rather than 'technocrats', argued a Green MEP (Lewis, 2004a). However, many politicians sought to avoid responsibility for such controversial decisions; the comitology rules provided a convenient way to do so.

3.3 National Bans on Trial

Alongside these legitimacy problems in approving new GM products, national bans were being imposed on GM products which already had EU approval. Bans by Austria and Luxembourg were succeeded by others such as Italy, Germany and Greece. These member states generally cited the 'safeguard' clause which permits a provisional ban, though EU scientific committees advised that the bans had no scientific basis.

The Commission lacked political support to put these governments on trial through the EU judicial system. In 1998 it proposed legislation requiring Austria and Luxembourg to lift their bans, but the Council gave no qualified majority. In later votes of the Council on similar proposals, not even a simple majority supported the Commission. So it had a weak basis to bring court proceedings against national bans: 'If there is a big group of member states in opposition . . . , then the Commission will hesitate to take steps against them', said a DG Environment official in 2003 (quoted in Toke, 2004: 173).

During the WTO dispute against the EU, the Commission raised the stakes. In November 2004 it proposed legislation requiring several governments to lift their bans, thus symbolically putting them on trial in the EU regulatory committee. The proposal gained fewer than one-third the total votes of member states, with many abstaining. This gained far less support than proposals to approve new GM products (FoEE, 2004a). Apparently some member states were defending national sovereignty against the Commission, not necessarily supporting the bans.

4 EFSA JUDGING—AND BEING JUDGED

Commission policy set high expectations for official expertise: 'science-based regulatory oversight' aimed 'to enable Community business to exploit the potential of biotechnology while taking account of the precautionary principle and addressing ethical and social concerns' (CEC, 2003d: 6, 17). As the EU regulatory procedure considered more GM products through 2003–04, national responses highlighted disagreements over expert judgements. EFSA had been set up to adjudicate such conflicts but was increasingly put onto the defensive.

4.1 Consulting Stakeholders

EFSA's opinions served a policy agenda of EU regulatory harmonisation, which often conflicted with national precautionary approaches. Each EFSA opinion briefly explains why it claims to accommodate, resolve or reject national objections, generally on grounds that they had no basis in scientific risk assessment. According to some Commission staff and expert advisors, national governments use scientific arguments to try to justify their dissent or delay of GM products:

> Some governments are running strong political agendas on agri-biotech, sometimes by using scientific arguments. The EU has an agenda to separate science from politics, as a step towards transparency about the political basis of objections to GM products. Some countries have difficulty in defending their stances in scientific terms. EFSA is like the High Court: after EFSA gives an opinion, it becomes more difficult for a country to return to its earlier risk assessment. (member of EFSA GMO Panel, interview, 2004)

On that basis, national objections could be dismissed as 'political' dissent by citing EFSA's advice.

However, any science/politics boundary became contentious. 'It may not always be straightforward to distinguish between scientific and political reasons for expert disagreements' (Commission staff member, interview, 2005). Environmental NGOs regularly criticised weaknesses in company risk assessments and attacked EFSA for accepting them as adequate science (e.g., FoEE, 2004b). With a more cooperative tone, consumer NGOs complained that EU regulatory procedures lacked clear criteria for necessary evidence to demonstrate substantial equivalence.

When EFSA held a stakeholder consultation on its risk-assessment guidance document, comments had divergent agendas. Some asked that data requirements be proportionate (EuropaBio, 2004a), while NGOs attacked EFSA for 'Failing Consumers and the Environment' (Greenpeace, 2004). Some CAs asked for a more explicit analysis of uncertainties in risk assessment. In the revised guidance document, applicants were asked for a 'risk characterisation' explaining uncertainties about data and assumptions. Moreover, applicants should explain 'the scientific basis for different options to be considered for risk management' (EFSA GMO Panel, 2004b: 46–50).

According to EU statutory guidelines, 'the overall uncertainty for each identified risk has to be described' (EC, 2002b: 32). The Panel's published opinions still did not characterise any 'uncertainty', consistent with the conclusion that no risk is identified for a GM product. Any uncertainties in risk assessment remained implicit in EFSA opinions.

Beyond public consultation on the guidance document, a participation procedure was also established for specific GM products. Any comments could be sent and posted on a special web forum, *Gmoinfo* (http://gmoinfo.jrc.it). This would also post any response by the national Competent Authority which had provided the risk assessment.

During 2003 many public comments had an ethical or common sense focus, e.g. by asking the applicant to justify why genetic modification was necessary; all such comments were dismissed as irrelevant to the statutory criteria. In response to any technical comments, EFSA said that these had already been considered. The average number of comments per product fell from twenty-one in 2003 to only four in 2004, sent mainly by specialised NGOs. Broader comments nearly disappeared in favour of technical ones (Consiglio dei Diritti Genetici, 2005). Prospective participants faced a dilemma: their wider comments would be predictably ignored, while technical comments would depend upon specialist expertise in risk issues. Thus participation could not be equated with more democracy (Ferretti, 2007).

4.2 Turning EFSA into a Defendant

During 2004–2005, each time a Commission proposal to approve a GM product went to the EU regulatory committee of member states, few gave support and many more voted against (FoEE, 2005a). A similar pattern arose when the same proposals went to the Council. Nevertheless the Commission granted approval for some products, citing the favourable advice of EFSA's GMO Panel. Environmental NGOs denounced this arrangement as 'profoundly undemocratic'.

EFSA's judgements were questioned as less than independent. As the Italian delegation stated, EFSA 'merely examines the scientific data supplied by applicants', but it should be able to arrange for further investigations, necessary for an 'independent assessment'. Rejecting this proposal, the DG-SANCO Commissioner said, 'Any change in the system would change the EU' s whole approach on GMO authorisations, and it would alter the burden of proof' (Smith, 2007). Indeed, member states were already putting a greater burden of evidence onto EFSA—to demonstrate its independence, as well as product safety.

DG-Trade pressed for faster approvals, especially as transatlantic pressure mounted to demonstrate that the EU no longer had a moratorium. Along with some member states, DG-Environment proposed that the decision procedure should give governments greater control. A former UK environment minister stated, 'Having a group of unelected bureaucrats deciding what food should be eaten is fundamentally undemocratic' (Lean, 2005).

Increasingly EFSA's GMO Panel was criticised for a pro-biotech bias, e.g. for simply accepting the data in company dossiers. Panel members made even greater efforts at performing consultation, yet they hardly accommodated the widespread criticisms—e.g., about weak evidence of safety, optimistic assumptions and implicit uncertainties. For the environmental risk assessment of Bt maize crops for cultivation purposes, some member states criticised EFSA for ignoring diverse environmental conditions across Europe (Levidow et al., 2005); the Commission had been warned about this prospect when centralising the risk assessment procedure.

At a special meeting set up with the GMO Panel, environmental NGOs emphasised its failure to carry out a quality check on scientific information produced by industry (Greenpeace Germany, 2006). Moreover, they attacked Panel opinions for bias towards the agbiotech industry. As extra evidence of guilt, three Panel members were overt supporters of an industry-funded pressure group lobbying for less stringent regulation (FoEE, 2006a). EFSA's response invoked the putative separation between science and policy, as if this body stood above politics. According to the Director, Herman Koeter, EFSA acted as a 'neutral risk assessment organisation' and was 'not prepared to enter into discussions of a political nature or on risk management areas, especially with regard to the societal debate on GMOs' (Herman Koeter, quoted in EFSA, 2006: 2).

By early 2006 member states were publicly attacking the EU-wide approvals procedure as undemocratic, given that the Commission's decisions had little support from member states, and that EFSA depended upon the applicant for scientific evidence. These conflicts came to a head at the March meeting of the EU Environment Council. According to the Danish Minister, 'It doesn' t do a great deal for democratic accountability' (Anon, 2006c). Some governments proposed that EFSA use scientific opinions available from national bodies; they warned EFSA that its own opinions must be seen to be 'scientifically objective'. Although the UK generally voted in favour of GM products, even its Environment Minister asked that EFSA's opinions be 'more robustly argued and more clearly explained' (Lewis, 2006a).

Similar proposals came from senior EU politicians. According to the Environment Commissioner Stavros Dimas, GM crops had poor consumer demand and low acceptance for food uses. Farmers would continue to grow conventional or organic varieties, which therefore should be improved. He emphasised that GM crops

> . . . raise a whole new series of possible risks to the environment, notably potential longer-term effects that could impact on biodiversity. Protected sites or areas, endangered or vulnerable species of plants and animals are of paramount importance in this respect. (Dimas, 2006)

Thus he echoed Austria's language justifying its ban on GM products—arguments rejected by EFSA as lacking a scientific basis. Moreover, he argued, the EU system should 'alleviate concerns regarding GMO products by improved risk assessment practices and making them more transparent' (Dimas, 2006).

After much internal debate, the Commission announced a policy shift in April 2006. Previously it had accepted EFSA's implicit role as a High Court, routinely judging and countering national objections to safety claims. Now it diagnosed the problem as expert disagreements arising from EFSA's procedures and advice, or perhaps even that advice itself. The Commission proposed practical improvements 'to improve the scientific consistency and transparency for decisions on GMOs and develop consensus between all interested parties'. Among other requests, EFSA would be asked 'to address more explicitly potential long-term effects and bio-diversity issues', and 'to provide more detailed justification for not accepting scientific objections raised by the national competent authorities'. Moreover, where a member state's observation raises important new scientific questions not properly or completely addressed by the EFSA opinion, 'the Commission may suspend the procedure and refer back the question for further consideration . . . ' (CEC, 2006a).

With these proposals, EFSA became a symbolic defendant—not simply a judge of dissenting member states. The Commission signalled that it would no longer routinely accept EFSA's safety claims. If EFSA did not adequately justify its advice, then the Commission might reject it or impose extra management measures to deal with uncertainties in risk assessment. The Commission proposals were pushing EFSA to go beyond an internal consensus.

The decision rules and procedures were in crisis, as recognised by DG-Environment. A representative reflected on the March 2006 Council meeting as follows: 'Many delegations voiced misgivings about the use of the comitology procedure for the authorisation of GMOs, with some delegations asking for the possibility of rejection by simple majority'. Towards a remedy within the current procedures, the Commission's April proposals could help clarify the broad role of precaution; however, this concept was absent from the Commission proposal (Tierney, 2006).

The new arrangements potentially used public comments to test safety claims. EFSA discussed how national views 'could be addressed in a more visible way'. When EFSA issued its next opinion on a GM product in May 2006, the Commission asked for public comments on that opinion (Lewis, 2006d).

Originating from the DG-Environment Commissioner, Stavros Dimas, those proposals had a double-edged character. As a practical reform, they could provide a way to accommodate more member states through greater rigour and thus potentially gain more votes for Commission proposals. On the other hand, if greater transparency meant a more explicit uncertainty, then this could undermine safety claims as 'objective' advice (Levidow and Carr, 2007a). Just such transparency was being provided by the Commission in the WTO agbiotech dispute.

5 COMMISSION DOUBLY ON TRIAL IN THE WTO DISPUTE

When the transatlantic trade conflict over agbiotech products began in 1999, the USA threatened to bring a WTO case against the EU. After the Bush Administration took office in January 2001, the US government intensified its rhetorical attacks on EU regulatory delays and blockages of agbiotech products. The Commission privately warned US officials that their overt threats were undermining its own efforts to establish a workable regulatory system, especially as NGOs attacked these efforts for promoting agbiotech and deferring to the US government (Murphy and Levidow, 2006: 161).

Indeed, using the transatlantic conflict, agbiotech critics pressed for more stringent regulations to defend EU sovereignty. In drafting new legislation in 2001, the Commission generally favoured less-restrictive criteria than some member states and MEPs, especially on grounds that EU rules must be workable and comply with international commitments. Critics denounced the Commission proposals as a surrender to US pressures (see Chapter 7). Thus the Commission was being put doubly on trial—for blocking trade from a US government standpoint, and for surrendering sovereignty from an NGO standpoint.

After the US decision to launch a WTO case against the EU in May 2003, agbiotech critics likewise portrayed any EU approval of a GM product as a surrender. The approval decision on Bt 11 sweet maize, as well as a subsequent approval of another GM product, was emphasised in a Commission statement to the WTO Disputes Panel: 'How else can you prove the absence of a moratorium if not through demonstrating that the approval process moves on and results in decisions?' (CEC-SJ, 2004a: 10–11).

In the WTO agbiotech case, arguments centred on the burden of evidence for many issues: whether or not EU procedures had 'insufficient information' to approve GM products in the late 1990s, and thus whether EU procedures had undergone an 'undue delay'. According to the US government, the delays and national bans were clearly an illegal barrier to trade. The EU had 'a politically motivated moratorium' lacking any scientific basis, especially given that the EU's own scientific committees had given favourable advice on the blocked products.

The Commission defence case linked regulatory delays with scientific uncertainty. In the earlier transatlantic dispute over hormone-treated beef, it had tried to broaden the interpretation of risk assessment in the WTO Agreement on Sanitary and Phytosanitary (SPS) Measures; it funded new risk research and cited expert disagreements to highlight scientific uncertainty. The Commission extended this strategy to the agbiotech dispute.

According to the Commission, the SPS Agreement should be interpreted in relation to the Biosafety Protocol, which could justify more flexible timescales. This argument was linked with jurisdictional sovereignty:

> This is, in the view of the European Communities, a case about regulators' choices of the appropriate level of protection of public health and the environment in the face of scientific complexity and uncertainty and in respect of which there is great public interest. It is a case essentially about time. The time allowed to a prudent government to set up and apply a process for effective risk assessment of products which are novel for its territory and ecosystems, and that have the potential of causing irreversible harm to public health and the environment. (CEC-SJ, 2004a: 1–2).

The Commission strategy sought to highlight divergent expert opinions as indicating scientific uncertainty about risk. It argued for the establishment of a WTO Expert Group, as a means for the WTO Disputes Panel to understand the broader context of scientific issues and disagreements. It appealed to the legitimacy problems of the WTO: 'The European Communities submits that a failure by the Panel to have regard to this broader context will risk undermining the legitimacy of the WTO system' (CEC-SJ, 2004b: 10). The USA argued that an Expert Group was unnecessary because the case could be judged on procedural grounds alone, and by drawing on the advice of the EU' s own scientific committees if necessary. In August 2004 the WTO Dispute Panel accepted the Commission's proposal.

Following that decision to establish a WTO Expert Group, the parties made further arguments about its appropriate composition and remit, especially the scientific questions to be answered. Several nominees were rejected by one of the parties; the expert group ultimately included members with diverse viewpoints, thus facilitating the Commission's strategy of highlighting expert disagreements. As a result of its proposals, the questions that were put to the experts did not simply impose a burden of evidence on the defendant to demonstrate risk; the overall list was relatively more balanced. The questions opened up greater opportunity to explore uncertainties about potential risks of agbiotech products in the late 1990s—the time when the delays in EU decisions first triggered the trade dispute. Some members of the Dispute Panel's Expert Group highlighted what was not known scientifically in the late 1990s.

In February 2006 the WTO Disputes Panel issued its interim report, which found the EU guilty. Upholding US complaints, the Panel agreed that there had been 'undue delays' in EU decisions on new GM products and unjustified national bans on products already approved by the EU. Green MEPs lamented that the ruling 'gives free trade precedence over the precautionary principle and the democratic right to regulate for the protection of either health or the environment'. In their view, moreover, the Commission should 'abandon its ambiguous policy on GMOs and side with its citizens' (Lewis, 2006a).

The ambiguous policy involved tensions between the Commission's domestic agenda and WTO submissions. Since 2003 the Commission had been asking all member states to lift their national bans and to support the approval of 'safe' GM products, thus accepting EFSA's safety judgements. In its WTO submissions, however, the Commission criticised evidence of safety in EFSA opinions as inadequate and argued that the EU regulatory procedure had legitimate alternatives to such advice. The equivalent expert committees in some member states had raised issues that were not adequately addressed (CEC-SJ, 2005).

After that submission was leaked, it was cited by NGOs as grounds for the EU to maintain its regulatory blockages and delays on agbiotech products. In the domestic arena the Commission was put metaphorically on trial for duplicity. According to environmental NGOs:

> The contrast between the public and private views of the European Commission on the risks of GMOs is staggering and will seriously dent public and member states' confidence in their ability to act fairly. All the evidence reveals a Commission policy of favouring the interests of the biotechnology industry over protecting the environment and human health. (FoEE/Greenpeace/Global 2000, 2006: 3)

This language was meant as a self-fulfilling prophecy: unlike consumer NGOs, environmental NGOs sought to undermine public confidence in the EU regulatory procedure and its safety claims for GM products.

As a greater difficulty for the Commission, it was being put increasingly on the defensive by some member states as well as NGOs. They were criticising political dependence upon inadequate risk assessments. Accommodating those demands in its April 2006 policy shift towards EFSA, the Commission incorporated precautionary features of its submissions to the WTO dispute.

EFSA's safety assessments were generally kept on the defensive in discussions among member states. Given those conflicts, the Italian delegation even proposed to revive the 1999 moratorium:

> . . . the EFSA reform process now underway is very positive, but it should do more to clarify the mandate of the European Authority concerning GMOs, guaranteeing at the same time a real and independent long-term impact assessment from the health viewpoint and a wide synergy with Member States on environmental impact assessments . . . During the reform process, and pending its finalization, all authorisations of GM crops should be suspended. (Italy, 2007: 7)

Italy's proposal came up at the October 2007 meeting of the EU Council of Environment Ministers, who predictably reached no common view. The official record emphasised the reform process, especially EFSA's role:

> Following interventions by the Italian and Austrian delegations, many delegations also stressed the need to continue this examination of the EFSA's role and the procedures for assessing any risks arising from GMOs (EU Council, 2007: 21).

Thus EU expert advice remained on trial.

6 CONCLUSIONS

At the start, this chapter posed the following questions, especially about the years after the EU Council's 1999 de facto moratorium:

- What aims drove EU reforms for regulating agri-food products?
- How did the agbiotech controversy test those reforms?
- How were state bodies put onto the defensive?
- What roles were played by expert advice in various trials?

A series of food safety scandals in the mid-1990s, especially the BSE crisis, stimulated EU reforms towards separating expert advice from policy decisions. These changes were designed to regain public confidence through an independent basis for regulatory decisions, so as to stabilise an EU internal market in agri-food products. Under new arrangements starting in 1997, expert bodies were meant to carry out only scientific risk assessment, functionally separated from regulatory decisions, for which the political authorities would take responsibility.

These general reform measures underwent a difficult test in the agbiotech sector, leading to disputes over the relation between science, policy and legitimate decisions. The Deliberate Release Directive was meant to link environmental protection with 'completion of the internal market' through harmonised standards. Conflicts arose over the standards that would shape a market for agbiotech products. By the late 1990s, member states were disagreeing more sharply about how to define the 'adverse effects' which warrant evaluation and preventive measures.

For GM crops that would be cultivated in the EU, official risk assessments focused on 'the natural environment', while ignoring or accepting undesirable effects in the agricultural environment. The European environment was conceptually homogenised—modelled according to intensive monoculture and its normal hazards, such as the pesticide treadmill, prospectively compounded by a genetic treadmill. This agri-industrial ordering of nature complemented a particular socio-political order: economic-competitive pressures to maximise agricultural productivity. In the late 1990s this potential future was being destabilised by greater conflict among member states, reflecting concerns of national expert bodies and citizens' groups. For defining 'adverse effects', diverse agri-environmental norms came from national concerns—e.g. organic agriculture in Austria, drinking water policy in Denmark, farmland biodiversity in the UK, weed-control issues in France, etc. These norms provided diverse pressures and criteria for translating public unease into uncertain risks which must be managed or prevented.

EU rules were changed to address these societal conflicts. With impetus from the 1999 *de facto* moratorium, the 1990 Deliberate Release Directive was revised to encompass a broader range of potential harms and explicit uncertainties. This legal reform provided greater scope for a more selective form of ecological modernisation, potentially favouring agbiotech products which are more environmentally beneficial. Through a managerialist frame, with publicly accountable risk assessments, societal polarisation over agbiotech could be potentially channelled into regulatory disagreements (cf. Gottweis, 1998: 319). As revised in 2001, the Deliberate Release Directive was implemented in the context of other EU-level changes.

The earlier policy on 'risk-based regulation' was recast as 'science-based regulation'. The new European Food Safety Authority (EFSA) was meant to provide cognitive authority for the science necessary to harmonise risk assessment and precaution. This strategy depended upon a science/policy distinction: risk assessment involves merely technical caution on scientific issues, while policy judgements about the precautionary principle await risk management as a later stage, according to the Commission's 2000 Communication (CEC, 2000). This distinction was formalised by the new EFSA procedure. For the agbiotech sector in particular, the GM Food and Feed Regulation provided greater public access to regulatory procedures and information, as means for potentially holding decisions accountable—indeed, so that regulatory procedures could potentially be kept on trial. At the same time, expert assessment was centralised in EFSA, on the basis that scientific neutrality would avoid or overcome conflicts among member states, at least according to the European Commission.

The Commission also appealed to concepts of sustainable agriculture. According to its key policy document on agbiotech, regulatory procedures should evaluate the risks of not taking action, e.g. of not approving a product, especially 'in areas where current agricultural practices are unsustainable'. As a remedy, GM crops could lead to 'more sustainable agricultural practices', e.g. pesticide reductions (CEC, 2002b: 14). In that account, some agrichemical practices were implicitly unsustainable and so provided an appropriate comparator for evaluating GM crops as environmentally preferable, even as essential. Such claims came under scrutiny for their assumptions about comparators and possible alternatives—first in the late 1990s, and again when the EU regulatory procedure resumed in 2003.

The procedure generated conflicts which tested official claims for expert objectivity. Safety claims were put onto the defensive, thus destabilising 'science-based regulation'. EFSA's GMO Panel generally advised that each GM product was as safe as its conventional non-GM counterpart; its advice framed any uncertainties as irrelevant or readily resolvable through extra data, thus internalising the Commission's policy on regulatory harmonisation. That policy sharply distinguished between science and policy, yet EFSA's advice encompassed policy judgements which came under challenge. The Commission had separated the precautionary principle from risk assessment and promoted this own account as a 'common understanding' among EU institutions. Yet EFSA's safety claims and data requirements corresponded to a more narrow account of precaution than objections from member states.

Through calls for greater precaution, agbiotech critics challenged optimistic assumptions in safety claims and held state authorities accountable for their decisions. Expert disagreements highlighted precautionary and policy issues in risk assessment, thus undermining EFSA's cognitive authority. The EU regulatory procedure extended the earlier dissonances from the late 1990s—rather than harmonising risk assessment, much less precaution.

From their apocalyptic frame, agbiotech opponents emphasised scientific weaknesses in safety claims, questioned EFSA's political independence and sought to intensify public distrust in the EU regulatory procedure. Product approval decisions were attacked as a Commission surrender to external political pressures, especially the WTO, symbolising the threat of globalisation to democratic sovereignty. In these ways, opponents circumvented the rules normally favouring specialist expertise in the regulatory arena.

Relative to the EU regulatory procedure of the 1990s, the new procedure from 2003 onwards had greater formal means for public comment, though this reform hardly enhanced state accountability. Proposals to authorise GM products drew numerous public comments in the formal consultation procedure, but many comments were rejected as 'ethics' and therefore irrelevant, while specialist technical comments were dismissed as raising no new issues. These participatory arrangements modelled the public as proxy risk assessors, as the only legitimate basis for comments that could be heard. Criticisms from specialist NGOs, like those from member states, had no apparent influence on EFSA's opinions or criteria, despite extending its consultative procedures.

Meanwhile in the WTO agbiotech dispute, the EU was being put formally on trial for blocking GM products, in turn creating multiple 'trials' for the European Commission. Its Legal Services sought to defend the EU's sovereignty to shape its own regulatory procedures and timetables. It proposed procedural arrangements which could open up expert disagreements and so question official safety assessments. Its submissions to the WTO emphasised scientific unknowns, inadequate information for risk assessment and thus weaknesses of safety claims from EU scientific committees. Those arguments were cited by NGOs as grounds to maintain the *de facto* EU moratorium. Moreover, NGOs denounced the Commission for duplicity about uncertain risks of GM products: its WTO submissions contradicted its public statements citing EFSA's advice as a basis for member states to lift national bans on GM products and to support the approval of new ones.

Any such approval posed legitimacy problems for the EU decision procedure. From 2004 onwards, Commission proposals to authorise new GM products lacked a qualified majority (or even a simple majority) in the Council; the Commission was left with legal authority to grant approval but a weak political authority. Senior politicians questioned whether the EU regulatory procedure had 'objective' expertise; some identified a 'democratic deficit' in the comitology procedure, which permitted the Commission to approve products without majority support from member states.

The Commission gained even less support in its efforts to remove national bans on GM products which already had EU authorisation. Member states voted overwhelmingly against such proposals in the EU regulatory committee and Council. When those governments were put symbolically on trial by the Commission, others supported their national sovereignty.

Facing a legitimacy problem, in early 2006 the Commission's policy shifted, now asking EFSA to address the demands of member states for more transparent, rigorous risk assessments—and even to re-do its risk assessments. In the name of 'scientific consistency', the Commission sought to accommodate diverse national norms and uncertainties in risk assessment, along lines of the Commission's submissions to the WTO dispute. Previously the Commission had depended upon EFSA to judge whether member states were guilty of unjustified objections to safety claims, but now the Commission would make its own judgements on expert advice.

EFSA's centralised expertise was originally designed to strengthen the political authority of the Commission vis à vis national authorities and their advisory bodies (Dratwa, 2004; Smith et al., 2004). Now EFSA too was being judged, almost being turned into a defendant. EU reforms were being tested for both precaution and democratic legitimacy. Similar trials also pervaded efforts at scaling up GM products to the commercial stage, as shown in the next chapter.

6 Scaling Up GM Crops, Testing Commercial Operators

INTRODUCTION

Since the 1990s GM crops have undergone experimental field trials for several purposes—to demonstrate that GM crops were being kept safe, while also testing them for evidence of safety if grown commercially. Often disputes arose over whether the experiments had adequate control measures to protect the environment, a rigorous design for yielding meaningful scientific results and a legitimate societal purpose (see Chapter 5). At the same time, GM crops underwent informal-symbolic trials of their status: agbiotech was variously cast as an environmentally friendly improvement in plant breeding, or as a risk-management problem, or as a pollutant per se (see Chapter 3).

Those multi-level trials converged in regulatory conflicts over the appropriate basis for commercial authorisation. From the late 1990s onwards, such conflicts intensified—over what potential effects must be tested and managed, what experiments could simulate them, what control measures could reliably prevent them and how responsibility would be assigned. This chapter analyses conflicts over whether or how GM products could be safely scaled up from field trials to commercial use.

In particular the chapter discusses the following questions:

- How did the agbiotech controversy extend the risks for which GM products should be kept on trial?
- How were commercial contexts anticipated, simulated and tested?
- How did regulatory procedures extend the responsibility of commercial operators?

This chapter first sketches divergent framings of risk issues around specific GM products, by continuing the story from Chapters 3 and 5. Then it analyses how the agro-food production chain was turned into a social laboratory for testing operator behaviour. GM crops were scaled up towards the market stage by designing commercial experiments. Although the story originates in the late 1990s, the chapter emphasises regulatory conflicts during 2003–2005, when commercial scale-up became more contentious.

1 EXPANDING THE REGULATORY SCOPE TOWARDS COMMERCIAL EXPERIMENTS

After agbiotech faced greater European controversy in the late 1990s, regulatory conflicts intensified over what potential effects must be tested and managed, what experiments could simulate them and what control measures could reliably prevent them. More risk issues became contentious whenever GM products were proposed for commercial authorisation. These issues challenged the original 'stepwise principle', whereby control measures on GM field trials would be abandoned at the commercial stage. This section analyses how public protest led to regulatory changes for managing extra uncertainties.

1.1 Moralising Operator Behaviour in the Agri-Food Chain

From the late 1990s onwards, agbiotech opponents elaborated discursive frames for gaining mass-media attention, mobilising activists and intensifying public suspicion towards GM products. Critics popularised a series of ominous risk metaphors—especially 'superweeds', 'sterilisation' and 'genetic pollution'—thus framing undesirable effects in apocalyptic terms (compare Tables 2.1 and 6.1). These metaphors associated GM crop cultivation with environmental dangers, even immoral activity; operator behaviour and regulatory irresponsibility were put symbolically on trial.

Early on, critics emphasised the prospect of a 'genetic treadmill', by analogy to the pesticide treadmill of pests developing resistance to chemical insecticides. For example, they claimed GM herbicide-tolerant crops would spread the tolerance trait to related plants, and increased herbicide usage would favour the resistance trait, leading to 'superweeds'; likewise that pest-resistant GM crops would generate resistant pests. Statutory regulations remained ambiguous about whether these scenarios would count as 'adverse effects', though they were eventually evaluated.

Since the mid-1990s UK conservationists warned about potential harm from the broad-spectrum herbicides for which herbicide-tolerant crops were designed. Their joint use could inflict 'sterility' upon farmland biodiversity, important for wildlife habitats. After the UK government accepted responsibility for these effects, the Deliberate Release Directive was revised accordingly, thus extending a national policy change to the entire EU-wide system (see Chapter 5). Under the revised Directive, a broader risk assessment included long-term and indirect effects of GM crops, as well as effects of any changes in management practices, e.g. any switch to broad-spectrum herbicides.

In the late 1990s agbiotech critics raised yet another risk issue: 'GM contamination'. This pollution metaphor had multiple meanings—e.g. 'unnatural' genetic combinations, 'filthy lucre' perverting science, globalisation corrupting national sovereignty, as well as GM pollen 'contaminating' non-GM crops (see Chapter 3). Friends of the Earth Europe adopted

Table 6.1 Three Contending Risk-Frames for GM Crops

	Eco-efficiency/ cornucopian (agbiotech business, e.g. Europabio, some farmers)	*Managerialist* (regulatory agencies, e.g. DG-Environment and national regulatory agencies)	*Apocalyptic* (environmental NGOs, *Coordination Paysanne*, Green MEPs)
Agricultural problem	inefficient farm inputs, uncompetitive outputs	uncertain biophysical effects of a new technology	intensive monoculture, farmer dependence on MNCs, pesticide treadmill
Nature seen as	cornucopian potential to be reaped	resources to be managed and protected	fragile resources under threat from uncontrollable, irreversible risks
GM crops seen as	safe eco-efficient tools to gain economic and environmental benefits	potential hazards to be evaluated and managed	pollutants threatening the environment, democracy and societal values
Solution	Apply routine management measures.	Design research and controls to manage uncertainty.	Block or deter GM products.
1) herbicide-tolerant weeds	Manage this agronomic problem through product stewardship, e.g. standard Good Agricultural Practices.	Evaluate control measures and their feasibility for this agro-environmental problem (which could affect herbicide usage).	Prevent 'genetic treadmill' and 'superweeds', which would perpetuate agrochemical dependence.
2) harm to farmland biodiversity from 'total' herbicides	Use herbicides more efficiently and so reduce environmental harm. Bayer: plan moral teaching for farmers.	Test relative harm of broad-spectrum and selective herbicides—effects contingent upon farmer practices.	Prevent herbicides from sterilising the countryside into green concrete.
3) mixture of GM and non-GM crops	Protect non-GM crops through standard isolation distances (for each crop).	Limit adventitious presence through national measures for segregation and coexistence.	Prevent 'genetic contamination'—but impossible or difficult.
4) Bt harm to non-target insects	Risk shown to be negligible, so no need for extra measures.	Investigate potential risks to biodiversity before and/or during commercial use.	Prevent threat to biodiversity by blocking commercial use.

Abbreviations
GAP: Good Agricultural Practices
MEP: Member of the European Parliament
MNC: multinational company
NGO: non-governmental organisation

a honeybee logo to symbolise 'unwitting agents of genetic pollution', i.e. long-distance unmanageable spread of pollen. Initially this problem was framed in two different ways. According to the eco-efficiency frame of the agbiotech industry, the unintentional presence of GM material is manageable through routine segregation measures, thus protecting the economic value of non-GM crops. According to the apocalyptic frame, 'GM contamination' threatens the environment, consumer choice and even democratic decision-making, by pre-empting alternative agricultural futures.

New regulatory language responded to that conflict. 'Adventitious presence' denoted levels of GM material which remain technically unavoidable, despite the operator's reasonable efforts to minimise its presence. This phrase implied a moral obligation to exercise and demonstrate such efforts through segregation measures (CEC, 2001a: 21). On this basis, GM labelling would not be required for GM material present below a threshold of 1% (see Chapter 7).

In sum, critics kept agbiotech on continuous trial through ominous risk metaphors, moralising an environment in danger from operator behaviour as well as from GM crops. Eventually EU regulatory frameworks translated these metaphors into technical-managerial criteria. To manage the extra uncertainties, the authorities revised and re-interpreted risk legislation.

1.2 Extending the Stepwise Procedure

From the start, EU agbiotech regulation had formalised 'the step-by-step principle'. According to early international guidelines, GMOs should be made predictable by progressively decreasing physical containment. Releases should follow 'a logical, incremental, step-wise process, whereby safety and performance data are collected' (OECD, 1986: 29). According to the original Deliberate Release Directive, the scale of release is increased gradually, 'but only if the evaluation of the earlier steps . . . indicates that the next step can be taken', i.e. safely (EC, 1990: 15). If justified by scientific uncertainty, a precautionary or proactive approach 'may be deemed necessary until a large and reassuring body of data has been accumulated and we can begin to treat the technology as familiar' (Tait and Levidow, 1992: 223).

Initially the stepwise principle was interpreted to mean that controls could be entirely relaxed at the commercial stage. Many expert advisors felt that risk assessment could not depend upon a particular manner of using a product, because such restrictions would be unfeasible at the commercial stage: 'If it needs special controls at that stage, then we have really lost it', according to a UK advisor (interview, ACRE member, 1991). For small-scale field trials in the mid-1990s, special measures were required to confine the GM material, e.g. by destroying the flowers and harvested plants, pending a regulatory decision on scale-up. From the empirical results of those trials, proponents cited 'no evidence of risk' as an argument for product safety, thus warranting no special conditions at the commercial stage.

More possibilities to test uncertainties at the commercial stage came from a legislative change. When the Deliberate Release Directive was revised in the late 1990s, it required two types of monitoring at the commercial stage. 'General surveillance' should monitor any effects which were not anticipated in the environmental risk assessment. 'Case-specific monitoring. . . . should confirm that any assumptions regarding the occurrence and impact of potential adverse effects of the GMO or its use in the environmental risk assessment (e.r.a.) are correct' (EC, 2001: Annex VII). Thus the stepwise principle was extended into the commercial stage, potentially extending the regulatory arena into operator behaviour in the agri-food chain. Monitoring also implied that GM crops remained 'on trial' in a literal, scientific sense at the commercial stage, since product approval could be revoked if an adverse effect occurred.

To clarify these requirements, DG-Environment drafted guidelines, which attracted many disagreements. The guidance notes underwent several drafts and discussion among CAs before being finalised. Changes proposed by DG-Enterprise and EuropaBio influenced later drafts along similar lines, potentially reducing the burden on companies for case-specific monitoring. For example, monitoring plans should take into account the 'cost-effectiveness' of any requirements for monitoring (EC, 2002c: 29). This criterion provided ways to avoid such a requirement—e.g. by imposing a prior burden of quantitative 'effectiveness' upon anyone who proposes to require monitoring, or by doubting that meaningful knowledge could be obtained, as happened for the risk of non-target harm (see later).

After enactment of the revised Directive, arguments continued over what 'assumptions' must be confirmed by monitoring commercial use, and whether some GM products would be entirely exempt from this requirement. There was a dilemma of circular reasoning, regarding how an environmental risk assessment could include 'assumptions' which simultaneously justify both commercial authorisation and a monitoring requirement—i.e. how product safety could depend upon assumptions, a term implying contingency or doubt. A remedy was proposed by industry:

> Monitoring should be linked to the conclusions of the risk assessment, i.e. from evidence regarding potential effects. The term 'assumptions' is too broad. How could we monitor to confirm all assumptions? (interview, company officer, 2002)

Accordingly, DG Environment replaced the term 'assumptions' with 'conclusions' in the final version: 'Where the conclusions of the risk assessment identify an absence of risk or negligible risk, case-specific monitoring may not be required' (EC, 2002b: 31).

For risk assessment and commercial-stage monitoring, another contentious issue was how to specify 'adverse effects' of herbicide-tolerant crops in relation to a comparator or baseline. After all, conventional crops too can generate herbicide-resistant weeds, and narrow-spectrum herbicides anyway harm farmland biodiversity. After industry proposed that 'conventional' practices provide a baseline, this was accepted in the final version: When considering the monitoring plan for herbicide-tolerant genetically modified crops, it may be appropriate 'to consider herbicide use for conventional crops as part of an appropriate baseline' (DG-Envt, 2002). This could be interpreted to mean that the most chemical-intensive cultivation of conventional crops would provide an adequate comparator. Environmental NGOs advocated a more stringent comparator (FoEE, 2002a).

National CAs later formed a working group on herbicide resistance, especially to consider issues of agri-environmental comparators and herbicide-tolerant weeds. According to its report, such weed problems should be regulated under the EC pesticide directive:

> If resistant weeds occur, other herbicides or mixes of herbicides will be applied and this can make it an environmental problem. The possible development of herbicide-tolerant weeds is mainly an agricultural problem, which falls under Directive 91/414/EC and /or any relevant Community legislation. (DG-Envt, 2003: 2)

As the report also noted, herbicide-tolerant crops could lead to more harmful use of herbicides, which in turn would come under the Deliberate Release Directive. This link later arose in evaluating specific GM products, even grain imports.

As another issue of how to define 'adverse effects', Bt insecticidal crops could generate resistance among insects, thus undermining the efficacy of the naturally occurring toxin too. Industry argued that such an effect should be excluded from the Deliberate Release Directive, on grounds similar to earlier risk assessments which had accepted insect resistance as a normal agricultural problem:

> Monitoring should not be required for insect resistance to Bt. Foliar Bt contains a variety of toxins effective against the same pest, so insect resistance to one toxin would not jeopardize efficacy of the sprays. (interview, industry officer)

Nevertheless the EC guidelines included this effect: 'The build-up of resistance by insects to the Bt-toxin through continued exposure is an example of a delayed effect' (EC, 2002c: 30).

All the above issues became contentious when regulatory procedures evaluated specific GM products. As public controversy broadened the range of potential effects to be prevented and managed, regulators faced a dilemma in scaling up GM crops from field trials to the commercial stage.

2 FARMLAND BIODIVERSITY: HERBICIDE MANAGEMENT ON TRIAL

Large-scale real-world experiments for GM crops were first developed for GM herbicide-tolerant crops. These crops allow farmers to replace selective herbicides with broad-spectrum ones, to use relatively small quantities, and to delay spraying until weed problems appear. In the USA the agbiotech industry promoted such crops as a tool for pesticide-reduction. This optimistic account was challenged there by environmental NGOs, mainly on grounds that weed resistance would necessitate a pesticide treadmill, leading farmers to use supplemental agrichemicals (BWG, 1990).

As an extra issue in the European context, critics there emphasised special risk of broad-spectrum herbicides, which kill all arable weeds. These herbicides could cause relatively more harm to biodiversity in and around fields, as compared to the selective herbicides previously used. This question was specially compelling in the UK, where two-thirds of land is agricultural, widely seen as 'the environment'. At issue was not simply the quantity of sprays, but also their timing. From a nature conservation perspective, earlier applications would more effectively eliminate weeds and associated insects on which birds depend; they could also cause more harm to the seed banks essential for farmland biodiversity (Schütte, 2002).

Prospective harm to farmland biodiversity was disputed from the standpoints of eco-efficiency versus apocalyptic frames (see Table 6.1). According to proponents, GMHT crops help farmers to control weeds more efficiently, by delaying sprays until the post-emergence phase and/or reducing the quantities sprayed, thus benefiting the environment. According to UK nature conservation agencies, however, such herbicides would damage agricultural habitats essential for wildlife, thus threatening biodiversity; also inadvertent hybridisation could lead to adverse changes in herbicide usage (JNCC, 1997). Moreover, greater efficiency could turn the countryside into 'green concrete'. Even if smaller quantities were sprayed, broad-spectrum herbicides could reproduce 'the sterility of the greenhouse in open fields', thus causing a drastic reduction in wildlife already depleted by modern agricultural practices (ENDS, 1998a).

A parallel disagreement concerned responsibility for any herbicide effects. According to agbiotech companies, any effects on farmland biodiversity would result from the herbicide, not the GM crop—as if it were inherently 'innocent'. Within this causal model, herbicide legislation would be appropriate means to protect farmland biodiversity, under the responsibility of the Agriculture Ministry. According to the Ministry, however, no new issues arose from herbicide-tolerant crops (MAFF, 1998); in practice, its remit covered only unintended effects, e.g. toxicological harm, not intentional effects. According to nature conservation agencies, the crop-herbicide distinction was false, given the integrated design for herbicide tolerance and farmers' motives for choosing GM herbicide-tolerant crops.

With their warnings about farmland 'sterility', conservation agencies aimed to delay a decision on commercial authorisation, pending efforts to obtain more scientific information. Similar language was adopted by anti-biotech campaigners in their effort to stop GM crops altogether. They made public appeals to save the countryside as 'the environment'. Amidst public protest, such appeals stimulated efforts to put GM crops on trial in large-scale field experiments managed by farmers, thus also putting their practices on trial.

2.1 Disputed Simulations

Some research projects translated the 'sterility' hypothesis into biophysical effects, by investigating wider environmental consequences of substituting broad-spectrum herbicides. In the UK and Denmark, two experiments demonstrated potential environmental benefits of reducing and delaying herbicide application on glyphosate-tolerant beet. Greater biodiversity resulted from spraying only in a narrow band along rows, not between them (Elmegaard and Bruus Pedersen, 2001; Strandberg and Bruus Pedersen, 2002). Funded by Monsanto, small-scale studies in the UK likewise delayed herbicide sprays; the results showed greater weed and insect biomass on agricultural fields cultivated with GM beets, without reducing yields relative to conventional sprays. However, as the researcher acknowledged, 'Farmers could achieve higher yields with early overall applications of glyphosate . . . ' (Dewar et al., 2003). Thus environmental benefits of glyphosate-tolerant beet depended upon farmer practices which sacrifice yield, relative to the potential maximum.

Partly on those grounds, the experimental designs themselves were put on trial. Arguments ensued over whether the tests had realistically simulated commercial practice. Sceptics questioned whether commercial farmers would accept less than maximum yield: 'this means they forfeit 10% of their potential yield. The proposed approach is clearly therefore not a realistic proposal', argued the organic farmers organisation (Soil Association, 2003). Those criticisms highlight methodological assumptions and weaknesses of effectively designing trials so as to enhance biodiversity, in ways which may not realistically simulate farmer behaviour in commercial agriculture. Methods for larger-scale trials had already been shaped by political pressures in the late 1990s.

Facing intense public controversy over GM herbicide-tolerant crops, the UK's Labour government was internally divided. The UK had been a prime site for agricultural biotechnology—i.e. for R&D investment, for planned marketing of the initial GM crops and for industry lobbying. Extending the policy of the previous Conservative government, the New Labour government still promoted biotechnology as essential for attracting

R&D investment and for enhancing economic competitiveness. The Cabinet Office sought to expedite commercialisation, but the two Ministers responsible for safety regulation sought to accommodate environmentalist criticism through more stringent regulation and tests.

After a long political impasse, in October 1998 the Environment Minister announced: 'We are effectively declaring a moratorium. We must take a precautionary approach' (quoted in Brown, 1998). This policy change concerned environmental uncertainty about effects of broad-spectrum herbicides sprayed on herbicide-tolerant GM crops. With the agreement of the agbiotech industry, there would be a 'managed development', with commercial introduction accompanied by farm-scale monitoring for ecological effects (DETR, 1998; see next section).

Henceforth the government' s advisory committee would assess 'indirect effects', e.g. any resultant changes in agronomic practice and the subsequent effects on biodiversity. UK's biotechnology regulators would now act as if their remit already included herbicide effects, thus re-interpreting the law and anticipating revision of the Deliberate Release Directive 90/220 along such lines (see Chapter 5). Even politicians began to say, 'GM crops are now on trial'.

As a rationale for the policy change, agrichemical usage had caused a long-term decline of farmland biodiversity, according to the Environment Ministry (DETR, 1999a). The Ministry was anyway seeking ways to reduce environmental harm from agricultural practices in general, so regulation of GM herbicide-tolerant crops offered an opportunity to pursue that general aim. Indeed, mass protest provided an extra incentive and opportunity to do so, despite Cabinet reluctance towards more precautionary scrutiny of GM crops.

Represented by SCIMAC, the agricultural supply industry had developed a Code for farmer discipline, in anticipation of commercial production. The industry accepted the plan for farm-scale monitoring as a way forward to validate the code and thus reassure the public.

> SCIMAC is pleased to collaborate in this programme, which will play a vital role in answering the questions raised by conservationists and environmental groups . . . We have always made it clear that the SCIMAC Code of Practice will be subject to regular review, based on experience of the technology in use. Indeed, this pioneering framework offers a unique opportunity to encourage the adoption of measures which will make a positive contribution to farmland wildlife and biodiversity. Through this initiative, the UK leads the way in promoting a responsible and measured approach to the uptake of new technology. (SCIMAC, 1999)

By contrast, environmentalist groups attacked the experimental plan as an abdication of the public debate that should guide such regulatory decisions. According to Genewatch, the government was 'misusing science to obstruct democratic questioning of GM crops; hiding behind a set of scientific experiments to avoid the debate and all its complexities is neither rigorous science nor good governance' (Mayer, 1999). Indeed, democratic accountability was being displaced in favour of regulatory experiments, whose design thereby became all the more contentious; their scientific rigour would be put on trial.

2.2 Farm-Scale Evaluations

As a central part of its 'managed development' strategy, the UK's Environment Ministry soon announced plans for large-scale, on-farm trials for a four-year period starting in 1999. These Farm-Scale Evaluations (FSEs) aimed mainly to test herbicide effects on farmland biodiversity. According to the Environment Ministry, 'The results of these farm-scale evaluations will ensure that the managed development of GM crops in the UK takes place safely' (DETR, 1999b).

The FSEs had several aims at once. In the short term, they gave the government a scientific rationale for postponing a politically awkward decision, especially regarding Bayer's herbicide-tolerant maize, which already had EU-wide commercial approval. In the medium term, the FSEs could address the concerns of nature conservation agencies, thereby potentially incorporating them into new institutions and policy processes. Ultimately the FSEs were intended to justify a decision about approving GM herbicide-tolerant crops, on the basis of more credible knowledge about commercial contexts. In that regard, the FSEs have been analysed as a 'regulatory experiment', which put prospective regulations on trial (Lezaun and Millo, 2006: 193). Indeed, they were designed to test a prospective policy: permitting commercial use with a maximum limit on herbicide usage, just as in the authorisation of other herbicides, but with no requirements on timing the sprays.

In testing such a policy, the empirical results and their credibility would depend upon the comparator chosen for GM crops. Represented on the Scientific Steering Committee, nature conservancy groups proposed that the experimental design should include some conventional fields where farmers spray relatively less herbicides, to provide a more stringent comparator for broad-spectrum herbicides. The request was accommodated: 'To fully represent the range of potential biodiversity effects, a full range of current conventional practices used in the UK should be included in the research programme' (UK SSC, 1999). The FSEs specifically included farmers who operated 'less intensive production systems' in the trial fields (UK SSC, 2000).

In each context, split fields provided an experimental 'control' or baseline for comparing biodiversity effects of the two different herbicide regimes: broad-spectrum and selective herbicides. In both cases the experiment was meant to design and simulate whatever herbicide regime would give farmers the greatest economic benefit. The Scientific Steering Committee discussed with SCIMAC the provision of advice to farmers managing GM herbicide-tolerant (GMHT) crops, 'to be assured that the aims of the procedure were to ensure cost-effective weed control' (UK SSC, 2000; Fairbank et al., 2003). From that principle, the supply companies recommended specific amounts to be sprayed in the FSEs. Environmental NGOs expressed suspicion that the experimental design could have a pro-biotech bias, e.g. by excessively spraying the conventional crops or under-spraying the GM crops, though they did not criticise the SCIMAC recommendations at that time. Instead they attacked the FSEs in principle—for posing risks of 'GM contamination' and evading democratic accountability for an eventual regulatory decision.

Despite the careful approach to methodological issues in the FSEs, government advisors signalled difficulties and limitations would still leave 'GM crops on trial' afterwards. The FSEs had design weaknesses, e.g. by monitoring only weed seeds and specific insect types; and the FSEs could detect only large changes in farmland biodiversity. Moreover, commercialisation would involve wider issues such as segregation from conventional crops (AEBC, 2001).

In 2003 preliminary results of the FSEs indicated large environmental differences between GM and non-GM crops. The overall plan had included four crops: spring-sown oilseed rape, winter-sown oilseed rape, sugar beet and maize. For two of the crops, spring-sown oilseed rape and sugar beet, much greater harm to wildlife resulted from farmers spraying broad-spectrum herbicides, by comparison to spraying selective herbicides on their conventional counterparts. For example, the GM fields had fewer weed seeds and insects important for bird diets (UK DEFRA, 2003b; UK SSC, 2003; Champion et al., 2003).

In the case of maize, relatively less harm resulted from spraying broad-spectrum herbicides than the selective ones in use at that time, e.g. atrazine. Since this herbicide was soon to be banned, NGOs criticised the comparison as invalid. They also questioned whether the experimental designs were realistic models for commercial practice, i.e. whether they really simulated farmers doing 'cost-effective weed control'.

In response to those mixed results, agbiotech companies defended GM crops from attack, mainly on grounds that environmental effects had been contingent upon specific management practices in the FSEs:

> Activist groups claim that GM crops were in effect 'green concrete' and would 'wipe out' wildlife. These studies show that this sort of scaremongering is not supported by the evidence. (Bayer regulatory manager, quoted in ENDS, 2003: 31)

Proponents argued that only the changes in herbicide use were 'on trial', implying that GM technology per se was 'innocent'. From their eco-efficiency frame, any harm was caused by weed-management measures, which could be flexibly adjusted in order to benefit wildlife. In a similar way, some scientists proposed that any environmental harm could be avoided by mitigation measures. For example, they cited agro-environmental schemes which provide financial incentives to minimise or delay pesticide usage. Thus the crop could be exonerated of guilt, but only by keeping its cultivation methods under suspicion.

In its eventual advice to government, the UK scientific advisory committee accepted the results of the trials as valid only for their specific conditions. According to that committee, adverse effects would not result 'if GMHT maize were to be grown and managed as in the FSEs'. Given that atrazine would be phased out soon, however, alternative herbicides could change the unfavourable comparison of conventional maize with GMHT maize, so there must be a scheme 'to monitor changes in conventional management practice' (ACRE, 2004: 10). For GMHT spring-sown oilseed rape, the committee would not support commercial approval from the available evidence; companies would need to submit proposals for how the glufosinate sprays could be managed to minimise harm (ibid.: 13).

That outcome further complicated the stepwise principle (OECD, 1986). Originally the FSEs had been intended to ensure that commercialisation proceeds safely (DETR, 1999b), thus implying that a specific crop-herbicide combination could be declared safe, as the finale of the stepwise procedure. Yet the 2003 expert advice implied the need for further experiments. Even for maize, the one GM crop that had favourable results, advisors anticipated commercial use as a larger experiment in disciplining and monitoring farmer behaviour. For two other crops which had unfavourable results in the FSEs, any judgement on commercial approval would depend upon more pre-commercial experiments, farmer discipline and environmental information. (For winter-sown oilseed rape, results were announced at a later stage; these likewise showed relatively more harm from broad-spectrum herbicides.)

3 DISPUTING COMMERCIAL EXPERIMENTS

Criteria for environmental harm and monitoring became generally more contentious in 2003–2004, when European governments were once more deciding whether or how to support commercial approval of specific GM products. Regulators considered ways to predict, test and/or prescribe the behaviour of agro-industrial operators who would handle them. Some arguments drew upon the results of the UK FSEs, which thereby acquired Europe-wide significance. In these ways, EU-wide commercial use was anticipated as a real-world experiment.

EC legislation provided a framework for such experiments. Under the Deliberate Release Directive, companies submitted each notification for commercial approval of a GM product to a national Competent Authority (CA). Then all CAs could express views on the risk assessment and on appropriate requirements. They judged what extra controls or monitoring might be warranted, regarding any 'assumptions' in the risk assessment.

Initially, the basis for those judgements remained unclear: 'Regulators have no clear criteria for what may or may not be considered an assumption for the purposes of requesting case-specific monitoring' (company regulatory manager, interview, April 2003). Regulatory judgements can be illustrated by the following four examples.

3.1 Maize Cultivation in UK

Farmland biodiversity issues became more specific for Bayer's herbicide-tolerant T25 maize, which had already gained EU authorisation in 1998. Each member state could set its own terms for cultivation, especially regarding herbicide usage (see Table 6.2, first product). Citing the FSE evidence of environmental benefits, the UK government ultimately announced that it would approve the crop for the 2005 season, but with extra conditions: First, herbicide spraying must follow the herbicide practices used in the FSEs or other practices 'that have been shown not to result in adverse effects'. Second, after 2006 further trials would be necessary to redo the environmental comparison with whatever herbicides replaced atrazine. Moreover, given that non-GM crops could lose economic value as a result of gene flow from a GM crop, there would be a compensation scheme 'to be funded by the GM sector' (DEFRA, 2004; ENDS, 2004b).

In those ways, commercial cultivation was anticipated and designed as a semi-controlled experiment in farmer practices. The FSEs were originally intended to simulate farmer behaviour and thus to predict environmental effects, as a realistic basis for relaxing control measures. Instead expert advice now cited the empirical results to prescribe farmer practices, which themselves would need monitoring for compliance at the commercial stage. For segregation measures to avoid admixture, the requirement for a compensation fund would give companies a financial incentive to enforce and monitor farmers' compliance with guidelines, though such a relationship had little precedent. Thus the commercial stage was made conditional upon real-world experiments to test farmer practices regarding the two risk issues—farmland biodiversity and admixtures from gene flow.

Having obtained a conditional go-ahead, Bayer Crop Science applauded the UK announcement—but withdrew its application:

> The Government has however placed a number of constraints on this conditional approval before the commercial cultivation of GM forage maize can proceed in the UK. The specific details of these conditions are still not available and thus will result in yet another 'open-ended' period of delay. These uncertainties and undefined timelines will make this five-year old variety economically non-viable. (Bayer Crop Science, 2004)
>
> New regulations should enable GM crops to be grown—not disable future attempts to grow them. (Bayer, cited in Hennessy, 2004; Vidal, 2004b)

Table 6.2 Managerialist Risk-Frames as Experimental Designs

	Bayer's H-T Maize: for Cultivation	*Bayer's H-T OSR: For Cultivation*	*Monsanto's H-T OSR: For Grain Import Only*
Regulatory proposals and decisions	EU-wide approval already granted in 1998, but each country can restrict herbicide usage. UK proposed approval with extra conditions in March 2004, but Bayer withdrew application.	Belgium decided not to support EU approval for cultivation in February 2004.	CAs demanded special conditions to monitor and control escape of grain. Commission proposal ignored those requests and gained little support in June 2004.
Risk issues			
1) **Herbicide-tolerant weeds:** Evaluate control measures and their feasibility for this agro-environmental problem	(not a major issue for maize)	Bayer planned product stewardship for farmers to eliminate volunteer weeds for 'clean' fields. Belgian advisors doubted the feasibility of Bayer's plan.	UK advisors proposed monitoring to verify the company's reassurances about seed spillage.
2) **Farmland biodiversity:** Test relative effects of total (broad-spectrum) and selective herbicides through Farm-Scale Evaluations (FSEs).	FSEs showed relatively less harm from GM maize. UK advisors supported authorisation only if farmers use herbicides as in the FSEs. UK regulators adopted that advice for their decision.	FSEs showed relatively greater harm from GM OSR. UK advisors did not support authorisation. Belgian advisors noted the unfavourable results of UK FSEs.	NGOs argued that it would be unacceptable to spray total herbicides on roadside areas.
3) **Mixture of GM and non-GM products:** Limit adventitious presence through coexistence measures set at the national level	UK would require companies to fund any compensation to non-GM farms whose crops lose economic value.	Belgian advisors noted the segregation problem.	Monsanto defended routine procedures as adequate for segregation. But UK advisors proposed case-specific monitoring for segregation.

Note: National Competent Authorities (CAs) and their expert advisory bodies attempted to impose extra conditions upon GM products under the Deliberate Release Directive. The managerialist risk-frame (in Table 6.1) was operationalised by designing commercial use as an experiment. This table summarises conflicts over specific herbicide-tolerant (H-T) crops, e.g. oilseed rape (OSR), regarding the three risk issues.

The company declined to take responsibility for a commercial-scale experiment under those 'economically non-viable' conditions, for several reasons. Government rules would impose extra burdens, e.g. post-market monitoring and financial liabilities of coexistence measures. Yet this five-year-old variety would give farmers no clear benefits relative to other options; in particular, T25 maize used an inferior germplasm, relative to other varieties more recently available (interview, Bayer regulatory manager, May 2005). Companies may normally take for granted some financial liability for their products potentially causing harm, but Bayer faced liability due to farmers' practices, which it could not readily predict or discipline.

Regardless of the company response to regulators, they encountered further difficulties in requiring a semi-controlled experiment at the commercial stage. According to UK legal advice, the Directive provided no clear basis for a member state to require a specific herbicide regime, so the UK authorities withdrew their original proposal. For various reasons, then, eventual regulatory approval would depend upon predicting realistic effects—rather than prescribing or prohibiting their behavioural causes.

3.2 Oilseed Rape Cultivation EU-Wide

Under the revised Directive, Bayer Crop Science sought commercial authorisation for a herbicide-tolerant oilseed rape which had been delayed by the 1999 EU Council moratorium. Within its new application, the company included cultivation guidelines for growers. Although not required *a priori*, these guidelines would have statutory force, thus incorporating aspects of a managerialist frame into an experimental design. Yet the company's plan was criticised for optimistic assumptions.

In its risk assessment for the application, Bayer translated the 'superweed' scenario into managerialist terms. According to its guidance, farmers would efficiently use herbicides to keep fields clean of weeds. Bayer planned multi-year field tests to evaluate weed-control methods, as well as any potential unexpected effects of this crop. For the prospect of generating herbicide-tolerant weeds, the overall risk was assessed as nil, 'taking into account the risk management strategies' (Bayer Crop Science, 2002).

Emphasising product efficiency, Bayer claimed that its stewardship programme would provide 'cohesive guidelines for field management' by farmers. 'Different networks of expertise are being consolidated in all countries by testing the efficiency of the herbicide and the performance of the varieties'. According to the results available so far, 'Standard Good Agricultural Practices provide adequate control of transgenic oilseed rape volunteers', i.e. seeds which germinate after harvest (ibid.).

Thus the company proposed to manage herbicide-tolerant weeds as if they could be an adverse effect. The case-specific monitoring plan was designed to confirm the company's assumptions about the occurrence, impact and management of such weeds in particular. The monitoring would compare matched pairs of GM and non-GM oilseed rape fields. The plan aimed to demonstrate that 'the potential adverse effects identified in small-scale field trials (volunteers and outcrossing) are fully manageable in a practical way in farmers' fields' (Bayer Crop Science, 2003). Bayer would licence the herbicide-tolerance system to seed companies, rather than own and market the seed directly, thus complicating the locus of responsibility: 'This means that stewardship becomes more difficult, requiring shared responsibility among stakeholders' (interview, Bayer, April 2003).

Moralising the environment in its own way, the company also undertook to share responsibility with farmers, who faced a cultural conflict between efficient weed control and farmland biodiversity:

> . . . where the farmer can adapt the [spraying] practice according to the real weed infestation (due to post-emergence application), it is possible to overcome this conflict . . . For example, we could leave part of each field unsprayed, which would provide an environmental benefit. We needn't kill all the weeds. Instead we can apply the herbicide only when and where necessary during the season. But it is not farmers' culture to leave weeds growing in the field. So we will need to do cultural teaching about the moral obligation of farmers to know the environmental consequences of their actions. (interview, Bayer regulatory manager, April 2003)

This cautious plan anticipated practical issues and regulatory difficulties which soon arose in 2004.

For Bayer's marketing application to cultivate herbicide-tolerant oilseed rape, the regulatory procedure considered all three risk issues from the public debate: herbicide-tolerant weeds, harm to farmland biodiversity and mixture of GM with non-GM crops (see again Table 6.1, lower half; and Table 6.2, second product). As rapporteur for the EU-wide procedure, the Belgium CA sought the views of its own advisors. They emphasised the problem that oilseed rape can readily become a weed, as well as spreading its genes through pollen flow to other *Brassica* plants.

For preventing this gene flow, the company's guidelines would be essential, but they are 'not all technically feasible', so vertical gene flow 'may not be controlled', argued Belgian advisors. Therefore post-market monitoring must assess farmer compliance with the guidelines. The advice also mentioned the other two problems: adventitious presence of GM material, with problems for coexistence; and broad-spectrum herbicides allowing better weed control and thus 'cleaner' fields, i.e. less biodiversity in farmland (BAC, 2004a).

Indeed, farmland biodiversity became a new regulatory issue through this case. Belgian advisors initially accepted the company argument that the issue lay within regulation of pesticides, not GM crops. However, Belgian anti-biotech campaigners circulated copies of the UK FSE results to the advisory body and held protest actions (Denys, 2005). As the FSE trials had demonstrated, this crop 'offers the opportunity to keep the fields cleaner' than fields cropped with conventional oilseed rape, thus resulting in a decline in organisms feeding on weeds. Ultimately Belgian experts reiterated the UK conclusion that 'cost-effective weed control' would result in adverse effects on farmland biodiversity, at least in the short term. Therefore the problem requires 'a continued monitoring of continuously evolving agricultural practices'. Indeed, such an effort would be worthwhile for all pesticide usage, since it could have anticipated the decay of farmland biodiversity in recent decades, argued Belgian advisors (BAC, 2004b).

Citing that advice, the Belgian government decided to reject cultivation uses of the Bayer crop. Its official rationale mentioned all three risk issues: herbicide-tolerant weeds, farmland biodiversity and admixture. Regarding herbicide-tolerant weeds, its advisors had regarded Bayer's plan as 'impractical, hardly workable and hard to control in current agricultural practices' (Belgium-FPS, 2004).

Its advisory body had suggested the need to impose and monitor extra control measures on crop cultivation, but the government declined to attempt such an experiment. Instead it advocated approval only for grain import, not cultivation. According to the Environment Minister, the broad-spectrum herbicide 'kills food for birds, bees and everything else that lives in nature', thus reclassifying farmland as nature (Reuters, 02.02.04). Such apocalyptic language supported a decision that would avoid domestic and EU-wide conflict.

Stakeholder groups disputed the issue of control from opposite standpoints. Bayer criticised the government decision as political: 'The experts raised some concerns but indicated that with proper controls it would be possible to cultivate this crop without impacting on the environment' (cited in Vidal, 2004a). On the other side, Belgian NGOs opposed even grain import, by arguing that the government should not encourage cultivation of such crops anywhere, 'in view of their uncontrollable environmental and agricultural consequences' (cited in Detaille, 2004).

Excluding cultivation uses, the Belgian government then proposed approval for import uses only, with this condition: 'The consent holder is required to provide a localized case-specific monitoring plan' for the presence of feral GM oilseed rape at ports and processing plants (Belgium-FPS, 2004). Those institutions would be turned into a social laboratory for a real-world experiment as the basis for any regulatory approval. However, this proposal did not get far. Regulatory arguments ensued about whether the proposal had procedural legitimacy, or whether the company must submit a new application for import uses only. Arguments also ensued over substantive issues, especially whether the risk assessment had adequately considered spillage of the rapeseed prior to processing. The Bayer rapeseed was eventually approved for import but not cultivation (EC, 2007a).

3.3 Rapeseed Import

Before the Bayer rapeseed proposal mentioned above, similar conflicts had arisen over other GM products proposed for grain import only. In such cases, companies had expected the environmental risk assessment to be straightforward, warranting no special requirements, since 'The grain handling system excludes grain from the environment' (company regulatory manager, interview, April 2003). For such products from Monsanto, the company claimed that any risk would be 'effectively zero'.

However, a proposal from Monsanto (2003) to import GM rapeseed provoked conflict, partly about the weediness scenario (see Table 6.2, third product). Environmental NGOs criticised the risk assessment for optimistic assumptions—e.g. that environmental release from grain imports can be prevented or managed, and that any escaped seeds that germinate would be readily displaced by other weeds. Moreover, argued such critics, it would be unacceptable to spray extra herbicides on roadsides to control volunteers, and GM crops should be prohibited in centres of biodiversity for *Brassica* species (FoEE, 2003b).

National experts and regulators anticipated that some grain could escape. Italy objected that any escaped rapeseed could contaminate related plants, especially land races, and thus undermine its national centres of diversity for *Brassica* crops. The UK advisory committee, concerned about 'the segregation of transgenic and non-transgenic material', asked that the company 'include plans for monitoring and controlling establishment of feral populations as a result of seed spill' (ACRE, 2003: 6). Expert advisors were asking the company to test nearby populations for the herbicide-tolerance trait, as a means to confirm its optimistic assumptions.

Taking up the spillage issue, governments proposed extra conditions that would test and even prescribe importers' behaviour. According to the UK, any EU authorisation should include 'acceptable procedures to minimise seed spills' and 'active monitoring', especially to confirm that populations of feral herbicide-tolerant oilseed rape do not emerge (UK DEFRA, 2003a). Likewise Danish experts proposed that EU approval should include control measures to prevent unintended dispersal during grain transport at harbours, as well as monitoring of any dispersal and gene transfer (DFNA, 2003).

Regulators disagreed about whether or how to impose such conditions—implicitly, about whether GM rapeseed import must be specially designed as a real-world experiment. According to the company, it could not feasibly take responsibility for extra control measures at ports or processing plants, which lay beyond its own authority. Eventually the European Commission proposed to authorise the GM rapeseed import without such a requirement, thus accepting the company's minimal plan as adequate. When that proposal went for a vote to the regulatory committee in June 2004, more member states voted against approval than in favour (FoEE, 2004a). This defeat for the Commission gave greater bargaining power to member states demanding a requirement for market-stage monitoring, though they succeeded only somewhat.

When the Commission ultimately granted approval a year later, this potentially held the company responsible for designing and monitoring commercial use as an experiment. The official decision required 'appropriate management measures to be taken in the case of accidental grain spillage' (EC, 2005). The Commission also announced a 'Recommendation concerning measures to be taken by the consent holder to prevent any damage to health and the environment' from accidental spillage (CEC, 2005b).

3.4 Bt Maize: Cultivation and Monitoring?

As new GM products went through the EU regulatory procedure, more disagreements emerged between member states, their national expert bodies and EFSA, which was meant to provide 'independent, objective and transparent advice' (see Chapter 5). Disagreements arose over uncertain risks that could not be resolved prior to commercial approval, in relation to the statutory criteria for monitoring requirements. EFSA's GMO Panel understood case-specific monitoring (CSM) as mandatory only for 'assumptions' that harm would happen—for which it saw no evidence. Indeed, according to its guidance document, CSM is necessary only when a risk assessment already has evidence of expected adverse effects: 'Where there is scientific evidence of a potential adverse effect linked to the genetic modification, then case-specific monitoring should be carried out . . . ' (EFSA GMO Panel, 2004b: 65).

For some CAs, however, such evidence of risk would warrant rejection of a product, while inadequate evidence would justify CSM. And they regarded general surveillance as less meaningful than CSM, which would experimentally test a specific cause-effect hypothesis. Some regulators wanted CSM to test safety assumptions based on inadequate evidence, not simply to explore evidence of risk. The EFSA GMO Panel invited CAs to attend workshops on scientific methods for general surveillance of commercial use, as a substitute for requiring CSM. But elaboration of general surveillance did not satisfy all member states (EFSA GMO Panel, 2005a). Conflicts continued over risk-assessment 'assumptions' relevant to uncertainties, control measures and monitoring for specific GM products.

These conflicts arose specially around proposals to authorise commercial cultivation of Bt insecticidal maize—the first such GM crop to be considered under the revised Deliberate Release Directive. In their applications to the EU, companies undertook to monitor commercial use for emergence of insect populations resistant to Bt, but not for harm to non-target insects, on grounds that any such risk was negligible. Their proposals were accepted by national rapporteurs for the EU-wide regulatory procedure—by France (for Bt 11) and Spain (for Bt 1507). EU-level experts likewise accepted the companies' safety claims (EFSA GMO Panel, 2005b).

Safety testing had originally been done for companies' applications to the US EPA, and their EU applications provided little more than the US results. Environmental NGOs and some national Competent Authorities criticised the scant testing of non-target organisms (e.g. Greenpeace Research Laboratories, 2003). Many CAs argued that the available test data did not adequately encompass European species and environments (FoEE, 2005b). As a way forward, some CAs proposed CSM for non-target harm as well as for insect resistance.

Moreover, Denmark proposed mandatory buffer zones to protect non-target species. According to its expert advisors, the Bt toxin Cry1F in Bt 1507 maize was relatively less well known, and there was less practical experience from field trials, so there was uncertainty about eventual long-term effects. Denmark emphasised rare butterflies whose existence there was already under threat. Pollen could disperse to plants on which their larvae feed, thus causing negative effects. Hence Denmark recommended a buffer zone of at least 1 metre between agricultural fields and natural habitats (Skov-og Naturstyrelsen, 2003; cited in Toft, 2005).

Beyond inadequate evidence of safety for Bt maize, evidence of potential risk became available. In 1998 the lacewing experiments had highlighted uncertainties about non-target harm and the weaknesses of safety assumptions. A later test indicated that Bt toxins could harm earthworms; Bt-fed larvae had lower weights after 200 days, much longer than previous experiments (Zwahlen et al., 2003a, 2003b). Anti-biotech NGOs publicised the results as indicating a threat to biodiversity, whose protection warranted more research.

However, the EFSA GMO Panel discounted the earthworm experiment, partly on grounds that its results came from simply 'a single worst-case laboratory study and in a single small scale field test'. EFSA's opinion raised uncertainties about whether the earthworm experiment had demonstrated a causal relation between the Bt toxin and harm—normally difficult to do in a lab experiment. More generally, its opinion emphasised methodological difficulties about detecting any non-target harm in the field, as extra grounds for why CSM would not be cost effective (EFSA GMO Panel, 2005b: 23):

> the recording of statistically sufficient data on the abundance of Lepidoptera would demand a high input of personnel and costs (Lang, 2004), especially if larvae, as the most susceptible and immobile development stage, are to be monitored. In addition, maize, a recently introduced species into Europe, is not a significant food source for endemic Lepidoptera and impacts due to pollen dispersal are likely to be transient and minor as demonstrated by studies on monarch butterflies in the USA. (Dively et al., 2004)

In its view, therefore any non-target risks should be relegated to general surveillance. Implicitly, similar arguments could dismiss evidence of risk and a CSM requirement for a wide range of cases.

In further individual comments, Panel members raised doubts about the likelihood and extent of harm in the field, as in these two views:

> The weight difference between the Bt and non-Bt treatment was not very strong in the lab, and was absent in the field experiment. It's also a matter of precaution to avoid publication of false-positive results. The difference found in the Zwahlen study could be a sampling error. (interview, Panel member, 2004)

> In the experiments by Zwahlen et al., the effect was apparently not dramatic in the sense that a whole population was lost from the habitat. And the system is resilient enough to return to normal after the Bt exposure. Even if the Bt caused the weight loss, apparently it has an ephemeral character. (interview, Panel member, 2004)

Going beyond arguments in EFSA's official opinions, those individual views had strong normative stances. For example, potential weaknesses in experimental method were equated with false positives, in turn seen as an anti-precautionary error; and true environmental harm would mean no less than an entire population being destroyed. On such grounds, the burden of evidence was readily shifted back to those who would suspect risk; companies would have no burden to conduct a commercial experiment testing non-target harm.

However, EFSA was put onto the defensive for safety claims such as this one (see Chapter 5). Evidence of both risk and safety was put on trial in the mass media. Disagreements about uncertain risks expanded into conflicts over whether or how commercial use should be designed as an experiment, including an obligation to monitor or prevent specific risks. National CAs were split along several lines—whether to grant approval on the basis requested, to grant approval only with mandatory CSM or to reject the proposal entirely.

With such weak support for commercial authorisation, a real-world experiment could not go ahead for the new Bt maize products, i.e. Bt 11 and Bt 1507. Meanwhile their non-approval was contributing to the EU blockage of US maize exports and thus EU non-compliance with the 2006 WTO ruling. US farmers were cultivating such GM varieties, and US-EU disagreements continued over how to ensure that US exports contain only varieties which had EU approval.

In October 2007 DG Environment proposed a way beyond the impasse. Under its internal proposal to the Commission, EU approval would permit food and feed uses, while excluding cultivation uses. As a rationale for that exclusion, it cited scientific studies of non-target risks (Prasifka et al., 2007; Rosi-Marshall et al., 2007; Nguyen and Jehle, 2007), as well as methodological difficulties of testing such risks in the lab or monitoring them in the field. On those grounds, DG Environment rejected EFSA's claims for scientific rigour:

> it is assessed that the degree of uncertainty attached to the results of the evaluation of the available scientific information as regards cultivation is high, and that, in accordance with much of the above evidence, the level of risk generated by the cultivation of this product for the environment is unacceptable. Adequate risk management measures cannot thus be taken at this stage and a more comprehensive risk assessment is necessary. (DG-Envt, 2007: 6)

The Commission did not adopt such a stance against cultivation uses, but it depended upon DG Environment to take the lead for authorisation of the products under the Deliberate Release Directive.

Given this new conflict within the Commission, agbiotech critics launched a campaign to block cultivation uses of new Bt maize products, while further putting EFSA's GMO Panel on the defensive. Earlier EFSA had cited methodological difficulties and high costs as grounds to reject requests from member states for case-specific monitoring. Yet now those difficulties were turned against EFSA's advice, as grounds to reject cultivation uses altogether.

With ongoing controversy over environmental risks of Bt maize, the French political conflict turned into a judicial trial. In February 2008 the new Sarkozy government banned MON 810, which had been legally cultivated since its EU-wide approval in 1998. The government cited advice about possible risks from the *Haute Autorité,* recently created as an alternative to the long-standing French advisory committee which had been favourable to GM products. A judicial appeal was launched by Monsanto and the *Assemblée Générale des Producteurs de Maïs* (AGPM), representing the relatively more industrial farmers. In March the appeal was rejected by the *Conseil d'Etat,* on grounds that the appellants had not demonstrated the illegality of the government's decision, rather than on substantive grounds.

Environmental NGOs celebrated the judicial decision, e.g. as 'a victory for agriculture and the environment'. According to *France Nature Environnement,* 'A judicial battle is difficult to win, but the war against GMOs continues' (Actu-Environnement, 2008, my translation). A prominent blogger raised the issue of state accountability: 'This decision shows the urgency of a new democracy applied to ecology' (Le Tourbillon Masqué, 2008, my translation).

Similar trials continued against Hungary's ban on MON 810 Bt maize. According to a research consortium headed by the Plant Protection Institute of the Hungarian Academy of Sciences, the Bt crop could endanger the high biodiversity and protected butterfly species of the 'Pannonian biogeographic region' of the Carpathian basin. In particular the long-term spread of pollen spreading to common weeds, such as nettle, could harm non-target species feeding on these plants (Székács et al., 2005; Bakonyi et al., 2006; Darvas et al, 2006). In Hungary, farmers do not regularly spray against European corn borer larvae, which is not a significant pest; so any risk had to be compared with organic maize.

After Hungary's moratorium began in January 2005, EFSA advised that it had no evidence of risk from MON 810. Citing EFSA's advice, the Commission proposed legislation requiring Hungary to lift the ban, but without success. In September 2006 the EU Environment Council rejected its proposal by a simple majority. In February 2007 the Council did so by a qualified majority. Although these procedures put Hungary on the defensive, the negative vote was a symbolic role-reversal, implicitly putting the Commission's proposal on trial.

In June 2008 EFSA asked Hungary (along with Austria and Greece) to appear at a hearing by the EFSA GMO Panel and defend the Hungarian moratorium on MON 810. On 11 June 2008 Ministry officials and Hungarian scientists presented their research results as a basis for the moratorium. In the discussion that followed, the Panel did not raise any point that would seriously question the significance and validity of the Hungarian results. Nevertheless, afterwards the EFSA GMO Panel (2008) stated again that data presented at the hearing are unpublished and so cannot be considered in their advice. The EFSA statement ignored the difficulty to obtain stronger evidence, given that Monsanto declined requests to supply its seed for biosafety research.

4 CONCLUSIONS

At the start, this chapter posed the following questions:

- How did the agbiotech controversy extend the risks for which GM products should be kept on trial?
- How were commercial contexts anticipated, simulated and tested?
- How did regulatory procedures extend the responsibility of commercial operators?

In the European agbiotech controversy, moral visions were attached to the proper social ordering of the agri-environment. From their eco-efficiency frame, agbiotech proponents diagnosed the agricultural problem as deficient inputs which can be remedied through molecular-level corrections. GM crops would control pests and bring environmental improvements, thus better ordering the agricultural environment. By contrast, agbiotech opponents stigmatised agbiotech as disordering nature, perpetuating intensive agricultural methods and imposing apocalyptic risks.

Agbiotech opponents framed such disorder in successively broader ways. Their discourses deployed ominous metaphors—e.g. 'superweeds' leading to a genetic treadmill, thus aggravating the familiar pesticide treadmill; broad-spectrum herbicides inflicting 'sterility' on farmland biodiversity; and pollen flow 'contaminating' non-GM crops. These discourses highlighted plausible harm beyond the 'adverse effects' being officially evaluated; they undermined official distinctions between environmental and agricultural harm.

Thus opponents expanded the charge-sheet of suspected crimes for keeping GM crops on trial. In earlier controversies, e.g. agrochemical effluent and mad cow, agricultural technologies and farmer practices had been stigmatised as immoral (Lowe et al., 1996, 1997; Buttel, 1998). In the case of agbiotech, a similar stigma was attached to practices of commercial operators, especially their efforts to maximise productivity. Eco-efficiency promises were cast as suspect or even as a threat (see Table 6.1, lower half).

Governments faced plausible scenarios of harmful effects not formally or adequately evaluated in official risk assessments. More member states cited special national contexts—concerning environmental diversity, vulnerability and uncertainty—as grounds to request extra information, monitoring requirements and/or control measures to prevent potential harm. Such issues could not be credibly resolved before the commercial stage, so regulatory decisions became more difficult.

Elaborating a managerialist frame, some regulatory authorities extended their responsibility; they translated the three ominous metaphors into scientifically measurable, manageable effects. For example, the 'sterility' metaphor was scientised through indices of biodiversity. 'GM contamination' was scientised through the new concept of 'technically avoidable' presence, and the means to detect GM material were improved (see also Chapter 7). As critics warned against 'uncontrollable risks' of GM products, regulatory procedures eventually devised control measures for anticipating or managing more effects than before; regulatory agencies scientised the extra risks and thus the protest. More national authorities took up those extra risk issues, which readily became Europe-wide ones.

Broader accounts of harm meant more uncertainties about whether GM crops could cause such harm in the agri-food chain. Uncertainties went beyond any testable characteristics of products per se, because the potential effects would depend upon operator practices. Regulators faced a further dilemma: how safety could be tested safely, while justifying a scale-up to commercial use.

Originally the 'stepwise principle' meant that any control measures could be relaxed at the commercial stage for GM products shown to be safe in laboratory and field trials (OECD, 1986). Product safety was seen as independent of the way the product was used, by analogy to conventional crops. A natural order of inherently safe product characteristics, symbolically normalising a GM crop, would complement a social order of market freedom for commercialising and circulating the product. By the late 1990s, however, this socio-natural order was becoming destabilised by public protest and expert disagreements.

As a more cautious way forward, the 'stepwise principle' was extended into the commercial stage to test assumptions about operator behaviour, beyond 'product safety'. Such measures were aimed mainly at reassuring or marginalising critics, yet the outcomes were not entirely predictable.

Commercial operators were to be put on trial—in moral and experimental terms. Regulators requested more knowledge about operator behaviour and diverse agri-environmental conditions; neither could be standardised in advance. Control measures assigned greater legal-moral responsibility to agro-industrial operators, including seed companies, grain importers, farmers, etc. Whenever risk assessments made assumptions about the operator behaviour necessary to avoid harm, these could be turned into prescriptions, which in turn could be tested for plausible compliance.

Commercial use was anticipated as a real-world experiment, testing assumptions about human practices as well as their environmental effects (cf. Krohn and Weyer, 1994). These steps towards commercialisation could be justified as cautious ways to generate knowledge for risk assessment, and thus as ways to gain moral licence for a technological scale-up. As strategies to prevent potential harm, commercial experiments underwent tensions between predicting, testing and/or prescribing operator behaviour. Agbiotech critics questioned whether preventive measures would be reliable, realistic or morally responsible.

Agricultural fields were being anticipated as a social laboratory for testing operator behaviour, which was also given moral meanings. As farmers were assigned a role as environmental stewards, their potential behaviour became a focus for greater monitoring, control and self-discipline. For example, farmer discipline was needed to control weeds, or to protect some weeds as biodiversity, or to do both at once—aims which could be mutually conflicting. For cultivating the most controversial GM crop (oilseed rape), the biotech company Bayer planned moral education for farmers, so that their use of herbicide sprays would minimise harm to farmland biodiversity. For the import of GM rapeseed, some member states sought extra measures to limit and monitor any spillage, which could lead to greater use of herbicide sprays.

For herbicide-tolerant oilseed rape, regulators disagreed about whether operator behaviour could be feasibly reorganised around the necessary social discipline to prevent environmental harm, and thus about how to design a technological scale-up (see again Table 6.2; cf. Wynne, 1995). For the prospect that gene flow would generate herbicide-tolerant weeds, Bayer proposed control measures for farmers to maintain 'clean' weed-free fields. However, Belgian government experts questioned whether such measures would be reliable; its national Competent Authority cited that doubt as grounds to reject Bayer's request for commercial authorisation.

For the prospect that broad-spectrum herbicides would harm farmland biodiversity, UK farm-scale trials turned agricultural fields into a social laboratory for testing cultivation practices. The trials were originally intended to simulate 'cost-effective weed control' with herbicides and thus predict their environmental effects, as a basis to make a regulatory decision on commercial use. After Bayer's GM maize crop showed favourable environmental results in the trials, however, the specific cultivation practices were made prescriptive, as rules to be enforced and monitored for compliance. Moreover, UK authorities also required agbiotech companies to compensate any farmers for economic loss from GM contamination. The financial burden would depend upon farmer behaviour, thus potentially involving companies in farmer discipline. Bayer declined the prospect of a real-life experiment with such uncertain burdens, so commercialisation did not proceed.

Also in the UK farm-scale trials, GM herbicide-tolerant oilseed rape caused relatively more harm to farmland biodiversity than its conventional counterparts, so this GM crop lost any prospect of commercial authorisation in the UK. Moreover, the unfavourable experimental results were circulated by anti-biotech activists in Belgium, whose Competent Authority was evaluating the same crop for EU-wide commercialisation. The UK trials became a factor in Belgium's decision to reject the product.

For new Bt insecticidal maize products, some member states disagreed with EFSA's safety claim regarding non-target risks. Regulatory conflicts ensued over whether or how commercial cultivation should be designed as an experiment, with a long-term requirement for market-stage monitoring. As a way around the impasse, in 2007 DG Environment sought to authorise the products for feed and food uses only. It cited uncertainties over non-target risks and monitoring methods, as grounds for why any commercial authorisation would need to await 'more comprehensive risk assessments'. EFSA's safety claims were being further put onto the defensive.

In sum, GM crops were kept continuously on trial in three related ways: by contending risk discourses which attributed moral meanings to the agricultural environment, by safety tests which simulated commercial use, and by special measures (or proposals) assigning greater responsibility to commercial operators. Through a managerialist frame, some governments broadened the policy problem by turning the agri-food production chain into an experiment. In so doing, national regulatory agencies had various policy agendas: ensuring that any approval decisions would be publicly defensible, assigning clear responsibility for any adverse effects and perhaps making any authorisation more difficult or burdensome. In the latter cases, governments could avoid politically unpopular decisions.

Regulatory conflicts ensued over how to anticipate and design commercial use as a real-world experiment (Levidow and Carr, 2007b). Scientific uncertainty about risks and experimental methods was cited in contrary ways—e.g. as grounds to authorise products within minimal control measures, or as grounds to require extra monitoring for commercial scale-up. Regulatory institutions were put 'on trial' for their trustworthiness in designing and evaluating such real-world experiments. State accountability for innovation choices was being further displaced onto regulatory issues, which thereby became more contentious. Within and among EU institutions, such conflicts continued over a legitimate basis for authorising GM products. Their commercial use was anyway blocked or deterred by food retailers in the late 1990s, as shown in the next chapter.

7 Labelling GM Products, Testing Free Choice

INTRODUCTION

When the first GM products first gained EU-wide approval in the mid-1990s, no GM label was required—consistent with the policy of the agbiotech and food industry. Widespread demands arose for GM labelling, based on consumers' right to choose non-GM food. Analogies were drawn between the BSE pandemic and uncertainties about GM food safety; official safety claims came under greater challenge, thus providing extra grounds for an informed choice by consumers.

These demands posed difficulties for commercial operators in the agri-food chain. US exports to the EU routinely mixed GM with non-GM grain, especially soya, which was widely used in processed food. Consequently, such products became vulnerable to public suspicion and market instability. At national and EU levels, GM labelling rules were sought as a means to stabilise the market for processed food. In the late 1990s the food industry became a new actor in the agbiotech issue by labelling GM food and then excluding GM ingredients.

This chapter discusses the following questions:

- How were industry and government put on the defensive for their no-labelling policy?
- How did broader GM labelling rules gain decisive support?
- How did agri-food markets test claims about agbiotech?

The chapter analyses EU-wide conflicts over how to set and extend GM labelling rules, amidst contending policy agendas around agbiotech. The chapter follows the sequence of developments. Under consumer pressure, retailers took the lead, following by food processors. Conflicts over GM labelling rules involved an interplay among regulatory, economic and cultural arenas.

1 DIS/ORDERING THE EUROPEAN MARKET

In the late 1990s, protest against GM food disrupted the UK market for processed food. The food industry came under pressure to apply a GM label or to exclude GM ingredients. Similar difficulties arose in other European countries. National variations and market instabilities eventually led European governments to support demands for mandatory GM labelling.

1.1 Stigmatising 'Frankenfoods'

In the mid-1990s most EU member states argued that no labelling could be required for safe products under European Union law. As an implicit motive for this stance, governments sought to avoid any symbolic association between GM products and uncertainty about risk. The UK acted as rapporteur for the first GM food product in the EU regulatory procedure, Monsanto's GM herbicide-tolerant soybean, on the basis of no requirement for a 'GM' label. The product gained approval on that basis; likewise the second major GM product, Ciba-Geigy's Bt maize (EC, 1996a, 1997a; see Chapter 3).

Soya and maize were already common ingredients in many processed foods, so these could contain GM ingredients, without any means for consumers to know their presence. When the first US shipments of GM grain reached Europe in 1996–97, activists held symbolic protests at ports, thus gaining mass-media attention for their message: Europe was being force-fed GM food (see Chapter 3). Some European food retailers asked grain traders for non-GM supplies but were told that segregation would not be feasible. From 1996 onwards UK NGOs and consumers protested at the failure to provide GM labelling and non-GM alternatives. Supermarkets were picketed and leafleted by Friends of the Earth, among other local groups. Critics denounced biotechnology companies for 'force-feeding us GM food', as if we were guinea pigs in an 'uncontrolled experiment'. Leaflets denounced GM soya as 'virtual food' from a 'test-tube harvest' (WEN, 1998).

In 1996 Greenpeace Europe launched a campaign against agbiotech by exploiting its market vulnerability, as a strategy to influence an entire industry. Encouraged by Greenpeace UK, some chefs and food writers called for a ban on GM foods. According to a well-known food writer, the Prime Minister 'must now make a decision between the public's understandable antipathy to meddling with the balance of nature and their future health and his other love, wealth creation' (quoted in Charles, 2001: 207–8). In public opinion surveys, half the people opposed the cultivation of GM crops and did not want to eat GM food, while only a small fraction expressed willingness to do so (ENDS, 1998c).

These protests provoked conflict between the food retail and agbiotech sectors. Companies had proposed and obtained market approval for their GM products without any requirement for GM labelling. Consequently, grain traders had no incentive to segregate GM grain; they also claimed that segregation would be unfeasible or very expensive. As consumer protest hit retailers, they in turn complained to Monsanto and grain traders.

Like UK retailers, the entire European food industry was dependent upon soya and maize imports from North America or Argentina, especially for processed food. So the UK's internal conflicts had analogies in other countries, especially in France, Germany and Austria. Environmental NGOs launched campaigns against GM food and for labelling on a Europe-wide scale. In a Greenpeace Germany ad, a family eats at the dinner table inside an animal cage, with the caption: 'Genetic experiments with foodstuffs turn people into guinea pigs'. As the campaign slogan said, 'Stop poison and gene technology in food'.

Consumer groups continued to demand GM labelling. Moreover, protest against GM food potentially disordered the overall food market. Major food retailers recognised a problem of ambiguous product identity: how to distinguish GM from non-GM products. As they realised, inadequate labelling would jeopardise their reputation and consumer confidence: 'If consumers react against this, then the image of the authorities, producers and retailers could be damaged, apart from the financial losses' (interview, EuroCommerce, 24.11.97).

Figure 7.1 'Genetic experiments with foodstuffs turn people into guinea pigs' (Credit: Greenpeace Germany, www.greenpeace.de).

1.2 Clarifying Product Identity

EU-wide conflicts over GM labelling began in the early 1990s, when the draft Novel Food Regulation required no GM labelling. Contentious among member states, this aspect gradually changed in response to the European Parliament (Toke, 2004: 154–55). Under a 1996 redraft, GM labelling would be required only if an analysis of 'existing data can demonstrate that the characteristics assessed are significantly different in comparison with a conventional food or food ingredient'. This wording had some resemblance to US FDA criteria, which meant no requirement to label GM foods being initially commercialised. Accommodating the European Parliament, the Commission and Council eventually agreed to delete the word 'significantly' in the final Regulation (EC, 1997b). So the new rules would apply to most foods derived from GM grain. However, the Novel Food Regulation did not apply retrospectively to GM products already approved, i.e. GM soya and maize; nor did it adequately specify the technical criteria towards harmonised EU-wide rules.

When the first GM grain shipments arrived in Europe in 1996–97, EU institutions were divided over the labelling issue. Several EU member states had already enacted their own rules for labelling GM food. They proposed that the EU do likewise, initially in the decisions to approve GM soya and maize. But most member states did not agree, so these decisions included no requirement for labelling (EC, 1996b, 1997a).

To fill the gap, retailers' organisations devised voluntary rules, which varied somewhat across countries as regards whether the criteria were detectable DNA and/or protein. Such national differences posed a problem for a European internal market. According to the Confederation de l'Industries Agro-Alimentaires, representing the European food industry, 'The EU has potentially 15 different [national] markets. If there are no EU rules, then there is no obligation to label in the same way in all member states' (interview, CIAA, 19.01.98). Companies and countries sought to avoid market competition for selling 'non-GM food', yet they were effectively competing according to different definitions of GM food.

The principle of mandatory GM labelling was soon widely accepted, but for diverse motives, which became difficult to accommodate in common European rules. Officially, labelling provides the information essential for free consumer choice among products whose safety is assured. Unofficially, the impetus for labelling arose from more awkward sources—public distrust of GM safety claims, public resentment at dependence upon official experts, and activists' efforts to block a market for GM ingredients. NGOs protested against inadequate GM labelling which gave consumers an *un*free choice; they demanded comprehensive labelling. Industry became more dependent upon state rules to structure a 'free choice' which could stabilise a market for processed food.

The stigma of genetic modification was difficult to overcome and even pervaded commercial discourses. From its apocalyptic frame, Greenpeace campaigned against GM 'contamination' and counterposed 'pure' food, i.e. organic products. Likewise the retail industry discussed tolerance limits of GM 'contamination', as a basis to label food as 'non-GM'. Some retailers planned ways to exclude GM ingredients and thus provide 'pure' or 'clean' food, thus implicitly linking GM grain with risk or impurity. Processed food had an ambiguous identity which left instability in the processed food market, especially in the face of consumer protest.

More national authorities proposed mandatory EU-wide labelling of GM food as an essential means to re-order a Europe-wide food market. Yet such rules were difficult to agree upon, formulate and implement. They were elaborated in three stages—in 1997, 1998 and 2000—each stage attempting to overcome ambiguities or gaps in the previous one, as described in the rest of this section.

The Monsanto soybean and Ciba-Geigy maize had gained safety approval under the Deliberate Release Directive 90/220 without any labelling requirement (EC, 1996a, 1997a). As the first gap-filler, Regulation 1813/97 required that the foods containing these GM products must be labelled if they are detectably 'not equivalent' to non-GM food. But the Regulation did not specify the technical criteria for detectability and thus non-equivalence, because member states could not agree. The Regulation had to be implemented by November 1997, but the deadline had no practical effect without agreed criteria.

Meanwhile a European market for processed food remained unstable, potentially in disorder. Without clear means to distinguish GM from non-GM food, product identity was uncertain, even in dispute, and could not be easily resolved. 'People are awaiting labelling requirements which will put order into the market', noted an official (interview, DGIII, 25.11.97). There were disagreements over how to interpret Regulation 1813/97 and how to clarify the criteria.

Meanwhile companies were being put on trial by NGO surveillance of food products for GM 'contamination', based on their own lab tests. Their publicity targeted retail chains and processors of any products which gave positive results. This campaign strengthened pressures on the food industry—to adopt voluntary rules, to adopt relatively broader criteria and to ensure that any GM content was duly labelled.

1.3 Extending 'GM' Labels

During the confusion over how to implement Regulation 1183/97, there were no harmonised EU-level rules for labelling food as GM. To fill the gap, in November 1997 the European food industry (CIAA) devised its own voluntary guidelines, which would apply a GM label to any food containing GM protein. According to the CIAA, tests would be done at the stage of bulk supplies at EU ports, i.e. before any further processing that could reduce detectability of GM protein. The test would be applied to all grain from countries where the GM crop is cultivated (interview, CIAA, 19.01.98). An implicit aim was to gain public trust, while avoiding competition among retailers according to different criteria for GM labelling.

Yet the criteria remained contentious within the food industry. On the one hand, the CIAA guidelines would result in labelling more products than required by a statutory criterion of GM protein in processed food, where the protein may no longer be detectable, e.g. because it is highly processed. On the other hand, the CIAA guidelines fell short of consumer demands for broader criteria which would encompass GM DNA and even rapeseed oil. The CIAA guidelines were criticised for 'discrimination' against GM products which have detectable differences, while bypassing other GM products. On those grounds, representing the European retail sector, EuroCommerce advocated mandatory labelling for all GM-derived products. However, the GM protein-only criterion was preferred by the UK and Belgian members of EuroCommerce. Thus divergent criteria continued within and across member states.

The November 1997 CIAA guidelines were set to be implemented in February 1998. However, they did not guide practice throughout Europe, partly because disagreements arose between the retail sector and food industry in some countries. Even where GM food was labelled as such, e.g. in the UK and Netherlands, this encompassed any products derived from US soya—a simpler course than trying to identify its precise origin and content. Some retailers sought certified 'non-GM' soya but could not obtain sufficient quantities to fulfil the consumer demand.

Those labelling criteria, differing both within and across EU member states, led to the second-stage EC rules. DG III drafted a regulation which would require GM soya and maize to be labelled if either GM protein *or* DNA is present. This criterion aimed to maximise the range of products which would require a 'GM' label—more products than being labelling under voluntary rules in some EU member states. Government stances on this proposal coincided somewhat with that of their respective retail sectors. The DG III proposal initially gained support from Sweden, Denmark, Germany and Austria; some other member states objected, e.g. by describing DNA tests as expensive and unreliable. Nevertheless the EU's Council of Ministers eventually approved Regulation 1139/98 with the broader criteria, i.e. GM protein *or* DNA; this became legally binding in September 1998 (EC, 1998b).

Another ambiguity remained, leading to the third-stage change in EC labelling rules. European retailers intended to apply a 'genetically modified' label to any soya or maize ingredients obtained from a country where genetically modified crops were cultivated—initially the USA and Canada, and later Argentina too. Exceptions would be highly processed products that contain no DNA or protein, e.g. rapeseed oil and lecithin. Yet consumer groups still demanded labelling of all GM-derived food, regardless of detectability. Some supermarket chains accommodated these demands, thus again generating different criteria across the EU. In late 1998 the European Commission announced an intention to require GM labelling for additives such as lecithin; this was eventually adopted (EC, 2000).

In sum, consumer protest and market instabilities led to voluntary labelling rules, whose diversity became an extra impetus for statutory EC rules. These were elaborated in three stages, each broadening the range of food products which must be labelled as GM. At each stage, current or proposed arrangements faced a multiple charge-sheet of suspicions: whether they neglected or adequately identified GM 'contamination'; whether they accommodated or denied 'free consumer choice'; and whether they facilitated or undermined the stability of the processed food market. Under these suspicions, alternative rules were being put on trial along with agbiotech. Through changes in labelling rules, often moving from voluntary to statutory status, 'GM' food was defined more broadly than before.

2 MARKET FORCES OUT GM GRAIN

Agbiotech had been initially promoted as essential for economic competitiveness in the face of market forces, as objective imperatives to be accommodated. By the late 1990s market forces gained more negative meanings from protest demanding non-GM food, backed by NGO surveillance and publicity on GM 'contamination' in food. Thus GM ingredients were being turned into a competitive disadvantage.

In May 1998 a UK supermarket chain, Iceland, found suppliers of non-GM soybeans in Brazil and Canada. This move reinforced a non-GM market, while undermining claims that segregated production would be impossible. Iceland also introduced negative labelling, i.e. 'contains no GM ingredients', while repeating Greenpeace claims about risks of GM food (ENDS, 1998b; Iceland, 1998). Its newspaper ads read: 'Like to be a human guinea pig in the largest food experiment of all time? Chances are you're one already'. Thus a supermarket chain denounced GM foods with a discourse similar to Greenpeace. 'Iceland doesn't test on humans', stated the company, echoing the Body Shop slogan, 'Not tested on animals'. The company portrayed the segregation issue as a 'David and Goliath' battle against Monsanto.

In this way, Iceland stimulated the development of a dual market for food products. To gain economy of scale, the company established an entire processing plant for its non-GM soya, which was then put onto the wholesale market. Facing public protest, more and more supermarket chains bought the non-GM soya supplies for their own-brand product lines, though they did not apply negative labelling. Reportedly the non-GM lines were sold out as fast as they could be supplied. The UK Agriculture Ministry published a list of companies which could supply non-GM soya.

Figure 7.2 Iceland versus Goliath/Monsanto (Credit: Russell Ford/Iceland).

To accommodate consumer concerns, retail chains soon excluded GM grain on a European scale. Labelling requirements deterred companies from using GM ingredients in their own-brand products, to avoid labelling them as GM. At least in northern Europe, most retailers excluded GM grain from their own-brand products; some gave public undertakings to do so. Increasingly their exclusion policy became process-based, i.e. independent of detectability. Such a policy required a documentary control system, potentially incurring extra costs. Food companies absorbed any extra costs, rather than charge a premium price for non-GM food (Levidow and Bijman, 2002).

Some major companies adopted even broader processed-based criteria, perhaps to avoid consumer criticism. They voluntarily labelled GM-derived additives and even oils, in which no GM ingredients would be detectable. Thus more and more companies went beyond EU requirements for GM labelling. In Germany and Austria, the entire industry moved towards negative labelling, e.g. 'GM-free' food. Some companies promoted organic meat as a way for customers to avoid GM animal feed (ibid.).

Alternative supply chains institutionalised the commercial blockage of GM soya and maize (ENDS, 1999). Major retailers established a Europe-wide consortium to obtain non-GM grain; consortium members soon included Sainsbury s and Marks & Spencer in the UK, Carrefour-Promodes in France, Effelunga in Italy, Migros in Switzerland, Delhaize in Belgium and Superquinn in Ireland. The largest food manufacturer in Europe, Nestlé, undertook to exclude GM-derived ingredients as far as practically possible, where the public demanded it. As a Nestlé company officer said to Monsanto in May 1999, 'Don't expect *us* to take a bullet for *your* GMO products' (cited in Charles, 2001: 242).

During 1999 GM grain was excluded by major processors too, e.g. Nestlé, Eridiana Béghin-Say, Frito-Lay (a subsidiary of Pepsico) and Gerber (a subsidiary of the biotech company Novartis), a baby food producer. As Gerber officials acknowledged, 'We have to act pre-emptively . . . The parents trust us. If they don't trust us, we are out of business' (quoted in Lagnado, 1999). As opposition activists emphasised, a biotech company was declining to sell its own product.

Unilever accommodated similar pressure, announcing that it would no longer use GM ingredients in its European production in May 2000. It left open such options for the future: 'We are continuing to research the use of biotechnology and genetic modification in the development of new products' .The company would retain the capability to include GM-derived ingredients 'if these are shown to be safe, approved by the relevant authorities and are wanted by consumers on a fully transparent basis' (Unilever, 2000).

Animal feed always was the major use of soya and maize, so much more grain would be needed for non-GM animal feed than for non-GM food. Segregation would be more difficult for these larger quantities (Wrong, 1999). Nevertheless some retailers undertook to sell meat only from suppliers which excluded GM-based animal feed. Other retailers said they would attempt to do likewise. By the late 1990s, non-GM animal feed had been established, mainly in the UK and France.

In all these ways, food companies were anticipating consumer pressures in advance. A major investment bank declared, 'GMOs are dead' (Deutsche Bank, 1999). Food companies had common Europe-wide sources of food materials and Europe-wide markets, so efforts to exclude GM ingredients operated across Europe, even in places which had no local protest. Those trends are exemplified by the following country-cases.

In the UK, by 1999 all the retail chains undertook to exclude GM ingredients from their own-brand products. Animal feed came under similar pressure. An extreme case was the UK's largest user of fresh produce, Tesco, which undertook to use only non-GM animal feed. A retailer reputed for high quality, Marks & Spencer, introduced a range of meat and eggs derived from livestock raised on non-GM diets. Likewise Sainsbury's sought suppliers of meat not derived from GM grain.

In France, both domestic and foreign pressure discouraged the use of GM grain. German food retailers indicated that they would not buy GM maize from French farmers (*Cultivar Actualité,* 05.05.99). French retail and processing companies found substitutes for GM soya or maize, e.g. non-GM or other grains (*L'Usine Nouvelle,* 27.05.99). The largest producer of animal feed in France (Glon-Sanders), as well as a Europe-wide producer of poultry (Bourgoin), declared that they would exclude GM grain; Bourgoin was also a partner of retail chains which import non-GM soya from the USA and Brazil (*Le Monde,* 02.09.99).

In the Netherlands, by mid-1999 the largest Dutch retailers (Albert Heijn and Laurus) were asking the suppliers of their own-brand products to label the presence of any GM ingredient. As a result, most producers changed their recipes to exclude any GM ingredients. According to the Dutch Dairy Organisation (NZO), it would decide whether GM feed crops for dairy cows will be grown in the Netherlands. Among other considerations, 'consumer acceptance in foreign markets are important signals for the Dutch dairy industry'. By taking this position, NZO effectively rejected Bayer's GM herbicide-tolerant maize as an ingredient in animal feed in the Netherlands, in a period when its cultivation was increasing in Spain.

In Spain, three of the largest retail chains were owned or co-owned by French retailers, which extended their own non-GM policy into the Spanish market. Local affiliates of foreign companies (Marks & Spencer, Unilever and Nestlé) also followed the latter's non-GM policy. By early 2000 Spanish food retailers adopted a policy of excluding GM ingredients from food. By 1998–99 Spain had the greatest cultivation of GM maize in Europe, but its use was limited mainly to animal feed.

Those pressures against GM ingredients in processed food also extended to the slow-ripening GM tomato from Zeneca/Calgene. The derived tomato paste had been marketed exclusively by Sainsbury's UK in 1997, at a lower price than comparable non-GM products, and had gained considerable sales. Clearly labelled as 'GM' from the start, the product had drawn no public criticism, even after Monsanto's GM soybean became a target of protest. Nevertheless the GM tomato paste was withdrawn when Sainsbury's moved to exclude GM ingredients from all its own-brand products in 1999. When a cartoon had portrayed Blair as a tomato head, the target was politicians rather than a specific product, yet the GM tomato became a casualty of larger-scale commercial decisions. Behind the scenes, some biotech companies such as Zeneca blamed Monsanto for antagonising the public and undermining any market for GM products.

Difficulties remained in finding reliable means to enforce the labelling rules; and companies were reluctant to sell GM-labelled products. In some countries, e.g. Germany and Denmark, 'GM' labels were rarely found on food. Perhaps some retailers were violating Regulation 1139/98, though many were avoiding GM ingredients altogether: either they found non-GM supplies or they substituted other ingredients for soya and maize.

In sum, activists took advantage of the industry structures whereby a market for GM products depended upon European food companies, in turn dependent upon consumer reaction (cf. Schurman, 2004). Companies were soon deterred from sourcing GM ingredients for their own-brand products. Reshaped by protest and new labelling rules, the European market forced out GM grain. Through this process, the public was no longer a passive supporter and consumer of technological 'progress'. Using political protest, financial power and cultural arguments, publics had gained a role of citizen-consumers who could influence markets and make their own judgements about safety claims.

3 PUTTING GM MATERIAL UNDER STATUTORY SUSPICION

Broader rules for GM labelling eventually helped to stabilise a market for processed food products, which were now better protected from the suspicion or stigma of GM content. But the new rules were overtaken by yet more demands. As a further link with the BSE crisis, in the late 1990s some NGOs proposed traceability system for all GM food and feed, so that any harm could be linked with a specific product. This proposal gained support from some member states as agbiotech became more controversial in the late 1990s.

In June 1999 several Environment Ministers issued a statement aiming to suspend any new authorisations of GM products, pending new regulatory arrangements. They demanded measures to ensure labelling and traceability of all GMOs and GMO-derived products for food and feed purposes; this proposal encompassed all products of GM techniques, i.e. regardless of detectability. As a rationale, the Environment Ministers cited 'the need to restore public and market confidence' (see Chapter 5).

As in earlier debates over GM labelling, the new proposals faced a charge-sheet of suspicions: whether comprehensive process-based labelling would enhance or deny meaningful consumer choice, whether or not traceability requirements could be feasibly implemented, whether such rules would help to resume the approvals procedure or instead provoke a trade war—and whether the moratorium should be lifted anyway. These disagreements followed patterns similar to the three contending frames of other issues around agbiotech (see Table 7.1). The rest of this section

Table 7.1 Principles for GM Labelling Rules: Contending Frames

Economic Individualism	*Market Completion and Function*	*Democratic Rights*
agbiotech developers: EuropaBio + EPP in EP	regulatory decision-makers: CEC, Environment Council, EP majority or compromise	agbiotech opponents: EEB, CPE, Eurocoop; Greens + United Left in EP
Consumers need relevant information to make individual choices in their interests.	Consumer choice is essential for gaining consumer confidence.	Consumer choice is needed as a democratic right—to eat safe food and to make an informed choice.
Approved GM products are identical in quality and safety to non-GM ones.	GM food and feed cannot be treated as equivalent to non-GM products.	GM products are inherently different from their conventional counterparts.
Need consumer freedom to choose between products which do/not contain detectable GMOs.	Need consumer freedom to choose between GM, conventional and organic food products (CEC, 2003a).	Need consumer freedom to avoid GM products as genetic pollution.
Recognise the realities of agricultural production, especially the unavoidable mixing in fields and agri-food chain.	Need rules which allow the internal market to function effectively in the real world (CEC, 2001d).	Recognise the threat of GMOs to non-GM products.

analyses those framings of the traceability issue and labelling thresholds, especially in relation to the US-EU trade conflict.

3.1 Symbolising Sovereignty

The blocking minority in the EU Council, also known as the *de facto* moratorium, pushed the Commission to accommodate its demands. New legislation was seen as a necessary means to resume the approvals procedure and thus make the internal market safe for agbiotech. Comprehensive GM labelling would apply to most US maize exports, including highly processed corn gluten, as well as vegetable oils—which previously lay outside EU labelling regulations.

The Commission soon took up the Ministers' demand in extra draft regulations on GM food and feed, as well as on traceability and labelling. (Chapter 5 already analysed how the former regulation changed the evaluation and approval procedure for GM products.) The new draft rules aimed to overcome market instability and national restrictions:

> The free movement of safe and wholesome food and feed is an essential aspect of the internal market . . . Differences between national laws, regulations and administrative provisions concerning the assessment and authorisation of genetically modified food and feed may hinder their free movement, creating conditions of unequal and unfair competition. (CEC, 2001d; EC, 2003b: 1)

According to the DG-Environment Commissioner, any extra economic burden would be a necessary price for societal progress:

> Certainly there is a cost for the producers and for trade, but what is at stake is our ability to build public confidence. European companies will only be able to seize the opportunities provided by bio-technology if this confidence is established. (CEC, 2001d)

As a way forward, the new rules had implicitly precautionary aims. Traceability and labelling can 'facilitate the withdrawal of products' in case of unforeseen adverse effects, as well as facilitate 'risk management measures in accordance with the precautionary principle', according to the Commission proposal. It would also allow consumers an informed choice about whether to buy food derived from GM crops. By providing legal certainty, these rules would facilitate the internal market (CEC, 2001f; EC, 2003b: 1).

Also along precautionary lines, consumer choice would allow consumers to consider any uncertainties about safety. NGOs made this aim more explicit as 'a basic consumer right to eat safe food and to make informed choices' (Eurocoop, cited in Agence Press Europe, 2002). This phrase implied that approved GM foods may not be safe, regardless of official expert claims, as grounds for consumers to make their own judgements on GM products.

The US-EU trade conflict over GM products loomed over the Commission proposals. According to many EU politicians, these legislative changes were essential preconditions for lifting the *de facto* moratorium on GM products and thus for avoiding trade war with the USA. Others made an opposite warning. After the UK Food Standards Authority denounced the Commission proposals as unworkable, a House of Lords committee issued a report along similar lines, while counterposing a 'GM-free' label: 'Otherwise we will end up in a dispute at the World Trade Organisation', stated its chairman (Coghlan, 2002).

The US government continued its threats to bring a WTO case against the EU—for its regulatory delays and blockages, as well as for its new labelling rules as a threat of unfair trade barriers. Some European politicians reinforced those warnings, while others advised the US government that such threats could be counter-productive for achieving a transatlantic market: 'I have been working very hard to try to explain to the Americans the enormous strength of feeling on food safety issues in Europe', stated the Environment Minister Margaret Beckett (Hinsliff, 2002). In its official arguments against the Commission proposals, the UK prudently did not mention the threat of a US-EU trade war.

Other arguments cited the US threats in the opposite way, by advocating more stringent rules as a political defiance of the US government. NGOs emphasised Europe's democratic right to decide on its own food supply. Broader process-based labelling rules were turned into a symbol of democracy versus globalisation.

Such discourses pervaded the debates in and around the European Parliament. In June 2002 its Environment Committee voted for more stringent criteria on GM labelling than in the Commission proposals. Afterwards FoEE sent out an action alert, asking supporters to send letters to MEPs before the Parliament's plenary session. According to the alert, 'More and more citizens are destroying their local field trials, as their democratic rights to veto the trials have been flagrantly denied over and over again'. The letter mentioned Jose Bové, who had been imprisoned in France for attacks on agbiotech, in parallel with Parliamentary speeches demanding his release (FoEE, 2002b).

Thus a letter-writing task was vicariously linked with glamorous direct action for democratic sovereignty. When Parliament's plenary session voted for more stringent labelling criteria, a Green MEP declared, 'It's a great victory for consumer choice and a clear message to Tony Blair and his American friends'. Indeed, the Commission's proposals for broader labelling criteria and traceability had been opposed by the UK as well as the US government (Agence Press Europe, 2002).

3.2 Disputing Feasibility

According to the Commission's proposed Regulations, all GM food and feed must be labelled according to its source, and the information must be transmitted throughout the agro-food chain. Such rules would 'allow the internal market to function effectively in the real world' (CEC, 2001d). This proposal generated sharp disagreements about the role of markets and consumers (see Table 7.2, top).

EC law had previously identified and regulated GM products according to their transgenes, but full traceability depended on information about the specific 'transformation event' from which the GMO was developed (CEC, 2001f; EC, 2003b: 25). As this requirement acknowledged, gene expression may be contingent on the specific event, i.e. the precise site where the transgene was inserted. Research was still investigating transformation events and their potential effects. To operationalise the new requirement, resources were needed to create detection methods for identifying the fragments of DNA specific to each 'event' (Lezaun, 2006).

The new rules would also depend on a documentary control system to follow a unique identifier throughout the agro-food chain—from seed to feed, and from farm to fork. This system was intended to minimise the need for sampling and testing grain shipments. The largest farmers' organisation, COPA, supported traceability of all agri-inputs as a general principle; it advocated solutions which combine documentary control and laboratory detection methods. Likewise the food retail industry supported the Commission proposal as necessary and feasible, given that their member companies already 'are dependent on correct and complete information from their suppliers as well as on appropriate documentation' (EuroCommerce, 2002).

Such a system was opposed by EuropaBio, the industry organisation. Agbiotech companies feared that consumers would perceive a 'GM' label as 'a skull and crossbones', thus leading retailers to exclude GM ingredients. Likewise the food industry warned that such a label would stigmatise their products and confuse consumers, leading to negative publicity campaigns by environmentalist groups.

Together with agbiotech companies, the food industry emphasised practical difficulties of comprehensive process-based labelling. The CIAA raised doubts about whether a control system would be reliable for processed food. For imports of processed grain, operators could sell unlabelled GM products without effective detection, thus circumventing the rules (e.g. Taeymans, 2001). According to EuropaBio, consumers would be denied a clear choice to buy either GM or non-GM products, especially because of unlabelled GM food. Such arguments came even more strongly from the US food industry.

Moreover, EuropaBio argued that process-based labelling would be unwarranted and unworkable, especially given the lack of reliable detection methods:

Table 7.2 Instruments for GM Labelling Rules: Contending Frames

Agbiotech Developers and Promoters	*Legislative and Regulatory Decision-makers*	*Agbiotech Opponents*
Traceability and Labelling Rules		
Process-based labelling cannot be enforced, may encourage fraud (especially from non-EU producers) and so may undermine consumer confidence & choice. [Also say CIAA + UK government]	Process-based labelling is needed to provide legal certainty for operators, to ensure consumer choice, to facilitate monitoring and to allow product recall if problems arise.	Process-based labelling is needed to trace any harmful effects and to provide maximum information to consumers.
Labelling rules should be based on detectability of GM material. 'GM' label may be perceived as a skull and crossbones.	T&L rules to encompass all products produced/derived from GMOs, regardless of detectability.	T&L rules should also cover animal products resulting from GM feed + products from GM enzymes.
Allow a 'may contain GM' label for imports, to ensure a workable system and to avoid problems at the WTO.	Require a 'may contain GM' label (CEC, 2001e). Require a 'contains GM' label for imports (EC, 2003a).	Oppose a 'may contain GM' label as a surrender to US pressure. Such a label would indicate inadequate control over supply chain.
A 'non-GM' or 'GM-free' label could provide consumer choice to avoid GM products.	A 'GM-free' label would unfairly transfer the burden of responsibility and costs from GM producers to non-GM ones.	Do not rely on a 'GM-free' label, which would assume that GM is the norm, and would reverse 'the polluter pays' principle.
Food: Threshold for 'GM' Label		
Threshold value is political but should be based on feasibility and science, i.e. technical capacity.	Set threshold for adventitious GM material, as a basis for labelling requirement.	Low threshold essential for precaution. Market will adapt to any EU rules; a high threshold would surrender to the USA.
Set threshold > 0.9% because it is on the margin of detectability, i.e. technically difficult and expensive [early stance]. 0.9% takes into account the need for consumer information + associated costs and practicability [later stance].	For EU-approved GM material, set 0.9% threshold; technical progress can improve detection methods and documentary control measures.	Set a threshold at the limit of detectability (e.g. 0.1%), to be lowered according to technical progress. Set zero tolerance for living GMOs.
Threshold should include 'pre-authorised' GM products; traces are unavoidable in the bulk commodity trade. Set no time-limit on this allowance.	Threshold should include GM not approved by the EU but evaluated as safe (to avoid disruption of food supply), for a three-year period.	Set zero tolerance for GM material not EU-approved. Any threshold would contradict the policy of separating risk assessment from risk management.

(continued)

Table 7.2 (continued)

Agbiotech Developers and Promoters	*Legislative and Regulatory Decision-makers*	*Agbiotech Opponents*
Seeds: Threshold for 'GM' Label		
Thresholds for conventional seed should take into account the growing of GMOs in the EU (EuropaBio, 2004b).	Recognise current reality of cross-pollination due to uncontrollable natural phenomena, while ensuring that food products do not exceed 0.9% GM content.	Considerable contamination in seeds would result in exceeding the 0.9% limit in food and/or imposing great costs (ZS-L, 2003).
Lowering the maize threshold to 0.3% would significantly increase costs (EuropaBio, 2004b).	Set adventitious threshold at 0.3% (maize) and 0.5% (rapeseed) for a requirement to label seed as GM. Set thresholds at 0.3% [DG-Envt and many national CAs].	Need purity standard of 0.1% for a precautionary approach, both economically and ecologically. Higher thresholds threaten new rules on approval (ZS-L, 2003).

Abbreviations
CA: national Competent Authority
CIAA: Confederation de l'Industries Agro-Alimentaires
CPE: Confederation Paysanne Européene
EEB: European Environmental Bureau
EPP: European People's Party in European Parliament
EuroCommerce: Retail, Wholesalers and International Trade Representative to the EU
Eurocoop: European Association of Consumer Coops
T&L: Traceability and labelling
ZS-L: Zukunftsstiftung Landwirtschaft ('Save our Seeds!')

> It is already difficult to detect the 35S promoter in a reliable way because it is ubiquitous in GM seeds, except in Monsanto's GA21. It is even more difficult to detect other transgenes in a reliable way. Each transformation event requires a reliable method, to ensure equal performance, which would increase overall costs. If testing is required at several stages—from inbred lines, to hybrids, to farmers, etc.—then costs will increase for that reason too. We should ask: is the cost proportionate to the harm to be prevented? (interview, company officer, 2002)

Such objections were voiced by the European People's Party, joined by some Socialist MEPs from the UK, thus following the policy of their New Labour government. Based in the EPP, the rapporteur for the draft Traceability Regulations warned that the Commission proposals 'open the way for fraud and deception, for misleading consumers, for distorting prices and competition, and for less favourable treatment of EU producers and companies' (Antonios Trakatellis at EP Plenary, 2002). Fellow EPP members added, 'So-called process-based labelling, with no facility to prove genetic modification in the end product, is simply not practicable and is inviting fraud. Europe would be inundated with a flood of supposed scandals . . . ' (Renate Sommer at EP Plenary, 2002). The rules 'threaten to break the bounds of practical feasibility' and may 'turn out to be a bureaucratic monster' (Horst Schnellhardt at EP Plenary, 2003a, 2003b).

Proponents of traceability rules turned those negative arguments against the agri-food industry, by highlighting its potential ignorance about the origin of food ingredients. Traceability was already being imposed by retailers and food companies across the food chain since the late 1990s. On those grounds, the DG-SANCO Commissioner ridiculed the objections: 'I have yet to meet a business operator who would admit that he or she does not know the origin of the ingredients he or she is using'. Indeed, this would be admitting that 'you do not know whether or not you are using a non-authorised GMO' (David Byrne at EP Plenary, 2002).

The Liberal Group had internal disagreements about whether the specific proposals would enhance or undermine consumer choice. Objectors were ridiculed by a United Left MEP: 'It is remarkable that those who usually defend consumers' freedom of choice in the market now do not wish to provide consumers with this freedom through the effective labelling of GM food' (Jonas Sjöstedt at EP Plenary, 2002). After the Regulation was enacted in July 2003, industry continued to question the feasibility and fairness of the new rules. In response, the EFSA chief Geoffrey Podger warned industry against trying to influence public opinion: 'If you go too far, you may stir up an opposition that was not there' (Lewis, 2004b).

For the Commission, technical limits of detection became an impetus to devise better methods. The EC's Joint Research Centre developed Polymerase Chain Reaction and biosensor methods, which could detect even a 0.1% presence. The JRC also provided scientific support for the European Network of GMO Laboratories (ENGL), which was set up in 2002 'to contribute more effectively to the uniform enforcement of legislation across the Community through the harmonisation and standardisation of means and methods of sampling'. Their tasks went beyond detection methods, e.g. by including the stability of plant transgenes (CEC, 2003d: 16). These efforts provided a technical basis to formalise the system of unique identifiers for transformation events (EC, 2004). Various technical advances undermined industry's arguments that traceability would be unfeasible.

In so far as any GM material is detectable, an effective control system depends upon access to reference material for all GM varieties which may be present, but national CAs had doubts about obtaining this material (interview, Belgian CA, 2003). Some national CAs had concerns about being bypassed altogether, unable to enforce control measures for traceability. To gain an adequate role, CAs for the Deliberate Release Directive proposed extra conditions when evaluating notifications for market authorisation. Authorisations required that the company include a unique identifier and provide reference material to CAs.

National access to this information was eventually mandated in the new Traceability and Labelling Regulation, but it specified no details (EC, 2003b: 27). According to some CAs, they would need adequate quantities of pure reference material at least a few months before commercialisation. National access is all the more important for member states which lack strong inspectorates to enforce the rules. But companies did not give permission for prior access, lest the reference material be found in places where it should not be, potentially leading to a 'contamination' scare.

Adequate implementation of the Traceability and Labelling Regulation remained in doubt:

> At national level we lack the right technical structure to implement traceability and labelling. Detection is easy with the right probe, but this is not always available. There are conflicts between biotech companies and public authorities, which do not have sufficient quantities of reference material. (interview, GMO expert of the Green Group of MEPs, 02.02.04)

Before the 2003 Traceability and Labelling Regulation came into force the following year, the European retail industry restated its understanding that traceability requires simply the capacity to withdraw a product from the market, not a specific means for doing so (EuroCommerce, 2004). The food co-ops federation sought to extend the regulations, e.g. by proposing post-market monitoring of food consumption (Euro-Coop, 2003).

Experts identified several problems in implementing these EU rules. The lower the threshold for statutory labelling, the more difficult and expensive the task of detecting such a level. For multi-transgenic crops, detecting the precise combination would require more complex methods. Potential for further transatlantic conflict arose not only from policy differences over the need for traceability, but also from methodological differences: the EU favoured DNA-based techniques, while the USA favoured protein-based techniques (Miraglia et al., 2004).

3.3 Setting GM Thresholds

The ensuing debate highlighted contending assumptions about the ab/normality of GM products and thus potential futures for European agriculture. In opposing the Commission's draft Regulation, the main alternative proposal was a 'GM-free' or 'non-GM' label, which some companies had already introduced. NGOs attacked this option for assuming that GM crops would become the norm and thus non-GM an exception. As a Green MEP stated, moreover, a 'GM-free' label would reverse 'the polluter pays' principle (Agence Press Europe, 2002). Perhaps adapting the NGO argument, the

Commission opposed a 'GM-free' label for transferring the responsibility from GM producers, where it properly lay (see Table 7.2, middle).

Within process-based rules, i.e. labelling all products of GM techniques, consumer choice would depend upon criteria for thresholds. According to the Commission's proposal for a Regulation on GM Food & Feed, labelling would not be required beneath a specified threshold of adventitious presence—i.e. at levels which remain technically unavoidable, despite the operator's reasonable measures to minimise that presence. It proposed a 1% threshold, controversially including GM material not authorised by the EU if it had safety approval elsewhere (CEC, 2001e: 21).

Such a threshold for non-authorised GM material generated disagreements among NGOs, Commission staff and MEPs. Early comments included the following:

> The Commission's proposal is the wrong reaction to increased pressure and threats from the US administration and GMO-producing companies . . . If the EU sets clear and uncompromising safety standards, the market will adapt to them. (Greenpeace, quoted in *Agence Europe*, 26.07.01)

> Many MEPs and the Committee report said Parliament must take a precautionary approach. (*AgraFood Biotech*, 07.05.02)

Here 'a precautionary approach' meant opposing the clause and thus any scope for non-authorised GM material.

In response to NGO and Green criticism, backed by some member states, the Commission altered its original proposal so that the 'non-authorised' allowance would apply only to GM products which had a favourable risk assessment from an EU expert body. The agbiotech industry now more easily promoted this allowance for 'pre-authorised GMOs', a phrase assuming that EFSA's safety advice would be straightforwardly turned into an approval decision. The proposal was supported by the European People's Party in similar terms: 'pre-authorised GM products . . . are ready to be licensed; they are in a holding pattern' (Renate Sommer at EP Plenary, 2003a, 2003b). This flight arrival metaphor implied great danger if the product were not ensured a safe landing, i.e. freedom to circulate freely in the internal market.

Environmental NGOs attacked the clause which would allow traces of non-approved GMOs. Among other reasons, to permit unauthorised GMOs would treat expert advisors as proxy decision-makers, thus violating the principle that risk assessment should be kept functionally separate from risk management (FoEE, 2002b). Some MEPs denounced such an allowance as contrary to precaution: a threshold should be allowed only for GM products which had EU approval.

Eventually a compromise was reached on the following thresholds:

- 0.9% for EU-approved GM material
- 0.5% (for a three-year transitional period) for GM material favourably evaluated as safe by an EU scientific committee but not EU-approved.

A temporary allowance for non-approved GM material gained support for many reasons. Such an allowance would ensure that the overall agro-food chain was not disrupted by illegal traces of GM material, while also accommodating a temporary gap between expert advice and the EU approval procedure. This compromise was agreed at the second reading (EP Plenary, 2003a, 2003b; EC, 2003a). For some MEPs, the threshold criteria were less important than the traceability principle, which would enhance the accountability of industry for the spread and potential effects of GM products.

The lower the GM threshold, however, the greater the difficulty to keep GM material below it, so the greater incentive for companies simply to label all processed food as GM. Consumer groups expressed concern that this might happen, thus precluding freedom of choice (BEUC, 2003). NGOs asked food companies for a commitment to continue excluding GM ingredients from their products; some major companies declined to do so, thus leaving open their options. If the 0.9% limit became too difficult to maintain, then GM labelling could become the norm, they warned.

The GM threshold in organic food had been contentious from the start and remained so, long after enactment of the 2003 Regulations. Without a statutory rule for organic food, it had no clear threshold for GM labelling, thus potentially jeopardising its product identity and market. In 2007 the issue came to a head, sharply dividing EU institutions. In April the Parliament voted to set the threshold for GM in organic food at the lowest level possible, widely understood as the detectability limit of 0.1%. This was supported by some member states—e.g. Italy, Greece and Austria.

Nevertheless the Agriculture Council accepted the European Commission proposal that organic products should be subject to the same 0.9% adventitious threshold as other food (EC, 2007b), thus over-riding the Parliament. This decision was denounced by NGOs and organic farmers' organisations for opening the way to GM contamination of organic food (Halliday, 2007). Other issues were contentious too. Under the same regulation, organic labelling must specify the place of production but not the producer's name. Organic producers must quantitatively analyse their use of conventional agricultural inputs and materials. According to the Coordination Paysanne Européenne, that costly requirement will penalise small-scale producers, while disfavouring natural agri-inputs and traditional seeds. Overall the EU policy was putting organic agriculture 'in danger of contamination and industrialisation', claimed the *paysan* critics (CPE, 2007). Afterwards the 0.9% threshold remained contentious, especially regarding the implications for measures to segregate GM material from organic crops (see Chapter 8).

4 LABELLING CONVENTIONAL SEEDS AS GM?

Relative to food or feed, seeds had much higher stakes for GM contamination. Such material would pose difficulties for farmers in keeping levels below the labelling threshold for food and feed. Such contamination could be more difficult to control if originating from a GM crop not approved by the EU. And its detectable presence would undermine industry claims for adequate control over GM crops. Moreover, opponents denounced any 'GM contamination' of seeds as a strategy to undermine efforts towards a non-GM agro-food chain, instead making GM crops the inevitable norm (FoEE, 2001). Proponents of relatively lax rules were put onto the defensive for threatening safety and consumer choice. This section analyses disputes over two issues: unauthorised GM material in seeds, and the statutory thresholds for labelling seeds as GM, with implications for economic burdens (see Table 7.2, bottom).

4.1 Unauthorised GM Material in Seeds

In early 2000 traces of unauthorised GM material were found in European conventional crops, cultivated from seeds imported from North America. Examples included Monsanto's GM cotton found in Greece, Novartis' GM maize found in France and Adventis' GM oilseed rape found in several countries. The GM material did not have EU approval, so its presence was illegal under the Deliberate Release Directive. Member states responded in various ways, e.g. by destroying the crops or by implying that the contamination was acceptable.

The Commission sought to harmonise national responses and avoid market disruption. Initially it proposed a Gentleman's Agreement for coordinated testing, with interim thresholds permissible for any unauthorised GM material in seeds. Towards a longer-term harmonisation, it proposed to set acceptable levels of 'adventitious' GM material, in cases where its presence was technically unavoidable (DG-SANCO, 2002).

Disagreements ensued over the practical meaning of 'adventitious', as well as over the feasibility and grounds to prohibit 'contamination' of conventional seed by GM seeds unauthorised in the EU (see Table 5.1). Industry advocated a permissible threshold, as had been done in Greece, thereby allowing cotton farmers to save their crop:

> In the case of cotton seeds in Greece, NGOs found contamination > 0.01%, while the government found no contamination > 0.1%, a criterion which allowed the system to salvage their cotton crop. Greece had no alternative source of seeds. (interview, company officer, 2002)

However, some regulators regarded any contamination as illegal. For example:

> GM contamination is a factual situation—not just an uncertainty . . . For seeds contamination by GM material not authorized by the EU, the seeds are unsafe, so such contamination should be (and is) prohibited within the detection limits, presently 0.1%. (interview, DG-SANCO, 2002)

In that view, illegal material must be treated as if it were unsafe.

The EU's scientific committee regarded a 'GM-free' standard as impossible, i.e. as incompatible with routine commercial activities. According to its advice:

> . . . a zero level of unauthorised GM seed is unobtainable in practice. A zero level would have severe consequences for Part B GM field releases, for biosafety research and for evaluation of new GM plant varieties. (SCP, 2001a: 16)

The expert advice of the scientific committee was criticised by NGOs as 'political'.

> The SCP advice is led by 'political and commercial assumptions, rather than scientific criteria.' (Greenpeace, quoted in *Agence Europe,* 26.07.01)

> Seed-producing farmers cannot be held responsible for the consequences arising from the presence of non-authorized GMOs in seed that is produced using supposedly GMO-free material. (COPA, 2002a)

> We are interested in the Commission's suggestion to look into the possibility of fixing thresholds for non-authorized GMOs that have undergone positive scientific testing. (COPA, 2002b)

As in the latter quotes, major farmers' organisations feared that unauthorised contamination could bring liability consequences, so they sought ways to normalise 'GM contamination'. An impasse continued over this issue. Meanwhile the presence of any non-authorised GM material would render a seed supply illegal, thus potentially disrupting the seed trade.

4.2 GM Thresholds in Seeds

A related contentious issue was the relative GM thresholds for labelling seeds and food. This relation would influence the relative burden of segregation measures along the agro-food chain—for seed breeders, farmers and food producers. A lower GM threshold in seeds would impose a relatively greater burden upon plant breeders but would help others to comply with the 0.9% limit for non-GM food. Seed producers sought a relatively high GM threshold in order to minimise their costs, which would be effectively transferred to farmers and grain traders.

In the Commission's proposal on GM material in conventional seeds, thresholds were set according to the outcrossing behaviour of each crop, and low enough to ensure that any GM presence in food would be kept below 1%—the threshold in the Commission's parallel proposal on food labelling. Even this level would be difficult to maintain over time, especially for male-sterile varieties of oilseed rape, which have a greater capacity for long-distance cross-hybridisation. According to expert advice, the proposed thresholds could be achieved only under ideal conditions:

> Achieving the 0.3 and 0.5% thresholds [for seeds] will become increasingly difficult as GM crop production increases in Europe. In due course the 1% threshold [for food & feed] set by the Commission may have to be increased. (SCP, 2001a)

> Varieties and lines containing male-sterile components will outcross with neighbouring fully fertile GM oilseed rape at higher frequencies and at greater distances than was previously thought (EEA, 2002b: 56).

> All farm types producing oilseed rape or conventional maize will need significant changes to meet their thresholds . . . changes may involve cooperation between neighbouring farms. (DG-JRC, 2002)

The seeds draft regulation was circulated in July 2002, on the day before the European Parliament voted for a 0.5% threshold in food—incompatible with the original thresholds for seeds, and much more difficult to achieve in the field. Farmers' organisations were concerned that lax thresholds would threaten the availability of non-GM seeds:

> 'Serious & irreversible damage' includes mixing with GM seeds if it eliminates conventional non-GM seeds, because this would go beyond a point of no return, into a grey zone. There would no longer be any choice. (interview, COPA, 2002)

A year later the Commission pressed forward with its seeds proposal. According to some critics, the seed thresholds were set too high to keep GM presence below 1% in food, thus requiring costly tests and management measures. In their view, it would be better to start from a strict purity regime, e.g. a 0.1% threshold in seeds. Another critical argument was that the gradual spread of GM material would undermine risk management and monitoring of GMO releases. 'A precautionary approach, both ecologically and economically, is therefore appropriate', argued an NGO (ZS-L, 2003). Following Austria's precedent, Italy announced a plan to set its own national threshold for GM presence at less than 0.1% (*Agra Europe,* 10.10.03).

As a more general concern, the Commission was maintaining its original threshold figures for GM material in seeds, even though the GM threshold for food had been lowered from 1% to 0.9%. The now smaller difference potentially put a greater burden on farmers' practices to segregate GM from non-GM crops. Alternatively, the burden could be put on seed breeders through a stricter standard for purity—as advocated by environmental NGOs, some member states and the food retail industry. With the originally proposed GM thresholds in seeds, it would become 'impossible to guarantee the recently adopted threshold of 0.9% for the adventitious presence of GMOs in food and feed', argued retailers (EuroCommerce, 2003).

Similar criticisms from member states led to proposals for lower seed thresholds. Initially the Commission sought to gain approval for its original 2002 proposal for thresholds in seeds. The decision could be agreed by a simply majority in the Standing Committee on Seed Propagation, hosted by DG-Agriculture. Any such decision faced a court challenge, however, given that the Deliberate Release Directive delegated authority to its national Competent Authorities (CAs) to set GM thresholds for labelling purposes. Partly through DG-Environment, those CAs asserted their authority to set lower figures than in the Commission proposal, which would need a qualified majority for approval in their CAs regulatory committee. Those two committees, under the DGs for Agriculture and Environment, had a rivalry for authority and their own specific proposals.

As a way forward, in October 2003 the Commission consulted them together. This move sought to avoid conflicts which could arise later, especially by different thresholds being set under different legislation. The Commission proposed 'thresholds for approval by qualified majority' by the CAs' Regulatory Committee under Directive 2001/18, hosted by DG-Environment (SCSP, 2003). But agreement could not be reached there.

Without any rule on thresholds, the seed industry faced a serious problem: any detectable level of GM material officially rendered a seed lot as 'GM', according to the Commission. Such seed would be potentially worthless in the EU seed market. Contamination arose especially in seed lots imported from the USA, though perhaps also from seed bred in Europe, where the contamination problem would be worsened by any commercial cultivation of GM crops. The seeds industry complained that conventional seed production was becoming more difficult in the absence of GM thresholds: companies have been 'making supplies of seed available to farmers under unrealistic, unclear and legally disputable national rules' (EuropaBio, 2004b).

In 2004 prospective rules were advocated or opposed according to divergent assumptions about future agricultural practices as a basis for 'realistic' thresholds. Environmental NGOs maintained their proposal for a 0.1%

threshold, i.e. the limit of detectability, as an implicit means to deter cultivation of GM crops. DG-Environment proposed that the GM threshold for maize seeds be set as low as for oilseed rape, i.e. at 0.3%. According to the agbiotech industry, such a low threshold would greatly increase costs: 'The thresholds must be fixed at a level which takes account of the growing of GMOs in the EU. To do otherwise is counter to established EU policies'. In support they cited expert advice that the Commission's earlier proposal could be achieved only 'under ideal seed production conditions', i.e. without nearby cultivation of GM crops (EuropaBio, 2004b, citing SCP, 2001a). Thus proponents of the lowest or highest thresholds deployed normative arguments based upon non-GM or GM futures, respectively.

Member states could not agree on the DG-Environment proposal. Amidst a regulatory impasse at the EU level, discussion continued about possibly delegating threshold rules to member states, though this would jeopardise an internal market in seeds. For harmonised rules, the new DG-Environment Commissioner advocated a seed threshold as low as possible. Arguments ensued over whether this meant a threshold which is technically feasible or economically viable. Meanwhile the Commission sought more expert advice on the costs of maintaining seed purity (Lewis, 2004b). The subsequent DG-Environment Commissioner suggested that the EU did not need to set a GM threshold for conventional seeds, yet only his DG could initiate such a proposal (Lewis, 2006c).

This issue still remained open for a long time, as statutory thresholds could not be agreed. Seed breeders faced difficulties without a mandatory GM label. Any particular threshold any particular threshold linked a cognitive view of gene flow with a normative stance towards commercialising GM crops. From the three contending frames, such commercialisation was variously seen as inevitable, or as a difficult management problem, or else as unacceptable. In these ways, proposals for relatively high or low thresholds were put on trial according to their normative models of the social order.

5 CONCLUSIONS

At the start, this chapter posed the following questions:

- How were industry and government put on the defensive for their no-labelling policy?
- How did broader GM labelling rules gain decisive support?
- How did agri-food markets test claims about agbiotech?

Early EU agbiotech policy treated 'safe' GM products as normal. In the mid-1990s GM soya and maize were approved for the EU internal market with no requirement for a special label. The Commission rejected demands for mandatory GM labelling, partly on grounds that such a requirement would have no scientific basis and would provoke a US challenge at the WTO. Thus GM grain was invisibly mixed in agri-food chains and processed food.

This policy did not successfully normalise GM products, instead provoking demands for comprehensive labelling. When the first US shipments of GM grain reached Europe in 1996–97, high-profile NGO protests turned these products into symbols of pollution and anti-democratic coercion, while gaining Europe-wide publicity. Local protests at supermarkets demanded GM labelling and non-GM alternatives. In rejecting GM food, many people were 'voting' as consumers, in lieu of a democratic procedure for a societal decision about a contentious technology.

In the ensuing debate, consumer choice was framed in contending ways which led to GM labelling rules. Agbiotech supporters adopted an economic individualism, which modelled consumers as rationally pursuing their individual interests in safe food (Klintman, 2002). According to EuropaBio, rules should be based upon intrinsic product characteristics which are scientifically verifiable and relevant to consumer interests. The European Commission initially adopted a similar policy, but it was put onto the defensive by a wide range of civil society groups which had diverse stances towards agbiotech. Consumer NGOs sought to ensure the consumer right to know and choose food according to its origin. From an anti-biotech standpoint, environmental NGOs demanded GM labelling as a democratic right and defence against both risks and globalisation; such rules could also be used to deter the commercial use of GM grain. Through these cultural discourses and boycott actions in the market arena, food companies were being pressed to use their economic power with grain traders.

As an extra pressure on industry, NGOs carried out surveillance of food products for GM 'contamination', based on their own lab tests. Their publicity targeted retail chains and processors of any products which gave positive results. Thus the industry was being put on trial for deception and/or denial of consumer choice. This campaign strengthened pressures on the food industry—to adopt voluntary rules, to adopt relatively broader criteria and to ensure that any GM content was duly labelled.

In the market arena, the food retail industry became a significant new actor which eventually redefined its interests along the lines of consumer rights. European retail chains had been building up their own-brand lines, designed to symbolise product quality, as a tool of competitive advantage; this strategy made retailers more vulnerable and responsive to consumer concerns. European retailers found themselves competing to sell processed food as 'non-GM', defined according to diverse, unstable criteria. Facing market instability, the European food industry sought common rules for distinguishing GM from non-GM products. Such rules were needed to clarify product identity, as a means to stabilise markets for processed food.

EU labelling rules redefined what is a 'GM' product according to detectability criteria which became successively broader, supported by some member states and the European Parliament. Stability of the internal market became extra grounds for successively tighter statutory EC rules, in three stages. At each stage, current or potential rules faced a multiple charge-sheet: whether they neglected or adequately identified GM 'contamination'; whether they accommodated or denied 'free consumer choice'; and whether they would stabilise or undermine the market for processed food. Labelling rules were being put 'on trial' along with agbiotech.

By 1999 European retail chains excluded GM grain altogether from their own-brand products, rather than apply a GM label, to avoid any market disadvantage. Market forces effectively tested and undermined claims that agbiotech enhances economic competitiveness. Commercial pressures against GM crops were extended across Europe and the agro-food chain. Given the strong consumer signals in some countries, food companies changed their ingredients or supply-chain sources across Europe. Consequently, by the late 1990s GM grain was being used only for animal feed. Farmers came under similar pressures from retail companies and faced uncertainty about a market for GM grain.

In the late 1990s some NGOs proposed a traceability system for GM products, so that any harm could be linked with a specific product, partly by analogy to beef being recalled in the BSE crisis. In 1999 many EU member states likewise demanded full labelling and traceability, beyond the detectability of GM material, thus applying to far more products than the current rules. Such demands cited precautionary reasons: consumers and regulators must have adequate information to deal with uncertain risks. The European Commission accommodated those demands: under its eventual proposal, labelling would be required for all products containing GM material above a 1% 'adventitious' threshold, i.e. a presence which is technically unavoidable. This term implied a burden to demonstrate efforts made to avoid such presence. The Commission also proposed a traceability system, linked with a unique identifier for each transformation event which produced the genetic modification. These proposals cited on several grounds—consumer choice, public confidence, and market-stage monitoring; all were implicitly precautionary grounds.

The Commission proposal was supported by European retail chains, but it was opposed by food processing companies, the agbiotech industry and the US government. In their view, full GM labelling would be unfair, impractical, misleading and thus contrary to consumer choice. Such disagreements also arose in the European Parliament, which eventually voted for more stringent criteria than the Commission had proposed. The new proposals were debated along several lines: whether comprehensive labelling would enhance or deny meaningful consumer choice, whether traceability requirements could be reliably implemented and whether these rules would intensify the transatlantic trade dispute, given that US maize exports were effectively blocked in the EU.

That trade dispute became a frequent reference point for the debate over EC rules on traceability and labelling. The Commission advocated such rules as a means to resume the EU approvals procedure and thus avoid a trade war. According to some opponents, however, such rules themselves would provoke a WTO dispute and would have no scientific basis. Amidst those disagreements, agbiotech critics denounced proposals for relatively less stringent rules as a surrender to the US government; they associated proposals for tighter rules with democratic sovereignty and political defiance, such as attacks on GM field trials.

After the Commission's labelling and traceability proposals were enacted in 2003, with a 0.9% 'adventitious' threshold, the new rules put GM products under permanent suspicion in market arenas as well as in regulatory ones. Despite their precautionary character, these rules did not much facilitate commercial authorisation of new GM products, as Commission proposals encountered extra obstacles (see Chapter 5). A GM label was now required for a broader range of products, so agbiotech opponents gained an extra weapon for pressurising food companies to exclude GM ingredients.

Also contentious were the GM labelling criteria for conventional seeds. The Commission advocated specific thresholds as 'realistic', given the eventual prospect that GM crops would be more widely cultivated, thus inevitably spreading GM material to conventional seeds far more than grain imports would do. Agbiotech critics attacked this rationale as a self-fulfilling prophecy, while counterposing lower thresholds. Any particular threshold linked a cognitive view of gene flow with a normative stance towards commercialising GM crops. In the three contending frames, such commercialisation was variously seen as inevitable (eco-efficiency frame), or as a difficult management problem (managerialist frame), or else as unacceptable (apocalyptic frame). In these ways, proposals for relatively high or low thresholds were put 'on trial' according to their normative models of the social order.

Through all those conflicts, agri-food markets tested models of consumer choice. Any criteria for GM labelling thus had a normative basis by favouring specific models of the market, public and legitimate grounds for consumer choices. After regulators were accused of denying consumers the right of free choice, new rules accommodated relatively more comprehensive criteria for GM labelling. With the 2003 traceability and labelling rules, the Commission aimed to ensure that 'the internal market could function effectively' in the real world. Such language took for granted an external objective reality—the prospect of GM crops being widely cultivated in Europe, and their necessity for European economic competitiveness, as affirmed in EC policy. Yet GM labelling rules were used by critics to block or deter that scenario.

In the market arena, GM products were symbolically put on trial—as 'contamination' and an economic liability. Since the late 1990s, a commercial boycott of GM grain contradicted claims that agbiotech would enhance the economic competitiveness of the European agri-food industry. Combined with public access to information, GM labelling and traceability rules facilitated a different socio-natural order: agbiotech as perpetually suspect, being continuously kept 'on trial' for potential risks, and its commercial prospects potentially influenced by consumer decisions. Moreover, EU labelling rules became a contentious factor in disputes over 'GM contamination', as shown in the next chapter.

8 Segregating GM Crops, Contesting Future Agricultures

INTRODUCTION

Beyond the 'adverse effects' of GM crops being evaluated in EU regulatory procedures in the late 1990s, an extra issue emerged—the prospect that GM material would become inadvertently mixed with non-GM products (see Chapter 6). The stakes were raised by new rules which required a 'GM' label for any product containing more than 0.9% of GM material (see Chapter 7). Such a label could reduce the market value of a crop, especially if intended for food uses.

The admixture issue involved diverse economic interests and political agendas. Farmers and food retailers sought to protect their products from economic loss. Agbiotech critics sought an extra means to restrict or block the technology; they warned that GM crops could irreversibly pervade the environment, thus 'contaminating' non-GM crops and the wider environment. The admixture problem generated further conflict about whether GM crops were simply an extra option—or rather a threat to other agricultural systems.

To address the admixture problem, there were various proposals for segregating GM crops and material from non-GM crops. The European Commission elaborated a policy for the 'coexistence' of GM with conventional and organic crops. Designed to mediate conflicts over admixture, 'coexistence' policy frameworks became contentious.

This chapter analyses such conflicts by discussing three questions:

- Why did the admixture (or 'contamination') problem become so contentious?
- How were various coexistence policies put on trial?
- How did 'coexistence' rules relate to different agricultural futures?

The chapter analyses 'coexistence' policy as a focus for contending accounts of biotechnological risk, agricultural futures and European democracy. As outlined in previous chapters, contending frames of agbiotech intersected with rural development paradigms (see Chapter 1).

1 SEGREGATION POLICY ON TRIAL

When the European Commission first discussed admixture as an extra issue beyond environmental risk, segregation measures had no clear statutory basis. In the market authorisation of a specific GM product, special conditions could be imposed—on grounds of uncertain risks or assumptions in the risk assessment—thus extending the experimental stage into commercial use (see Chapter 6). However, admixture concerned GM products which already had (or would soon obtain) market approval; they could freely circulate within and across EU member states. Some national authorities devised or demanded segregation measures to prevent admixture, as a general pre-condition for supporting approval of GM products or for permitting their use.

European policy actors framed the admixture problem according to their general stance towards agbiotech, corresponding to agricultural development models (as shown in Table 8.1; see also Chapter 1). Agbiotech opponents warned that 'GM contamination' would aggravate uncertainties about environmental and health risks, as well as jeopardising non-GM alternatives, as grounds for a ban on GM crops. Originating in an apocalyptic frame, this stance eventually linked with an agrarian-based rural development paradigm, promoting 'quality' agriculture. By contrast, favouring an agri-industrial paradigm, agbiotech proponents argued that good agricultural practices would suffice to limit admixture, which would be only an economic problem, for which non-GM farmers should have the main responsibility.

The European Commission initially adopted a similar argument, which came under widespread attack for denying farmers and consumers the right to a free choice. Eventually the Commission accommodated pressures for greater restrictions on GM crops, as ways to ensure that GM crops would coexist with organic and non-GM crops. Responsibility was shifted to GM farmers, followed by various proposals to increase the burden upon them. In this process the Commission elaborated an agri-diversity frame. This section analyses the policy conflicts and shifts.

1.1 'Adventitious' Criteria as a Moral-Legal Burden

New developments raised the stakes and mass-media profile of the admixture problem. By the late 1990s the major food retail chains had excluded GM grain from their own-brand products, as a measure to accommodate consumer preferences (Levidow and Bijman, 2001). In early 2000 regulatory authorities disclosed that European farmers were cultivating seeds which contained GM material lacking EU approval. This discovery of illegal planting' intensified debate on inadvertent mixtures of legal GM material in conventional seeds and in agri-food products more generally. Such 'admixture' could have many possible sources—e.g. GM grain imports

Table 8.1 Coexistence Policy Frames

Agbiotech Promoters **Agri-industrial paradigm** (EuropaBio, EPP Group of MEPs)	*Legislators and Regulators* **Agri-diversity policy** (European Commission and some governments)	*Agbiotech Opponents* **Agrarian-based rural development paradigm** (Greenpeace, FoEE, CPE, Green Group of MEPs, AER)
Technological innovation = progress. GM crops can safely sustain intensive agriculture. Don't stigmatise GMOs: protect the environment!	Consider the risks of not developing technologies to remedy unsustainable agriculture (CEC, 2002b: 15).	De-intensify and relocalise agriculture; support multifunctional farms and sustainable family farming (CPE). Promote a community-territorial basis for skilled, quality agriculture
Support aims of Lisbon summit (implying that competitiveness depends on greater productive efficiency).	Develop the knowledge-based economy for international competitiveness of European industry (EU Council, 2000b).	Oppose GM agriculture as non-competitive, e.g. losing markets and jeopardising non-GM markets.
Consumers need a choice to buy GM or non-GM products.	Consumer choice is essential for gaining consumer confidence.	Consumer choice is paramount (in its own right).
Freedom of choice for economic operators (consumers and farmers)	Freedom of choice for economic operators.	Freedom of choice to buy non-GM products
Coexistence is necessary and feasible.	Coexistence is necessary and feasible. No form of agriculture should be excluded; national rules should allow market forces to operate freely.	A coexistence system appears to be practically impossible and unnecessary (2001). Coexistence needs stronger rules, even 'GM-free zones' (2003)
Admixture is about economics—distinct from safety issues.	Admixture is an economic issue about how to limit adventitious presence of safe GM material.	GM pollution (admixture) would aggravate uncertainties about environmental and health risks.
Normal management measures will be adequate to segregate different forms of agriculture. Disproportionate burdens on GM crops would deny choice to farmers and consumers.	Management measures should be proportionate, based on specific knowledge about agricultural systems and gene flow.	Management measures may never be adequate to protect non-GM crops.

(continued)

Table 8.1. (continued)

Agbiotech Promoters **Agri-industrial paradigm** (EuropaBio, EPP Group of MEPs)	*Legislators and Regulators* **Agri-diversity policy** (European Commission and some governments)	*Agbiotech Opponents* **Agrarian-based rural development paradigm** (Greenpeace, FoEE, CPE, Green Group of MEPs, AER)
Growers who meet a quality standard that provides a higher value product should not expect their neighbours to bear their management costs of meeting that standard.	The burden should fall on economic operators who intend to gain a benefit from a specific cultivation model (Fischler, March 2003). Farmers who introduce a new production type should bear the responsibility (CEC, July 2003)	The polluter should pay: GM growers should have full responsibility and strict liability.
National law on civil liability offers abundant possibilities to seek compensation for any economic damage.	Liability rules are a matter for each member state. Environmental Liability Directive covers only (significant) environmental harm from GMOs.	Need new legislation to guarantee compensation for GM pollution—not covered by the Environmental Liability Directive.

EU-level organisations: abbreviations
AER = Assembly of European Regions
COPA = Committee of Agricultural Organisations of the EU
CPE = Coordination Paysanne Européenne
EPP = European People's Party, especially Christian Democrats
FoEE = Friends of the Earth Europe

spilling around ports, pollen flowing from GM to non-GM crops, or the same farm machinery being used for both types, or grain storage silos mixing the two types, especially in countries which export grain to the EU. Stakes were also raised by the pressure for full labelling and traceability of GM products, as demanded by EU member states in 1999 (see Chapter 7).

To address those dual problems, the European Commission developed a new policy framework which went beyond simply accidental mixtures. The new term 'adventitious' meant a 'low-level, technically-unavoidable and unintended presence' (CEC, 2001a: 29). Reasonable efforts should be made to prevent mixing: if only traces of GM material occurred and were technically unavoidable, then their presence would not trigger a requirement for a 'GM' label. For food and feed products, the threshold figure was eventually set at 0.9% adventitious presence of GM material (EC, 2003a).

The term 'adventitious' implied a moral-legal burden to demonstrate that the presence was truly unavoidable. According to the new law, 'operators must be in a position to supply evidence that they have taken appropriate steps to avoid the presence of such material' (EC, 2003a: 11). This condition would later become controversial. Demands for lower thresholds of GM labelling converged with demands for segregation measures to minimise the presence of GM material (see Chapter 7). 'Adventitious' originally denoted GM material present in grain or seed imports into Europe. Later the term also denoted gene flow from GM crops cultivated in Europe, which would require preventive measures to isolate them in a farming context.

Through such measures, government sought to ensure farmers' freedom of choice to cultivate GM, conventional or organic crops. According to this 'coexistence' policy, no agricultural option would be excluded. This new concept evoked the Cold War slogan, 'peaceful coexistence among nations of different political-economic systems', which aimed to avoid direct military conflict. By analogy, a European 'coexistence' policy sought to avoid or manage political-economic conflict over agbiotech.

Beyond GM thresholds in food and feed, the European Commission also attempted to set thresholds in conventional seed. Legislative proposals were discussed for a long time but remained at an impasse. By default, meanwhile seeds had to be labelled as 'containing GM' if such presence was detectable (see Chapter 7). Itself contentious, this requirement raised the stakes for any GM material present, thus intensifying conflict over segregation measures.

1.2 Coexistence: Routine Measures or 'Mission Impossible'?

Conflicts over GM labelling and environmental issues intensified debate over the feasibility of coexistence. Environmental NGOs argued that coexistence would be difficult—even 'mission impossible', especially for oilseed rape, which would easily spread and readily contaminate its conventional counterpart (FoEE, 2002c). Going beyond economic implications, moreover, such NGOs and even some regional authorities portrayed any GM material present as an environmental risk issue, given uncertainties and expert disagreements about risk assessment (see Chapter 5 and 6).

Admixture was foreseen by farmers as more than just an economic problem in another sense. Representing large-scale industrial-type farmers, the Committee of Agricultural Organisations of the EU (COPA) favoured free access to GM crops in principle but anticipated strong public reactions:

> Farmers fear they will become the centre of a public dispute. If the thresholds are exceeded, despite farmers' best efforts, then what? . . . If admixture eliminated conventional non-GM seeds, then this would go beyond a point of no return, thus ending freedom of choice. (interview, COPA, 2002)

In the latter scenario, GM material could spread irreversibly, to the point where non-GM seeds become more difficult or impossible to obtain.

Organic crops further raised the stakes for 'contamination', economic loss and thus conflict among farmers. According to the EU Organic

Standards Regulation 1804/1999, organic farmers could not 'use' GM crops, but their food products had no statutory threshold for any GM material which may be inadvertently present. According to the Commission, the labelling threshold therefore would be 0.9%—i.e. the same as for conventional crops, by default of any separate rule. The prospect of allowing such amounts of GM material in 'organic'-labelled crops became contentious, given that GM maize had EU approval for commercial cultivation and so could plausibly spread its pollen to organic sweet maize.

Consequently, the out-crossing and admixture problem posed a threat to European organic producers. Their business depended upon a strict, feasible separation of agricultural systems: 'We need to make sure we protect a promising growth sector', they argued (BÖLW, 2003). In their view, organic crops must remain strictly GM-free.

Responding to the controversy, the Commission eventually changed its initial policy framework on coexistence. Originally this foresaw agri-biotechnological techniques being assimilated into conventional and organic agriculture (CEC, 2001c: 7). From this standpoint, admixture would pose no special problem. Later its policy emphasised segregation from GM crops, to achieve several aims:

> In order to fully apply the principle of freedom of choice for economic operators and to safeguard sustainability and diversity of agriculture in Europe, public authorities in partnership with farmers and other private operators need to develop agronomical and other measures to facilitate the coexistence of different agricultural practices without excluding GM crops. (CEC, 2002b: 14)

The Commission also undertook to protect seeds from GM material. It would explore possible options

> for agronomic and other measures to ensure the viability of conventional and organic farming and their sustainable coexistence with genetically modified crops. Moreover, the Commission recognises the importance of safeguarding the existing genetic resources in agriculture. (CEC, 2002b: 30)

Thus the Commission began to elaborate its own policy frame for promoting agri-diversity, while managing the societal conflict over GM admixture.

1.3 Responsibility for Segregation?

At issue was the burden of responsibility for segregating GM crops and thus clear accountability for any admixture. In Commission policy, the responsibility was initially assigned to non-GM farmers. This issue could not be avoided by expert reports seeking to advise the authorities on admixture problems and possible remedies.

In January 2003 a Danish report implied that significant responsibility lay with GM farmers, though the issue remained contentious. The report based its scenarios upon the GM labelling thresholds in new draft laws. According to its advice, 'co-existence at the proposed threshold values is possible, as long as the measures suggested are adopted'. The report did not assume that such measures would be straightforward:

> It is evident that co-existence in many cases will require good farm management and great care in the production. The group therefore suggests that a course in GM crop cultivation and handling is made compulsory for farmers—possibly as part of the farmer's education. (Tolstrup et al., 2003: 6)

In March 2003 the Agriculture Commissioner Franz Fischler issued advice so that each national authority could specify coexistence measures appropriate to local conditions. Authorities could also develop or clarify legislation to provide liability for economic damage from adventitious presence of GM material. The Deliberate Release Directive, the main risk regulatory framework on GM crops, could not be used to regulate adventitious GM presence because it did not qualify as environmental harm, Fischler argued. For similar reasons, local authorities could not simply impose a blanket ban to declare entire areas 'GM-free', as proposed by some regions (see later sections). Moreover, the burden of coexistence measures should 'fall on the economic operators (farmers, seed suppliers, etc.) who intend to gain a benefit from a specific cultivation model they have chosen' (Fischler, 2003: 5). Under this criterion, the main burden would fall upon non-GM farmers, especially if 'a benefit' meant a premium price for their products, as was already the case in European grain markets. The Fischler criterion provoked sharp disagreement over the burden of responsibility.

Concerns about economic loss and liability were spreading from organic to agri-industrial farmers, represented by the Committee of Agricultural Organisations of the EU (COPA). This organisation welcomed the Fischler proposal to provide financial liability in cases of economic loss to non-GM farmers. But they wanted clearer management rules so that all producers would have 'legal security' and so that all agricultural options would be economically viable (COPA, 2003).

Agbiotech opponents attacked the Fischler document, especially for reversing the 'polluter pays' principle; instead, polluted farmers would pay for preventive measures or lower prices. Several environmental NGOs attacked the Commission for 'dodging its responsibility' to prevent genetic 'contamination', given its foreseeable consequences: 'With no hard legislation in this area, genetic contamination will soon become a *fait accompli* in EU agriculture, depriving European consumers and farmers of the right to choose'. Therefore clear, effective legislation was needed 'to protect the agricultural assets of Europe' (FoEE/GP/EEB, 2003; also FoEE, 2003a). Along with the Coordination Paysanne Européene, environmental NGOs demanded clear statutory rules assigning responsibility and liability entirely to GM farmers.

That demand was echoed by some politicians. According to Germany's Agriculture Minister, Renate Künast, 'those who want to produce without GMOs in the future should under no condition be confronted with extra costs'. According to a press release by three Socialist members of the European Parliament's Environment Committee, the Commission should launch a 'new, more ambitious proposal' on coexistence (quoted in FoEE, 2003c).

As some critics conflated economic and other risks, Commissioners again distinguished between them. At a European round table on coexistence, the DG-Research Commissioner agreed with Franz Fischler that it would make no sense to transform the coexistence issue into a debate about the environmental risks of GM agriculture: 'We must concentrate on the economic risks . . . and recognise that coexistence is inevitable and essential to ensuring freedom of choice for both farmers and consumers' (Cordis News, 2003). He emphasised the choice of GM as well as non-GM products.

Like Fischler, the agbiotech industry advocated that all responsibility lay with non-GM farmers, especially those doing 'quality' agriculture. EuropaBio favoured the market freedom of efficient industrial production.

> EuropaBio considers that growers who will benefit from a specific quality standard should not expect their neighbours to bear the special management costs of meeting that standard; to do so would reverse fundamental concepts of freedom of economic activity and would establish a dangerous precedent. To allow specialty operators to formulate unrealistic standards for GM in their own produce would impose impossibly high standards on other activities and would effectively bar competition and impose a ban on the choice of other producers. (EuropaBio, 2003)

However, that 'freedom of economic activity' would be restricted when the Fischler criterion was soon changed by fellow Commissioners.

1.4 Shifting Responsibility to GM Farmers

In July 2003 new Commission recommendations reiterated the need for national segregation measures 'to manage the possible accidental mixing (admixture) of GM and non-GM crops'. Such measures would depend upon the specific characteristics of the agri-environment and the crop. Member states should examine their liability laws for adaptation to any economic damage from adventitious presence of GM material (CEC, 2003a).

The Commission somewhat accommodated demands to shift the responsibility, as follows: 'farmers who introduce the new production type should bear the responsibility of implementing the actions necessary to limit admixture' (CEC, 2003a: 9). GM crops would be a new production type nearly everywhere in Europe, so this framework provided a basis to place the major burden on GM farmers, at least in the short term. Why hadn't the Fischler document allowed this option?

> It would be difficult to justify why a specific technology, and its safe products, should be punished with extra burdens. But it is defensible to burden those farmers who create the coexistence problem. This is a neutral way to handle the issue. (Commission staff, interview, 2004)

This 'neutral' way meant defeat for the agbiotech industry. Its main lobby reiterated the earlier stance about responsibility for a 'quality standard', though it acknowledged the responsibility of GM farmers to follow segregation guidelines: 'On the other hand, the question of compensation for economic loss in case of admixture must be considered where growers do not follow the guidelines for good agricultural practices' (EuropaBio, 2003).

The Commission recommendations implicitly extended a risk-management model from environmental to economic risk. They also extended earlier concepts of farm stewardship, especially concerning seed and crop purity: 'developing stewardship schemes and best practices for coexistence is a dynamic process that should leave room for improvement . . . ' (CEC, 2003a). Measures should be 'efficient and cost-effective and proportionate' to the aim: namely, how to maintain any GM presence below the statutory 'tolerance threshold' for GM labelling.

That labelling criterion was contentious. According to the Commission recommendations, GM farmers had a burden to keep any GM material below the 0.9% threshold for GM labelling and thus avoid any economic loss to non-GM crops. However, EU labelling law specified a threshold of 'adventitious presence', i.e. 'technically unavoidable' amounts. This phrase could mean that inherent difficulties should be pragmatically accepted as a basis for admixture; or else it could mean a duty to overcome the difficulties. According to the Commission, small amounts would be acceptable, provided they do not exceed the 0.9% threshold for GM labelling.

At around the same time, a statutory change potentially blurred the official distinction between economic and environmental issues. Under pressure from the European Parliament, the Commission finally agreed to amend the Deliberate Release Directive, so that 'Member states may take appropriate measures to avoid the unintended presence of GMOs in other products', under new Article 26a (EC, 2003a: 20). Article 22 requires member states to permit any GM product which has EU approval, yet the new Article could justify restrictions on grounds beyond environmental risk. Commission staff saw the outcome as an awkwardly 'mixed Directive', which could complicate the task of policing the distinction between economic and environmental issues: 'These articles are potentially contradictory, unless regulation is done in a balanced way' (interview, Commission staff, July 2005).

Indeed, various aims for segregation were potentially contradictory, as seen in responses to the Commission's coexistence recommendations. The agbiotech industry welcomed these for ensuring 'true choice' (EuropaBio, 2003). By contrast, environmental NGOs criticised the recommendations for unduly limiting stricter national rules, thus allowing some 'GM contamination'. There were calls to limit admixture as much as possible and to clarify who holds responsibility. According to consumer groups, thresholds for GM material are acceptable only within a system that 'guarantees that the contamination is really adventitious', i.e. a system that minimises admixture. Farmers would need special training for measures to achieve this aim (BEUC, 2003: 2).

Likewise, according to a European Parliament report: 'In contrast to the Commission's view, coexistence is certainly not limited to "economic aspects associated with the admixture of GM and non-GM crops"'. Coexistence measures should aim to exclude GMOs 'as far as is technically possible', not simply below the statutory threshold for a GM label. Liability rules should protect non-GM farmers, as an element of a coexistence regime (EP, 2003: 14, 15). Moreover, member states must have 'the right to prohibit completely the cultivation of GMOs in geographically restricted areas' (ibid.: 8). Similar views came from expert advice solicited by environmental NGOs (Lasok and Haynes, 2005; cited in FoEE, 2005c).

With this emphasis on lower thresholds and liability rules, some environmentalists implicitly changed their strategy. Earlier, they criticised coexistence as impossible, perhaps even undesirable. In a strategic shift, a Green MEP acted as rapporteur for the 2003 European Parliament report. Meanwhile the Green Group softened its overt opposition to GM crops:

> Without legislation for liability and coexistence, we oppose GMOs. Of course, if we accept coexistence, then we accept GMOs in some form, provided that all procedures are in place and respected. We have undergone a political evolution, from opposing GM crops to demanding stringent rules. . . . In the political evolution of the Green Group, MEPs from some countries—Netherlands, Spain, Germany—have relatively greater influence. This is important from a political viewpoint, so we propose the most stringent regulations possible. (EP Green Group advisor, interview, February 2004)

Thus 'coexistence' rules were being devised to make GM cultivation difficult, perhaps even impossible—in the name of protecting the freedom to choose conventional or organic crops.

2 EU-NATIONAL-REGIONAL CONFLICTS AS TEST CASES

For all the above reasons, 'coexistence' rules generated conflicts within member states, as well as between them and the Commission. Various organisations and authorities sought to restrict or deter GM crop cultivation, while promoting alternatives to agri-industrial production. Mainstream farmers' organisations had originally advocated market access to

GM seeds, but later they became more cautious—partly in response to pressures from their own organic sections, as well as concerns about economic liability and public suspicion.

According to Commission recommendations, appropriate coexistence rules depend on variations in agri-environmental conditions, e.g. the size and shape of farms in a region. In practice, however, political variation among countries seemed a more significant reason for diverse policies across Europe (interview, Commission staff member, July 2005). Indeed, political variations remained a major impetus for differences in coexistence rules, which shape the terms of competition among agricultural systems.

Coexistence rules were being justified, scrutinised or criticised according to different EC laws. Government authorities could cite the new Article 26a of the Deliberate Release Directive, authorising member states 'to avoid the unintended presence of GMOs'. Such draft rules became test cases of compatibility with EU law, especially Directive 98/34, which requires that any technical standards be proportionate to their purpose. It aims to avoid 'the creation of new barriers to the smooth functioning of the internal market' (EC, 1998c). On the basis of those criteria, the Commission accepted some national frameworks for coexistence, such as a Danish draft law (Toft, 2005).

However, the Commission rejected some other frameworks for disproportionately burdening GM farmers, especially by aiming to keep any admixture far below the EU threshold for GM labelling. Likewise, according to the agbiotech industry, such constraints 'would reverse fundamental freedoms of economic activity and would establish a dangerous precedent'; unrealistic standards for non-GM products 'would effectively pose a ban on the choice of other producers' (EuropaBio, 2005). Reinforcing this criticism, experts advised that such an excessive burden could mean substantial income loss for maize farmers and seed producers (DG-JRC, 2006).

According to the Commission, coexistence rules may not distort the internal market for products approved as safe; such rules must 'allow market forces to operate freely in compliance with Community legislation' (CEC, 2006b: 3). Yet any rules restrict some freedoms more than others, thus favouring one agri-paradigm over others. National and regional conflicts with Commission policy express contending agendas for future agriculture, as illustrated by the following examples.

2.1 Wales: Statutory Isolation Distances

In Wales the neoliberal agro-industrial paradigm has been seriously challenged by an alliance promoting 'quality' agri-production for rural development (Marsden and Sonnino, 2005). This alliance has sought to exclude GM crops. With its newly devolved powers, the National Assembly for Wales has taken as restrictive an approach as is possible within EU legislation, with the aim of keeping its products distinctive for marketing purposes. The Assembly expressed concerns that the Organic Standards Regulation 1804/1999, which sets minimum standards for organic production, was inconsistent with gene flow from GM crops (Oreszczyn, 2005).

This conflict took the form of segregation rules. In the UK an industry-wide body had already set standard isolation distances to limit gene flow to non-GM crops (SCIMAC, 1999). In 2001 Wales decided to put those standards onto a statutory basis, as a measure necessary to provide 'an environment where non-GM crops can be grown'. On its behalf, the UK invoked the safeguard clause of the Deliberate Release Directive. In response, the Commission sought the advice of its scientific advisory committee, which stated that the rules had no basis in environmental or health risks (SCP, 2001b). The Commission initially regarded Wales' rules as invalid but took no formal action, at a time when anyway no GM crops were approved for cultivation in the UK.

By 2003 the political context had changed in favour of Wales' rules:

> Under Article 26a [of the Directive], the UK could impose isolation distances. In fact, the July 2003 Commission recommendations considered the SCIMAC [1999] guidelines together with other input. (interview, Commission staff member, February 2004)

Those isolation distances varied according to gene flow from each crop, so the Commission could accept Wales' rules as proportionate to the admixture problem. At the same time, these rules aimed to deter GM crop cultivation, while favouring an agrarian-based rural development paradigm.

2.2 Upper Austria: GM-Free Areas?

In Austria agbiotech has symbolised a threat to quality agriculture, especially organic products. Even before GM crops became a high-profile issue there in the late 1990s, the government was promoting organic farming—as ecologically sound, as 'quality' products, and as an economically feasible market-niche for an endangered national agriculture. This 'competitiveness' scenario undermined the putative EU-wide imperative to increase agricultural productivity through agbiotech. Austrian regulators unfavourably compared potential environmental effects of GM crops to methods which use no agrochemicals, among other grounds to oppose commercial approval (Torgersen and Seifert, 2000).

In 2003 the regional jurisdiction of Upper Austria decided to establish 'GM-free agricultural areas' and notified its draft Act to the Commission under EU Treaty Article 95(5), the safeguard clause allowing national restrictions for environmental and health protection. Its draft Act aimed to protect agricultural products and natural biodiversity from 'GMO contamination', especially in 'sensitive ecological areas'. As a rationale, a 2002 study had noted the small size of farms and the 7% organic production there, which together would make coexistence nearly impossible, e.g. by requiring a 4km isolation distance from any GM cultivation.

In response, the Commission requested advice from the European Food Safety Authority. According to its experts, Austrian documents contained no new information regarding risks to human health or the environment, nor evidence to show that admixture concerns such risk issues; its advice explicitly excluded issues such as the management of coexistence (EFSA GMO Panel, 2004a). Citing that advice, the Commission rejected the restrictions on several grounds: that the Austrian concerns 'relate more to a socio-economic problem' than to environmental protection, and that a blanket ban would not be a proportionate measure (EC, 2003c).

The region refused to back down, thus defending its economic strategy within an agrarian-based rural development paradigm. According to the regional premier, the EC proposals 'may be suitable for large-scale agri-business, but they are certainly not for Upper Austria's small-scale farming' (*Agra Europe,* 05.09.03). Upper Austria brought a case to the European Court of First Instance, whose October 2005 decision supported the Commission stance. In December 2005 Upper Austria appealed to the European Court of Justice, which eventually reiterated the earlier judicial ruling.

2.3 Carinthia: Licencing System

In late 2003 Austria's regional jurisdiction of Carinthia drafted a 'precaution law on gene technology', especially to address coexistence. According to the draft, a farmer must request authorisation, describing the structure of local fields and their own isolation measures, before cultivating a GM crop. Then the local authority would decide whether to grant permission. If granted, then the authority would accept liability for any damage, provided that the farmer implements the agreed measures. The rules would allow the region to prohibit GM crops in 'ecologically sensitive areas', e.g. near nature reserves.

Carinthia notified its draft law to the Commission, which raised several concerns about future burdens on GM farmers and consequent restrictions on GM crops. According to the Commission, such restrictions must be justified by the characteristics of a specific crop, as well as by the aim to comply with Community legislation on GM thresholds, i.e. purity levels of GM content in non-GM crops in neighbouring farms, rather than thresholds lower than 0.9% (CEC, 2003b). Those criteria were unclear in the Carinthian proposal.

Carinthia redrafted the rules to accommodate some of the Commission's criticisms and gained a somewhat positive response. 'Although appreciating the quite significant improvements made to the draft initially notified, it was not yet possible to assess the proportionality of practical coexistence measures' (CEC, 2003b). 'Nevertheless the Commission accepted the proposal in principle, providing that additional conditions are fulfilled and that a balanced approach would be taken which would make GM farming possible' (interview, Commission staff, February 2004).

Carinthia became a test case for how stringent burdens upon GM farmers could be made possible, in ways compatible with EU law. GM crops were potentially deterred, in favour of an agrarian-based rural development paradigm. As policy advisors there noted, moreover, 'GMO-free zones might create a specific image for marketing regional products and services', such as tourism, thus broadening their economic relevance (Jank et al., 2006).

2.4 Germany: 'Peaceful Coexistence'?

In Germany the stakes for coexistence were raised by more intense conflict between the earlier pro-biotech policy versus proposals for alternative agricultures. In 2000 the BSE crisis in Germany provided a new opportunity for critics of the *Agrarfabriken,* or factory farming, a phrase pejoratively linking intensive agriculture with animal diseases. As a challenge to the agro-industrial paradigm, the government initiated the *Agrarwende,* (literally) turning agricultural policy towards consumer interests, informed choices and sustainable methods. Led by a Green Party politician, a new Ministry promoted organic agriculture as a model for more sustainable farming and aimed to increase its share to 20% within ten years. As a result, agbiotech came under increased pressure, as it was perceived as threatening future choices within the *Agrarwende* (Boschert and Gill, 2004).

Germany initially debated coexistence as a necessary means of safeguarding consumer choice. Later the emphasis shifted to producer choice, especially the need to avoid a 'war in the villages', e.g. court cases claiming economic damage from other farmers. In 2002 a German study sought ways to reconcile opposing interests, through a system for preventing and mediating conflicts among farmers. It regarded the Deliberate Release Directive as an appropriate statutory basis for such measures: 'permission to market a GMO may include an order to take measures to avoid property damage through pollination', as a specific condition of using and handling GM products (Öko-Institut, 2002a).

In transposing the Deliberate Release Directive, Germany built upon its new Article 26a, which said: 'Member states may take appropriate measures to avoid the unintended presence of GMOs in other products'. According to the new German law which took effect in 2005, coexistence would have three instruments: an obligation to take precautionary action to prevent 'negative material effects' of GMOs, in particular through 'good farming practice'; a site register informing local farmers about any GM cultivation; and a compensation scheme for conventional and organic farmers. As a government rationale for such measures, 'there will be little hope of arriving at a state of peaceful coexistence' if farmers remain under threat of economic loss and legal liability (cited in Boschert and Gill, 2004).

'Precaution' was also cited as grounds to minimize out-crossing, to safeguard agricultural and consumer choices, and to manage the 'peaceful coexistence' of different forms of agriculture and consumption. Under the new law, environmental parameters such as out-crossing would serve as dual indicators of environmental and economic damage, thus implicitly conflating them. Financial compensation would be due if GM presence meant that a food product could not be labelled in a way that it otherwise could be. The law establishes 'joint and several liability' for economic loss due to GM presence in non-GM crops, so that all local farmers could be held liable. (Boschert and Gill, 2005)

In the drafting stage the Commission requested clarification on possible conflicts with EU law. For example, if Germany later adopted a 'GM-free' label, then this could mean overly stringent measures to ensure no detectable GM presence, thus going beyond the scope of the July 2003 guidelines. Nevertheless, the Commission provisionally accepted the final law as proportionate, pending future developments.

The new German law intensified domestic conflicts between advocates of contending agri-paradigms. The law was attacked by the main farmers' union and the agbiotech industry for unduly constraining cultivation of GM crops. After the SPD-Green coalition lost power in 2005, the Christian-Democrat/Social Democrat coalition government sought to lighten potential burdens upon GM farmers.

2.5 Italy: *De Facto* Ban

Italian agbiotech opponents have sought to protect the agro-food chain as an environment for craft methods and local specialty products, known as *prodotti tipici*. In the 1990s the Italian government allocated subsidies to promote specialty products but foresaw these being displaced by GM crops. According to a Parliamentary report, the government must 'prevent Italian agriculture from becoming dependent on multinational companies due to the introduction of genetically manipulated seeds' (Terragni and Recchia, 1999).

Moreover, when local administrations apply EU legislation on sustainable agriculture, they should link these criteria with a requirement to use only non-GM materials, argued the Parliament. It adopted such arguments from Coltivatori Diretti, a million-strong union of mainly small-scale farmers, who saw GM crops as a threat (ibid.). Since then environmental NGOs, farmers' organisations and regional authorities have built a network seeking to exclude GM products from Italian agriculture.

Italy's policy favours non-GM and alternative agricultures, for reasons beyond the admixture problem. Under the 2003 Italian law implementing the EC Deliberate Release Directive 2001/18, GM crops must be kept compatible 'with the need to safeguard the agro-biodiversity of agricultural systems and the agricultural production chain, with particular reference to typical [local], biological, and high-quality products'. A 2005 decree linked experimental GMO field trials with potential negative impacts on the image of local products. The decree also regarded any shift of resources towards more efficient new production (e.g. GM crops) as a risk that should be evaluated, regarding any loss of competitiveness and markets for pre-existing crops (cited in Niespolo, 2005).

The admixture issue was used for a *de facto* blockage of GM crops. In 2005 the Agriculture Ministry issued an administrative decree that no GM crop cultivation would be permitted until each region established technical rules to ensure coexistence of conventional and organic crops. According to the Agriculture Minister, the decree would avert 'the risk of diffuse and uncontrolled contamination by GMOs' (FoEE, 2005d). In October 2005 the Commission sent Italy a warning that its law breached EC treaty obligations, though Italy took no steps to comply. By late 2006 there were still no technical rules that could lift the ban, which gained broad support. Overt dissent came mainly from Futuragra, which demands 'freedom of research and of GM cultivation' (*Agro-Food Biotech,* 12.12.05); favouring the agri-industrial paradigm, this cultural association represents relatively more industrialised farmers in northeast Italy.

Developments in 2006 illustrate how an agbiotech blockage gained support from most political parties during the Berlusconi government, despite their sharp divisions on other issues. Together they proposed to protect Italy's traditional agro-food heritage from genetic modification. The Italian 'GM-Free Coalition' met with officials from relevant Ministries, who all agreed on several proposals, especially to protect the identity of organic products from GM contamination, and to request Commission support for restrictive measures (Monitoraggio normativo OGM, 2006).

2.6 Spain: Contending Alliances

Spain has been the only EU member state with significant commercial cultivation of GM crops. GM maize is cultivated in regions where grain cooperatives store all maize together for sale as animal feed within Spain, with no segregation, so GM material does not lower the market value. From an economic standpoint, segregation measures would be needed only near organic maize cultivation. Exemplifying the agri-industrial paradigm, a Right-wing farmers' group has promoted GM crops as a means to enhance productivity, economic competitiveness and environmental benefits.

In 2004 the government adopted guidance on coexistence, based on expert advice about pollen flow. The seed producers association accepted these recommendations as compatible with their own guidelines on good practices for cultivating GM maize. A government Commission of Biovigilance was set up to ensure coexistence of GM crops with other types (Tàbara et al., 2004).

The government guidance came under criticism, especially for accepting any 'GM contamination' below the EU labelling threshold and for assigning inadequate responsibility to GM farmers. According to some farmers' groups, coexistence is impossible because segregation cannot be guaranteed. Some organic maize farmers claimed that GM contamination had already lowered the value of their crops (Greenpeace España, 2004). A later report documented ineffective control measures and widespread 'GM contamination' of organic maize (Greenpeace, 2006).

When the government drafted statutory rules on coexistence, more stringent ones were demanded by the Rural Platform, representing Left-wing farmers' unions, environmental groups, organic farming associations and rural development organisations. They proposed that GM 'contamination' be kept below 0.1%, i.e. the level of detectability, not simply below the 0.9% EU labelling threshold. They also proposed greater separation distances, an obligation upon GM farmers to inform neighbouring farms well in advance, and clear liability for GM contamination (Anon, 2005).

In that regard, organic farmers wanted statutory protection, not dependent upon bringing other farmers to court. EU and Spanish policy assumed that organic farmers could gain protection by such means if necessary. Yet they were generally unwilling to do so, lest this aggravate conflicts with neighbours. Most withdrew from organic farming in the eight years since Bt maize cultivation began in 1998. In practice, then, they had no 'free choice' (Binimelis, 2008).

Spain served as a special test case of Commission policy in its progress report. Spain had uniform national coexistence rules, no differential price for GM and non-GM animal feed, considerable cultivation of GM crops, yet with uneven adoption across Spanish regions. On that basis, coexistence rules were being put on trial for whether they impede the internal market:

> An uneven adoption rate of GM crop cultivation across Member States and regions does, therefore, not necessarily imply a distortion of the market. At present, the impact of differences in the approaches to coexistence on the internal market can not be sufficiently assessed. (CEC, 2006b: 3)

As tested by Spain, then, coexistence rules would not distort the internal market or impede safe GM crops—the main suspected crime, according to Commission policy. In complying with that policy, Spain's practices provoked domestic conflict with alternative agricultures.

2.7 Contending Market Freedoms

Given the various rules contradicting its recommendations, the Commission challenged some authorities for disproportionately burdening GM farmers. According to its first progress report on national plans for coexistence, 'National coexistence legislation should allow market forces to operate freely in compliance with [EU] legislation'. However, some 'proposed measures . . . appear to entail greater efforts for GM crop growers than necessary, which raises questions about the proportionality of certain measures'. The report warned against 'overly restrictive measures that go beyond the objective of ensuring coexistence, and which may make the cultivation of GM crops practically impossible' (CEC, 2006b: 6, 9).

In such cases, where national measures impede the internal market, the Commission could invalidate them:

> By the end of 2005, 20 items of draft legislation from seven Member States had been notified under Directive 98/34/EC laying down a procedure for the provision of information in the field of technical standards and regulations. In 10 of these cases the Commission considered that the notified measures could create obstacles to the free movement of goods. . . . (CEC, 2006b: 4)

In other cases, where governments did not notify their measures to the Commission, these would not be enforceable if challenged: 'failure to notify national measures under the appropriate procedure means that the measures cannot be invoked against third parties' (ibid.: 7). These warnings sought to gain compliance with Commission policy but had little apparent effect on national or regional measures.

The coexistence debate featured contending accounts of market freedom. The agbiotech industry distinguished among different segregation rules: 'while some member states have set in place reasonable science-based rules to achieve a fair co-existence regime, others have clearly developed disabling rules that are aimed at denying choice to farmers and consumers' (EuropaBio, 2006a). This conflict erupted starkly at an EU Presidency conference entitled 'Co-existence between GM and Non-GM Crops: Freedom to choose'. Some speakers defended Commission policy for ensuring that farmers would have all options, while others regarded GM crops per se as a threat. In the conference workshop on stakeholder viewpoints, health and environmental impacts featured prominently (Gaskell, 2006). After the conference, the agbiotech industry criticised the views of some participants, including the rapporteur of a European Parliament report:

> . . . comments by Mr Graefe Zu Baringdorf MEP that farmers are being 'forced' to use GM technology are simply wrong. It does however appear there are many here today who wish to deny any choice to farmers. (EuropaBio, 2006b)

In other words, they sought to deny farmers the choice of GM crops, by claiming to protect non-GM farmers from contamination and coercion.

Environmental NGOs invoked consumer choice in a different way. They sought to discredit the Commission for denying consumers the freedom to choose non-GM crops, especially GM-free organic products: Under the Commission policy, 'consumer confidence in organic food will be seriously undermined' and so would 'EU policy to expand this sector of agriculture' (FoEE, 2006b: 5).

Moreover, the Commission was trying to bully 'any country or region that wants to defend the right of farmers and consumers not to plant GMOs or eat GM food', argued Greenpeace (Lewis, 2006b). By allowing GMOs in quantities up to 0.9%, the Commission had an irresponsible, coercive policy—'permitted contamination', as environmental NGOs called it (FoEE, 2006b). They advocated stringent measures to segregate GM crops and thus prevent any unavoidable 'contamination': agro-industrial operators must make any effort for minimisation, they have a moral responsibility to demonstrate such efforts, and segregation rules should keep GM admixture as low as possible.

Thus the 'freedom to choose' debate had contending accounts of market freedom versus coercion, each justifying different measures. The overall debate was framed by the EU internal market. According to a critic, this project seeks a single universally valid environmental risk assessment for each GMO, and expert advice tends to incorporate that policy: 'this artificial uniformity imposed by science is the unacknowledged consequence of the entrenched commitment to a single market for all member states, with a corresponding single regulatory system, thus uniform EU risk assessment' (Wynne, 2006: 31). In the Commission's coexistence policy, however, that political logic also ran in the other direction: any restrictions must be justified as proportionate to the case, e.g. specific characteristics of a crop and its commercial context. Thus anti-agbiotech attitudes generated claims about local crops or environments being specially vulnerable to 'GM contamination.'

3 'GM FREE' AS A REGIONAL BRAND

As the debate intensified over which segregation measures would be legitimate for 'coexistence', greater restrictions on GM crops were driven by an alternative agricultural model, which soon became more explicit. Since the 1990s agbiotech had been promoted as 'sustainable agriculture' through greater eco-efficiency, as well as being attacked as 'unsustainable', especially by contrast to 'quality' agri-production. By 2005 more policy actors were framing the admixture issue along the latter lines, especially to deter GM crop cultivation.

At the 2006 EU Presidency conference on coexistence, mentioned above, the rapporteur contrasted traditional quality agriculture with agbiotech innovation:

> The competitiveness of EU agriculture is becoming increasingly important for the economic well-being of farmers and for the development of rural areas. Many participants have stressed the importance of quality production, including products linked to traditional practices

> and geographical origin in order to safeguard the European model of agriculture, with its balance of socio-economic, environmental and territorial aspects. Others have emphasized the need to create a culture of innovation and to put science, including biotechnology, at the service of agriculture. (Austrian Presidency, 2006, Conclusions)

Yet alternative agricultures too depend on an innovation culture, as the following story shows.

After Commission policy assigned responsibility for segregation to the 'new production type' in mid-2003, demands came for greater protection of non-GM crops. In November 2003 ten regional governments asked the Commission to agree that they could define their own territory as a 'GM-free zone' and to assign liability according to 'the polluter-pays principle' (Anon, 2003; FoEE, 2003c: 15; 2003d). In response, Franz Fischler warned that regions cannot prohibit GM crops unless they provide scientific evidence that contamination of conventional crops is unavoidable (Spinart, 2003).

The regional plans provoked arguments about whether they were politically motivated—indeed, about how to distinguish between political, economic and scientific reasons. According to DG-Environment, 'the request for establishing GM-free zones is very often driven more by ideological concerns than by an objective assessment of the risks involved' (ibid.). By contrast, 'This discussion [on coexistence] is not about ideology but about practice and economics', according to Graefe zu Baringdorf, a Green MEP and vice-chair of the European Parliament's Agriculture Committee (Anon, 2003).

His more restrictive proposals gained support from mainstream politicians and farmer organisations, seeking to protect the market value of non-GM crops. Mainstream farmers originally advocated market access for GM seeds, but some became more cautious about market insecurities, given the threat of legal liability for GM contamination and consumer suspicion. Recognising this economic uncertainty, a UK advisory body stated that crop cultivation must go with the grain of the future direction for farming: reconnecting farmers to the national and international marketplace and a strong shift in the direction of enhancing the farmland environment' (AEBC, 2003: 12).

Within an agrarian-based rural development paradigm, alternative agricultures were increasingly counterposed to agbiotech. At their founding conference in Berlin in January 2005, several regional authorities linked 'GMO-free zones' with food sovereignty, 'quality' labels on food products and regional-cultural biodiversity. With the slogan, 'Our Land, our Future, our Europe', their charter identified GM crops as a threat to 'sustainable and organic farming and regional marketing priorities for their rural development' (FFA, 2005). In particular:

> Most European regions have made the promotion of sustainable and organic farming and regional marketing priorities for their rural development [. . .]. Most Europeans don't want GM-food. To serve

> this demand is part of a region's food sovereignty and an important economic chance. Regional authorities must be able to protect quality labels, purity standards, organic production and designations of origin at competitive prices. (FFA, 2005)

At a subsequent Florence conference, speakers more explicitly promoted a geopolitical alliance for a 'sustainable' future agriculture. As conference host, the Tuscany Regional President linked the precautionary principle, zero tolerance for the presence of GMOs, and uncertainty about their compatibility with environmental protection:

> We wish to avoid any standardisation of products which no longer have anything to do with their place of production. In Europe there must be room for a model of agriculture which is based on a genuine identity, cultural characteristics, high-quality GMO-free products. (AER/FoEE, 2005)

The Florence conference resulted in 'The charter of regions and local authorities of Europe on the coexistence of GMOs and conventional and organic crops', which in turn started the Network of GMO-free Regions. According to the charter, specific 'coexistence' plans would be based on in-depth feasibility studies examining the environmental, socio-economic and cultural impact of GMOs. Areas could be designated as 'GMO free' in order to protect any added value of certified quality products.

A larger conference broadened the network for alternative futures—now counterposed to agri-industrial methods, including GM crops. Entitled 'Safeguarding Sustainable European Agriculture: Coexistence, GMO free zones and the promotion of quality food produce in Europe', the conference was sponsored by the Assembly of European Regions and Friends of the Earth Europe. It aimed 'to define the most appropriate EU legal framework for an efficient coexistence regime' (AER/FoEE, 2005). Moreover, local environments were framed as cultural-economic assets under threat from GMOs. In their declaration, the organisers sought:

> To allow regions to determine their own agricultural development strategy, including the preservation and development of regionally adapted genetic resources and the right to prohibit GMO cultivation. (ibid.)

Numerous regional representatives elaborated that agri-development theme, by describing their 'natural' environment or special cultivation methods as a basis to market local products and services. According to a speaker from southwest England, for example, their local authority is committed to 'treating the environment as a highly valuable capital asset to be managed intelligently for long-term economic benefit' (FoEE, 2005e). According to a report on the conference, the speakers had explained 'how their local specialised agriculture was a precious resource that plays a vital role in marketing their region' (ibid.: 15).

An even larger conference promoted 'GMO-free regions' to a broader public during International Green Week in January 2006 in Berlin. By this time, 'GMO-free' declarations had come from more than 160 regions, 3500 municipalities and local authorities, and tens of thousands of farmers in Europe (see Figure 8.1). According to the GMO-free Network, regions were reclaiming their rights to local and regional self-determination—with regard to their landscapes, eco-systems, agricultural practices, food

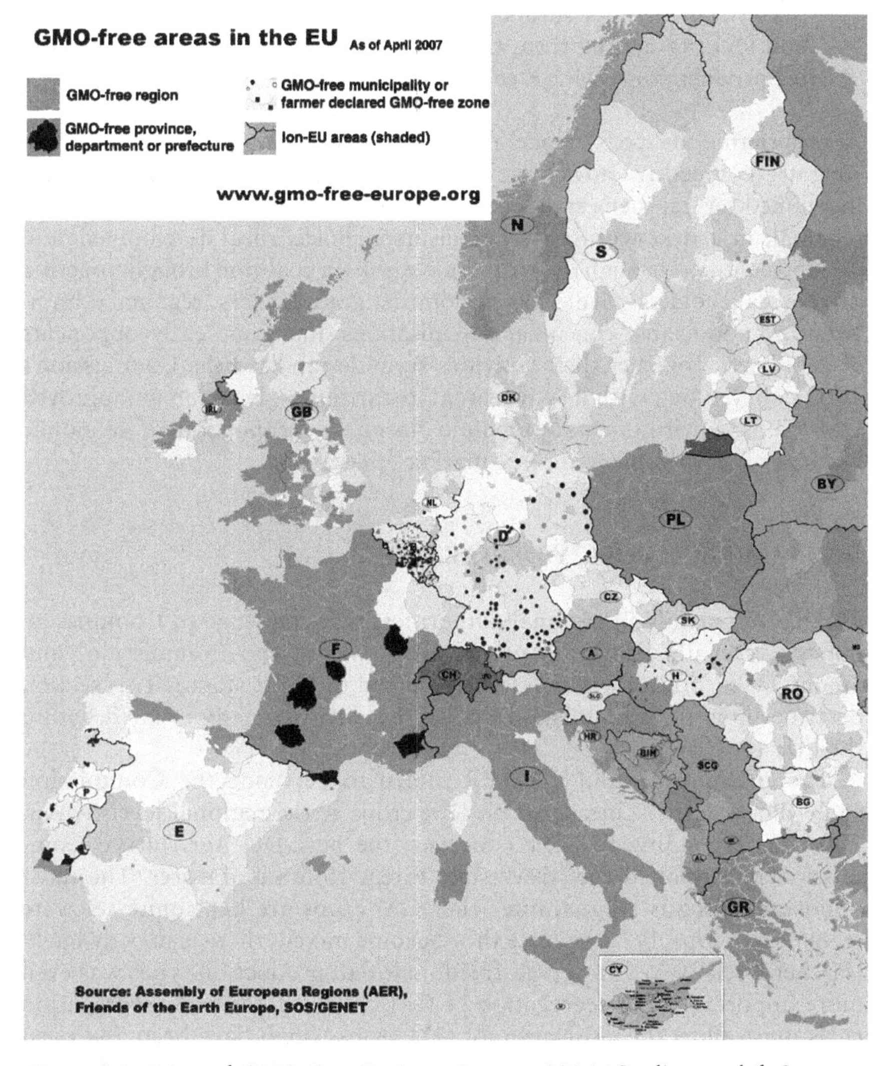

Figure 8.1 Map of GMO-Free Regions, January 2006 (Credit: copyleft Save our Seeds/GENET).

traditions and future economic development. Moreover, farmers and food processors cooperated to find sources of non-GM animal feed for their GM-free animal products, thus increasing the pressures to segregate distinctive markets for grain (AER, 2006).

The previous 'GMO-free' conferences were followed up with a stronger emphasis on alternatives for rural development, in particular:

> The erosion of biodiversity, independent farming, and regional quality food production arising from the exclusive control of seed by fewer and fewer companies.
>
> The challenges for rural development, biodiversity, food culture and food security arising from the global competition between food and fuel production. (GMO-Free Regions, 2007)

Appropriating the 'coexistence' metaphor, then, a new discourse-coalition sought a competitive advantage for alternative agricultures. Their storyline linked several themes: precaution, environmental risk, socio-economic regionalism, market competition, consumer choice, rural development and *paysan* identity. Through these discursive links, a coalition brought together diverse stakeholders: farmers, agronomists, grain traders, regional administrators, politicians, consumer organisations and some early opponents of agbiotech. Together their activities soon destabilised the Commission's 2003 policy, along with its conceptual distinctions—between un/approved GM crops, environmental/economic harm, dis/proportionate economic burdens on GM farmers, un/free market forces, etc.

4 COMMISSION POLICY DESTABILISED

Originating mainly in regional authorities, that challenge to Commission policy eventually generated overt splits—within the Commission, and among EU institutions, which largely echoed the AER stance. This section describes how the Commission's policy became gradually isolated within the EU.

The coalition around the AER aimed to persuade the Commission to permit restrictive measures on GM crops at the regional level (Anon, 2005a). Responding to their concerns, the new DG-Agriculture Commissioner mentioned the irreversible threat from admixture: 'The most important thing is to guarantee that GM crops are kept quite separate from other crops, because once they become mixed, there is no way back' (Fischer Boel, 2005). However, freedom for all production types was even more important than segregation of GM crops; in her view, segregation rules must allow the cultivation of GM crops which have been approved as safe:

> Where a product has been shown not to be harmful, in principle the rules of the free internal EU market apply. So, also, do WTO rules, as we have seen. The debate on co-existence must be about ensuring co-existence, not preventing it . . . co-existence policy is not about the safety of people, animals or the environment. (Fischer Boel, 2006)

For such reasons, authorities must respect the GM labelling threshold as the basis for segregation measures: 'The use of GMOs in organic produce is not acceptable, but we have a standard contamination threshold of 0.9% that is in line with the real world', argued Fischer Boel (Anon, 2006a). She dismissed demands for lower thresholds, especially zero tolerance for organic crops, by invoking objective reality, thus implicitly promoting a particular future.

However, a fellow Commissioner acknowledged uncertainties in risk assessment of GM products, as grounds to blur the official distinction between environmental risk and economic loss:

> As an Environment Commissioner, I am keen to ensure that the environment is protected from potential risks arising from the cultivation of GMOs. Coexistence measures, on top of the benefits they provide in purely commercial terms, can play a role in this respect. (Dimas, 2006)

On such grounds coexistence policy should also protect the environment 'from the potential risks arising from the cultivation of GMOs' (Anon, 2006b).

Thus, when a Commissioner attempted to put regional authorities on the defensive for disproportionate restrictions on GM crops, her effort was undermined. The issues could not readily be contained within the Commission's sharp distinction between legislative regimes for environmental risk and economic loss, especially given that some regional authorities had wider reasons for deterring GM crops. Some national authorities were considering environmental issues within coexistence rules, as well as including segregation issues within risk regulation procedures (Levidow and Carr, 2007b).

Sooner or later, other EU institutions criticised Commission policy and destabilised its key distinctions. Early on, the European Parliament had challenged the official distinction between environmental versus economic issues, as well as between approved and unapproved GM crops (EP, 2003, cited earlier). In 2004 the Economic and Social Committee likewise did so, as grounds for 'coexistence' measures to exclude GM crops from specific areas, thus echoing Austria's stance:

> . . . measures to protect nature conservation areas in line with Directive 92/43/EEC on habitats, flora and fauna and Directive 79/409/EEC on the protection of birds, and other ecologically sensitive areas, should be regulated at national and local level. . . . (CESE, 2004)

It also justified bans in cases where GM crops makes traditional production of plants 'impossible or unduly difficult' (ibid.).

The EU's Committee of Regions (CoR), which advises the Commission on subsidiarity policy, eventually took up the coexistence issue with a conference in 2005. The following year its statement echoed national and regional criticisms of Commission policy:

> Conventional and organic crops, too, are part of the environment and are therefore to be protected under the precautionary principle. . . .
>
> European legislation consistently aims to impose constant vigilance regarding the potential risk to human health and the environment, and so an approach which bases the method of coexistence exclusively on economic aspects appears contradictory. . . .
>
> The precautionary principle has its origins in that of sustainable development . . .
>
> . . . the European Economic and Social Committee recommended that the growing of GM crops be banned when it makes traditional production of plants of the same or related cultures impossible or unduly difficult . . .
>
> . . . a voluntary or regionally restricted renunciation of GMO cultivation may be the most effective and least costly measure for guaranteeing coexistence. (CoR, 2006: 2, 3, 10)

Thus the Commission's official separation of economic from environmental aspects was criticised as contrary to precaution. With this framing, the CoR justified product bans and 'GM-free zones'.

With its general remit for subsidiarity, the CoR had been long seen as a key pillar of EU democracy. In its submission to the EU's 1996 Institutional Review, the CoR had proposed that subsidiarity policy should designate local and regional bodies as primary units of European democratic culture. As the CoR's rationale for a role in decision-making, subsidiarity 'avoids the creation of an excessively centralized European power disconnected from the problems of ordinary citizens, the closeness of the Union to its citizens being one of the basic components of this legitimacy' (CoR, 1995). Under its proposal, the CoR would have an institutional status equivalent to the Parliament, though this proposal was not accepted (McGrew, 1997: 181–82).

According to the CoR's self-description:

> Established in 1994, the CoR was set up to address two main issues. Firstly, about three quarters of EU legislation is implemented at local or regional level, so it makes sense for local and regional representatives to have a say in the development of new EU laws. Secondly, there were concerns that the public was being left behind as the EU steamed ahead. Involving the elected level of government closest to the citizens was one way of closing the gap. (CoR, 2007)

The EU's Committee of Regions had always maintained a distance from pro-agbiotech policies. When the Commission published its *Strategy for Life Sciences and Biotechnology* (CEC, 2002b), the Committee of Regions took no position. Nevertheless, according to the Commission, all the European institutions supported its policy (CEC, 2003d: 8).

As the Commission steamed ahead with an internal market for GM products, citizens were indeed left behind, seeking alternative fora to represent their views. As a defiant response, since 2003 more regional authorities were collectively declaring 'GM-free zones', promoting 'quality' agriculture, creating an alternative European identity. This *de facto* subsidiarity was being led by an autonomous non-EU body, the Assembly of European Regions; its alternative project overtly challenged the internal market, its decision procedures, its legitimacy and thus the official form of European democracy.

5 CONCLUSIONS

At the start, this chapter posed the following questions:

- Why did the admixture (or 'contamination') problem become so contentious?
- How were various coexistence policies put on trial?
- How did 'coexistence' rules relate to different agricultural futures?

Since the late 1990s, inadvertent mixing of GM material with non-GM products has posed an economic problem: GM contamination potentially lowers the price of 'GM' grain or food, so commercial operators could lose income. From their apocalyptic frame, agbiotech opponents turned the admixture problem into an extra risk issue: GM crops would spread 'genetic pollution', thus endangering the environment and non-GM crops. The stakes were raised by actual mixtures of illegal GM material with conventional seeds in 2000, as well as by 2001 draft EU rules for 'GM' labelling, which covered a broader range of products than before. These rules introduced a new policy concept: 'adventitious' presence meant 'technically unavoidable', not simply an accidental presence of GM material. It implied a moral-legal duty upon commercial operators to demonstrate their efforts to exclude any GM material, as grounds for why a 'GM' label should not be required (see Chapter 7).

As member states were devising or demanding segregation measures, the European Commission addressed the admixture problem with a new policy: 'coexistence' of GM with conventional and organic crops. This aimed to ensure that farmers could freely choose among those production systems, which would develop side by side. Segregation measures would be justified to ensure that any 'adventitious presence' of GM material remains below the 0.9% threshold for labelling products as GM. Practical measures to achieve segregation were delegated to member states. Their rules could assign a burden of responsibility for segregation, as well as liability measures to compensate non-GM farmers for any economic loss.

Commission policy favoured a specific form of market freedom, by justifying only those restrictions necessary to maintain GM material below the labelling threshold and thus to stabilise an internal market. On that basis, coexistence rules would ensure that 'market forces can operate freely' for different crop systems, especially with free circulation of GM products authorised for commercial use. With this coexistence policy, the Commission extended its earlier managerialist frame to an agri-diversity policy, protecting a range of agricultural options.

Yet 'freedom to choose' was framed in contending ways, especially for assigning responsibility for segregation measures (see Table 8.1). Accommodating widespread demands, the Commission eventually accepted that the extra economic burden of segregation measures should fall upon 'the new production type', generally meaning GM farmers. Any extra burden had to be justified as necessary to maintain GM material below the statutory labelling threshold and thus to protect the EU internal market; otherwise such burdens would be illegitimately interfering in the market.

For agbiotech opponents, however, free choice meant total protection from GM material. In their view, GM products posed an uncontrollable risk of 'genetic contamination', as well as an existential threat to alternative agricultures. They denounced 'coexistence' as impossible, though this overt opposition was later softened, in favour of more stringent rules that would effectively preclude GM crops.

Proposed segregation rules were criticised for denying free choice, from two opposite standpoints. In some regions and member states, draft rules sought to keep GM levels as low as possible, on grounds that 'adventitious' should mean only those admixtures which are technically unavoidable. Uncertain environmental risks were cited as precautionary grounds for such measures. The Commission rejected such rules for imposing a disproportionate burden and thus denying market freedom to GM crops. The Commission in turn was accused of coercion, by forcing farmers and consumers to accept 'GM contamination'.

Within the Commission's coexistence policy, agbiotech would be co-produced along with specific forms of nature and society. Its policy framework linked a safe biotechnologised nature, EU expert authority for product safety, a European internal market for safe products, a dominant agri-industrial production system, and segregation measures to keep GM material below the 0.9% labelling threshold. This policy framework depended upon clear, stable distinctions: between GM products approved as safe and those not yet approved; between environmental risk within agbiotech risk regulation and merely economic loss within coexistence policy; between 'proportionate' segregation measures and unjustifiably more stringent ones.

However, those distinctions were blurred by alternative policies. Some regional and national authorities proposed extra restrictions on GM crops; as a rationale, they cited special vulnerabilities of local agri-environments or indigenous land races of crop varieties. As the first overt challenge in the legislative arena, in 2003 Upper Austria drafted a law establishing 'GM-free agricultural areas'; this would protect agricultural products and natural biodiversity from 'GMO contamination', especially in 'sensitive ecological areas'. When the Commission rejected the draft law as incompatible with the internal market, Upper Austria brought a court case but received an adverse ruling.

Some other regional authorities devised coexistence rules that could deter GM crops, albeit in the name of 'coexistence'. These rules effectively put GM crops 'on trial'—e.g. by requiring prior evidence that their cultivation would not contaminate other crops, or by strongly penalising any contamination through strict liability. The Commission scrutinised these rules for possible incompatibility with operator freedom in the internal market. The same draft rules were defended as market freedom or else denounced as coercion, thus undergoing informal trials from two opposite standpoints.

A more explicit challenge to Commission policy came from a 'GM-free network'. This was led by an autonomous non-EU body, the Assembly of European Regions, in alliance with environmentalist and consumer groups. They aimed to protect local genetic resources, other environmental assets and regional marketing strategies for high-quality alternative agricultural products. Agbiotech was framed as a threat, due to economic competition or simply the stigma of any nearby GM crops.

Quality alternatives illustrate a general strategy of territoriality, whereby a locality builds upon a distinctive history and identity. This involves 'positive relationships between the three components whose reciprocal interactions produce territory: the natural environment, the urban environment, and the social and human environment'. In this strategy, agricultural producers expand their roles towards the production of common goods, e.g. hydro-geological conservation, land reclamation, rural tourism, craftsmanship, etc. (Magnaghi, 2005: 83, 87).

Dissociation from agbiotech symbolically reinforced the high-quality status of local food, while promoting an agrarian-based rural development (Levidow and Boschert, 2008). This created alternative European identities through local agri-food and environmental diversity—in conflict with the European integration project of regulatory harmonisation for standard commodities. 'GM-free zones' implicitly meant 'EU-free zones', defying the internal market for agbiotech products. Those initiatives challenged the European integration project, its decision procedures and its legitimacy, at least for GM products.

The Commission attempted to put regional authorities on the defensive for 'disproportionate' restrictions on GM crops. However, such measures were eventually defended by other EU institutions—the European

Parliament, the Economic and Social Committee, and eventually the Committee of Regions, which officially advises the EU on subsidiarity issues. The Commission eventually faced overt internal dissent over its coexistence policy. Legally speaking, it could have brought court cases against several member states for breaching their treaty obligations, but they were not put formally on trial.

Rather, 'GM contamination' was further put on trial, along with the Commission's coexistence policy, for denying a free choice of non-GM products, especially quality agriculture. This multi-level trial expressed a wider European contest over possible future agricultures (see Table 8.1). Any coexistence rules would limit the choices of some farmers more than others. Coexistence became a new battlefield for long-standing rivalry between an agrarian-based rural development versus a neoliberal agri-industrial paradigm (cf. Marsden and Sonnino, 2005). With its 'coexistence' policy, the Commission had appropriated a Cold War metaphor as a means to avoid, contain or mediate conflicts among farming systems. Yet contending policy agendas shaped the official coexistence framework and kept it contentious.

9 Conclusion
Testing European Democracy

The European agbiotech controversy has been played out through multi-level, multi-arena, overlapping trials. Together these undermined policies favourable to the commercialisation of GM products. Agbiotech became a focus for contesting claims about European 'democracy'. Europe was told that it had no choice but to accept agbiotech, yet this imperative was turned into a test of democratic accountability for societal choices. This conflict expressed legitimacy problems of European integration.

To explore those dynamics, the Introduction posed several questions under four headings—trial roles, policies, democratic legitimacy, and broader lessons learned—which structure this concluding chapter.

1 TRIAL ROLES AND TYPES

Questions

- Who and what have been put on trial?
- What accusations have been made against the defendants?
- What types of trial have they undergone?

Since the 1980s agbiotech has been promoted as a symbol of European progress and political-economic integration. According to proponents, agbiotech provided a clean technology for enhancing eco-efficient agri-production; this would fulfil the beneficent promise of a European Biosociety, by analogy to the 'Information Society', likewise based on technological progress. By the early 1990s biotech further epitomised promises of a 'knowledge-based society', promoting capital-intensive innovation as essential for economic competitiveness and thus European prosperity. Through such efforts, Europe would become 'the most competitive and dynamic, knowledge-based economy in the world', declared the EU Council at its 2000 Lisbon summit.

By then, however, agbiotech was being turned into a symbol of anxiety about multiple threats: the food chain, agri-industrial methods, their inherent hazards, global market competition, state irresponsibility and political unaccountability through globalisation. By referring to various food scandals, especially the BSE crisis, 'GM food' was negatively associated with factory farming, its health hazards and unsustainable agriculture. Since the late 1990s the phrase 'GM-free' has become a common reference point for positive social identities; these link appeals to European cultural diversity, agri-food quality and democratic sovereignty.

An opposition movement grew by encompassing diverse issues, arenas and constituencies. New alliances linked agricultural, environmental, consumer and trade issues, thus going beyond the normal remits of individual NGOs. Expert bodies are generally limited to specialist scientists, but agbiotech critics successfully entered this arena by challenging official expert advice—for scientific ignorance, political bias and narrow accounts of environmental harm. Opposition strategies mobilised cultural resources across all relevant arenas—judicial, regulatory, expert, experimental and commercial—in ways bypassing the implicit rules which normally limit access to those arenas.

The agbiotech controversy often gained large public audiences through the mass media, as well as active involvement of many civil society groups. They took up ideas from small activist groups as well as from high-profile campaigns of large NGOs. Together these activities had several roles: testing models of citizen roles and public representation, developing citizens' capacities to challenge official claims, creating civil society networks to which governments could be held accountable through various trials.

These dynamics continuously expanded trials, defendants and arenas—what was put on trial, how, where and by whom. Measures promoting, facilitating or restricting agbiotech have undergone trials of many kinds. In a familiar sense of the word, formal trials tested evidence so that an official agency could reach a decision, e.g. in regulatory, expert or judicial procedures. Informal trials symbolically tested meanings—e.g. through discourses of blame, irresponsibility, threats, etc. Such informal-symbolic trials drove, shaped and permeated formal trials. Each issue featured role-reversals: an accusation or accuser was often put on the defensive. Such reversals happened around many issues and trials, as shown in Table 9.1. Arrows indicate a response to another text: x → y means that y results from (or responds to) x, while x ← y means the reverse.

As an early example, a high-profile controversy arose over the 1988 Commission proposal for an EC Directive authorising broader patent rights for 'biotechnological inventions'. This accommodated industry complaints that inadequate patent protection was deterring R&D investment. For proponents of the Directive, the free exchange and commercial use of GM material would be biopiracy, depriving inventors of their rewards. Critics reversed the accusation, so that biopiracy meant 'Patents on Life', i.e. privatisation of mere discoveries or common resources. After the Patents Directive was finally enacted in 1998, it was challenged by some member states—and was not transposed into national law by many more, so the European Commission brought court actions against them. This formal

Table 9.1 Trials, Tests and Role-Reversals

Issue	*Formal Trial or Test* (Judicial, experimental, expert, regulatory)	*Role-Reversal* (Reversal of accusation, accuser or defendant)
Patent rights (Ch. 2)	After EC Patent Directive is enacted in 1998, it is challenged by some member states and not transposed by many more, so the Commission brings court cases against them.	Agbiotech industry demands protection from biopiracy through patents for 'biotechnological inventions'. → NGOs denounce the 1998 Directive for promoting biopiracy through 'Patents on Life'.
Government ir/responsibility (Ch. 3)	Activists are prosecuted for sabotage of agbiotech development, e.g. GM field trials, seed stocks, lab experiments, etc.	→ NGOs use the opportunity to put the state on the defensive for irresponsibility and undemocratic decisions. Greenpeace UK publishes *GM on Trial*, a critique of safety science.
1990 EC Directive (Chs 2 and 5)	In public hearings in the early 1990s, the 1990 EC Directive is criticised for undermining European economic competitiveness, as grounds for easing approval procedures and shifting GMOs to product legislation (Ch. 2).	In the late 1990s the Directive is criticised (by NGOs and some member states) for defining risks too narrowly, thus warranting a revision along broader, more precautionary lines, as subsequently done in 2001 (Ch. 5).
Regulatory approval decisions and procedure (Chs 3, 5 & 6)	NGOs bring case against French government for failure to apply precautionary principle in approving Bt 176. *Conseil d'Etat* initially accepts that plaintiffs may have a valid case (Ch. 3). When French government bans MON 810, agbiotech proponents challenge the decision at the *Conseil d'Etat*, but the challenge fails (Ch. 6).	NGOs are criticised for politicising scientific issues. → NGOs accuse the Commission of approving GM products for political motives, without adequate scientific evidence or a democratic mandate. Politicians criticise EU comitology procedure for a 'democratic deficit' (Ch. 5).
Product safety: health risks (Ch. 5)	Novel transgenes are routinely tested on animals in the lab; results are cited as evidence of safety. After testing GM potatoes on lab rats, however, Pusztai obtains abnormal results and publicises them as grounds to question the safety of GM food.	→ Pusztai is dismissed by his employer and is subjected to character assassination. His experimental methods are criticised by a Royal Society report. → NGOs accuse Pusztai's employer (and critics) of commercial and political motives.

(continued)

Table 9.1 (continued)

Issue	*Formal Trial or Test* (Judicial, experimental, expert, regulatory)	*Role-Reversal* (Reversal of accusation, accuser or defendant)
Substantial equivalence (Ch. 5)	After Italy bans several foods from GM maize, on grounds that they do not qualify as substantially equivalent, the Commission asks Italy to lift its ban and requests support from other member states.	→ Member states side with Italy against the Commission, while also challenging the regulatory short-cut for safety approval based on substantial equivalence.
Lay expertise (Ch. 4)	Participatory TA exercises consider agbiotech by evaluating, anticipating or simulating advisory expertise, while marginalising questions about innovation pathways and alternatives.	These exercises test models of public representation and citizenship, especially the capacity of ordinary people for lay expertise.
Product safety: environmental risks (Ch. 5)	Bt toxins are routinely tested on non-target insects in the lab, whose results are cited as evidence of safety in the field. In a special experiment, however, harm is found when lacewing are fed on insects which had ingested Bt maize.	→ Lacewing experiment is criticised (especially by biotech proponents) for failing to simulate realistic conditions. → In response, routine 'safety' evidence is criticised for similar weaknesses and poor methods.
Environmental effects beyond product safety (Ch. 6)	UK's Farm-Scale Evaluations compare environmental effects of farmers' herbicide usage on GM and non-GM crops. Belgian authority questions the plausibility of company plans for farmers to eliminate herbicide-tolerant weeds.	EU initially approves GM crops as safe, as if farmer practices were irrelevant to any environmental risks. → ← Through ominous metaphors (treadmill, sterilisation, pollution), NGOs generate controversy over broader hazards which depend on operator behaviour.
Expert advice (Ch. 5)	EFSA opinions generally deny that member states have any grounds for objecting to safety claims or for demanding more rigorous evidence of safety. Thus EFSA acts as if it were a High Court, putting member states on trial.	→ Member states increasingly criticise EFSA for inadequate evidence for safety claims. Facing isolation, the Commission asks EFSA to justify its rejection of national criticisms, and even to re-do its own risk assessments.
National bans or restrictions (Ch. 5)	Commission proposals are put to a vote in EU regulatory committee. Few member states vote in favour, with even two-thirds voting against the Commission, thus weakening its political authority to bring court cases against national bans.	Several member states ban GM products after EU approval. ← Commission repeatedly demands that member states lift their bans on GM products.

WTO versus EU (Ch. 5)	US government leads a WTO dispute against the EU for an 'illegal moratorium' and for national bans on GM products. Although the EU is formally on trial, it proposes arrangements which could open up expert disagreements and so question official safety assessments.	→ NGOs attack the WTO case as a threat to democracy and consumer rights. The Commission reminds WTO that its legitimacy will depend upon respect for the democratic sovereignty of governments to establish prudent regulation.
Labelling rules (Ch. 7)	Commission initially dismisses demands for mandatory labelling—on grounds that such a requirement would have no scientific basis, would jeopardise the internal market and would provoke a US challenge via the WTO.	→ NGOs criticise retailers and governments for denying consumers a free, informed choice. → Food retailers exclude GM grain from their own-brand products in the late 1990s, thus testing (and undermining) claims for economic competitiveness of agbiotech.
Non/labelling practices (Ch. 7)	European retail-food industry devises its own criteria for whether to label food products as 'GM', in lieu of clear statutory rules in the late 1990s.	→ NGOs do surveillance of food products for GM material and alert consumers about non-labelled products, thus increasing the pressure on the industry to use broader criteria or to exclude GM ingredients.
Coexistence policy (Ch. 8)	After the Commission disqualifies Upper Austria's draft law for blocking GM crops, the region brings a court challenge but obtains an adverse ruling.	Under some draft coexistence laws, GM crops would be put on trial, e.g. by requiring prior assurance that they would not contaminate other crops. → The Commission challenges such rules as a 'disproportionate' burden on GM crops.

Note: Various measures—for promoting, facilitating or restricting agbiotech—have undergone trials of many types. 'Formal trial' means an organisational procedure which tests claims in order to reach a judgement. In parallel, informal trials symbolically tested meanings, e.g. through discourses of promise versus threat. For each issue, trial roles were often reversed, as shown in the right-hand column. Arrows indicate a response to another test: x → y means that y results from (or responds to) x, while x ← y means the reverse.

trial could not resolve the legitimacy problems of patents. Meanwhile broader patent rights raised suspicion among civil society groups, e.g. trade unions of professional workers, who were otherwise inclined to regard agbiotech as societal progress. Thus agbiotech was kept symbolically on trial for biopiracy (Chapter 2; Table 9.1, row 2).

Other controversial issues likewise featured informal-symbolic trials, formal trials and role-reversals. Many of those trials can be grouped around three main themes—safety versus precaution, eco-efficiency versus agri-industrial hazards and globalisation versus democratic sovereignty—as elaborated in the rest of this section.

1.1 Safety Claims Versus Precaution

Lab and field trials were intended to generate evidence of product safety, thus demonstrating a scientific basis for expert risk assessments, which in turn could justify commercial authorisation. Yet safety science became contentious. Expert safety claims underwent criticism for bias, ignorance and optimistic assumptions, leading to multiple trials (Chapter 5; Table 9.1, rows 5, 6 and 9).

When France led the EU-wide approval of Bt maize 176, its favourable risk assessment was widely criticised by member states as well as NGOs. When France further proposed to approve maize varieties derived from Bt 176 in 1998, Ecoropa and Greenpeace filed a challenge at the *Conseil d'Etat,* the French administrative high court, on several grounds—that the risks had not been properly assessed, that the correct administrative procedures had not been followed and that the Precautionary Principle had not been properly applied. These NGO arguments gained some support in the court's interim ruling. Thus a government was judicially put on trial for failing to put a GM product on trial.

When UK lab experiments found harm to rats from GM potatoes, the disclosure led to trials of other kinds. The project leader, Arpad Pusztai, questioned the safety of GM foods on a television programme. He was soon dismissed from his post and was then subjected to character assassination by other scientists. His experimental methods were criticised by a Royal Society report. International networks of scientists took opposite sides on that issue. NGOs put his employers and other critics symbolically on trial, by attributing their actions to political and commercial motives. The controversy reinforced consumer anxiety about GM food, as well as retailers' efforts to exclude GM grain from their own-brand products.

When a Swiss lab experiment found that an insecticidal Bt maize harmed a beneficial insect (lacewing), expert authority was put on trial. Criticising the experiment, other scientists cast doubt on its methodological rigour and its relevance to commercial farming, as grounds to discount the results in the regulatory arena. In response, agbiotech critics reversed the accusation: they raised similar doubts about the rigour of routine experiments that had supposedly demonstrated safety. Potential harm to non-target insects remained a high-profile issue, attracting further research and expert disagreements. Citing scientific uncertainties, some regulatory authorities rejected Bt maize or demanded that its cultivation be subject to special monitoring at the commercial stage, thus further testing safety claims and monitoring protocols.

In the latter two risk issues, surprising experimental results were deployed to challenge safety claims, optimistic assumptions and expert safety advice. Regulatory authorities were put symbolically on trial for failure to develop adequate scientific knowledge for risk assessment. When new evidence of risk was criticised for inadequate rigour or relevance to realistic commercial contexts, similar criticisms were raised against safety claims and their methodological basis.

For the safety assessment of GM food, EU regulatory procedures were eventually put on trial. After Italy banned several foods derived from GM maize, the Commission sought to lift the ban (Table 9.1, row 7). In 2000 the Commission requested support from the EU regulatory committee of member states, thus putting Italy on trial by its peers. But they instead sided with the defendant, while criticising the regulatory short-cut for safety approval based on substantial equivalence. After this role-reversal, the Commission soon abandoned substantial equivalence as a basis for easier approval in the GM Food and Feed Regulation. This retreat opened up more methodological issues, scientific uncertainties and assessment criteria for expert deliberation; such judgements were kept on trial by member states as well as civil society groups.

1.2 Eco-Efficiency Versus Agri-Industrial Hazards

Agbiotech began with a cornucopian promise, attributing beneficent properties to genetic modification. Thanks to precisely controlled genetic changes, GM crops would provide smart seeds, as eco-efficient tools for sustainably intensifying industrial agriculture. These promises were extended by the 'Life Sciences' project, featuring mergers between agri-supply and pharmaceutical companies, in search of synergies between their R&D efforts. It promised health and environmental benefits as solutions to general societal problems.

Critics turned agbiotech into a symbol of multiple threats; productive efficiency was pejoratively linked with agri-industrial hazards. Biotech companies were accused of turning consumers into human guinea pigs. In France, critics cast agbiotech as *malbouffe* (junk food), as threats to high-quality *produits du terroir*. In Italy GM crops were cast as agri-industrial competition and 'uncontrolled genetic contamination', threatening diverse, local quality agriculture. Using the term *Agrarfabriken* (factory farm), German critics linked agbiotech with intensive industrial methods, threatening human health, the environment and agri-ecological alternatives (Chapter 3).

Through these cultural meanings, agbiotech was put symbolically on trial as an unsustainable, dangerous, misguided path. Institutions faced greater pressure to test claims that GM crops would provide agri-environmental improvements as well as safety.

Those informal trials shaped conflicts over regulatory criteria from the mid-1990s onwards. When EU procedures initially evaluated GM crops for cultivation purposes, they were deemed safe by accepting the normal hazards of intensive monoculture; this normative stance was portrayed as a scientific judgement, while casting any criticism as irrelevant or political (Chapter 2). Yet such hazards were being highlighted by critics, framing risks in successively broader ways. Their discourses emphasised three ominous metaphors: 'superweeds' leading to a genetic treadmill, thus aggravating the familiar pesticide treadmill; broad-spectrum herbicides inflicting 'sterility' upon farmland biodiversity; and pollen flow 'contaminating' non-GM crops (Chapter 6).

These ominous metaphors expanded the charge-sheet of hazards for which GM products could be kept on trial. Moreover, these broader hazards would depend on the behaviour of agri-industrial operators, which consequently became a focus of prediction, discipline and testing. Regulatory procedures came under pressure to translate the extra hazards into risk assessments. In its risk assessment for GM herbicide-tolerant oilseed rape, Bayer claimed that farmers would eliminate any resulting herbicide-tolerant weeds, but Belgian experts questioned the feasibility of such measures. Citing that advice, the Belgian national authority rejected the product, rather than invite a commercial-stage experiment for testing the extra hazards (Chapter 6; Table 9.1, row 10).

GM herbicide-tolerant crops had been promoted as means to reduce herbicide usage and thus to protect the environment. But UK critics portrayed more efficient weed-control as a hazard: broad-spectrum herbicides could readily extend the 'sterility' of greenhouses to the wider countryside, which would be turned into 'green concrete'. The UK government was denounced for ignoring the agri-environmental implications. The Environment Ministry eventually took responsibility and funded large-scale field experiments, where farmer behaviour was put on trial along with the GM crops. The experimental results led to a regulatory impasse for the GM crops that could have been approved (Chapter 6). In those ways, putative societal benefits from agri-industrial efficiency were cast as threats and thus as extra accusations to be adjudicated.

More generally, broader accounts of harm meant more uncertainties about whether GM crops could generate such harm in the agri-food chain. Uncertainties went beyond any testable characteristics of products per se, because the potential effects would depend upon operator practices. As a more cautious way forward, the commercial stage was anticipated as a real-world experiment, testing assumptions about human practices as well as their environmental effects (cf. Krohn and Weyer, 1994).

1.3 Globalisation Versus Democratic Sovereignty

Field trials were meant to demonstrate the agronomic efficacy and safety of GM crops, as well as the diligent responsibility of the authorities. But the fields were turned into theatrical stages for protest. They used an 'X' or biohazard symbol to cast agbiotech as pollutants and unknown dangers, thus justifying sabotage as environmental protection. When facing prosecution, they used the opportunity to put the state symbolically on trial for inadequately evaluating or controlling GM crops, and thus for political irresponsibility (Chapter 5; Table 9.1, row 3).

Agbiotech had been promoted as an essential response to global market competition, but critics represented agbiotech as a threat and agent of 'globalisation'. The originator of GM soya, Monsanto, was targeted as a global bully 'force-feeding us GM food'. Before the European Commission approved GM soya in 1996, NGOs and some member states demanded mandatory labelling for all GM foods. However, this demand was rejected, with warnings that any such requirement would provoke a WTO case against the EU (Chapter 5 and 7; Table 9.1, row 11).

On this basis, the no-labelling policy became vulnerable to attack as globalisation undermining consumer choice and democratic sovereignty (Chapter 3). Pressure mounted from local protests at supermarkets, linked with labelling demands from Europe-wide consumer and environmentalist groups. By 1998 European retail chains adopted voluntary labelling of their own-brand products with GM ingredients. But different rules across EU member states, alongside NGO surveillance of GM material in food products, potentially destabilised the internal market. So the EU soon established more comprehensive, standard criteria for the EU-wide market (Chapter 7). Thus a commercial trial undermined the original EU policy, which then changed along lines accommodating public demands.

Activists appealed to democratic sovereignty when carrying out and defending sabotage actions. The UK government implied that decisions about GM crops lay elsewhere, beyond its political control; this claim was denounced as an irresponsible, undemocratic surrender to globalisation, thus legitimising sabotage. Further to the French example above, in 1998 the WTO approved higher US tariffs against several specialty foods including Roquefort cheese, as compensation for lost exports of US beef. *Paysans* attacked McDonald's as a symbol of WTO rules forcing the world to accept hazardous *malbouffe* such as hormone-treated beef and GM food. As defendants in court, the *paysans* sought to put 'globalisation' on trial, represented by the French government as well as biotech companies.

Democratic sovereignty also became an explicit theme in judicial trials. When some EU member states blocked the authorisation procedure for new GM products from 1999 onwards, they were demanding precautionary reforms in EU rules. At the same time, their *de facto* EU moratorium was turned into a public symbol of sovereignty versus globalisation. Agbiotech opponents discursively linked the moratorium with attacks on GM crops and support for the *paysan* activist Jose Bové against his prosecution by the French authorities.

When the US government brought a WTO case against the 'illegal moratorium' and national bans on GM products, the EU was formally put on trial. In its defence case, the Commission warned the WTO about its legitimacy problems, as extra reasons for the WTO to respect the sovereign right of governments to regulate products in a prudent manner (Chapter 5). At the national or regional level, democratic sovereignty became general grounds to legitimise measures or actions restricting GM products.

In sum, returning to the earlier questions about what has been on trial and how: Agbiotech was turned into a symbol, object and catalyst for multiple overlapping trials. Beneficent claims were challenged along several lines: safety versus precaution, eco-efficiency versus agri-industrial hazards and globalisation versus democratic sovereignty. Activists' accusations were taken up by large environmentalist and farmers' organisations, as well as by wider groups in civil society, acting as a pervasive mobile 'jury'. Through a cultural rationality, questioning the drivers and aims of an innovation, protest challenged the technical rationality of official risk assessment (cf. Krimsky and Plough, 1988; Fischer, 2005). The defendant symbolically on trial was expanded—from product safety, to biotech companies, their innovation trajectory, regulatory decision-making, expert advisors, government policy and its democratic legitimacy. Informal-symbolic trials shaped formal trials, as well as role-reversals between accused and accusers.

2 POLICIES ON TRIAL

Questions

- What policies have been on trial?
- How and why have they changed since the late 1990s?
- How have these policies promoted or marginalised potential European futures?

Agbiotech opponents were often criticised for exploiting the controversy—for example, for encouraging irrational fears, turning agbiotech into a proxy for extraneous issues, and thus politicising a technology. Yet politics were always involved in agbiotech. Its promotion incorporated specific models—of the market, the public, the environment, agriculture, regulatory expertise, scientific knowledge and European integration.

Neoliberal policies informed these models, while depoliticising them as objective imperatives: they became necessary adaptations to external reality rather than open societal choices. Since the 1980s the EU has elaborated a commitment to 'clean' technological innovations which promise to reconcile economic competitiveness with environmental protection. Conversely, such technological imperatives became an instrument for shaping European integration as a 'competition state', i.e. directing resources towards the domestic capacity for global competitive advantage (cf. Cerny, 1999). Thus the public good was displaced—or was even redefined along those lines.

Epitomising the competition state, by the late 1980s agbiotech was being promoted along with trade liberalisation, broader patent rights and marketisation of public-sector research. Together these policies were pushing European further towards the US agri-industrial model for which GM crops were designed. From the mid-1990s onwards, the EU internal market project was also being designed for transatlantic trade liberalisation. This future was promoted by the Transatlantic Business Dialogue, whose main slogan was 'Approved once, accepted everywhere'. Its Agriculture working group featured agrichemical companies which were undergoing transatlantic integration, as well as incorporation of seeds companies. In policy documents, however, market competition among those companies was represented as competition between US versus European economies, as an imperative for Europe to catch up in the race (Chapter 2).

Those policy frameworks complemented the neoliberal construction of Europe as a smooth homogeneous space for circulating standard products (cf. Barry, 2001). Through a technicist harmonisation agenda, national regulatory differences were treated as merely technical issues to be standardised, as if standards could lie above cultural values. Within this model, expert assessments could legitimately differ only because of geographical differences in resources or environments at risk.

Environmental contexts across Europe were homogenised in the policy framework of 'risk-based regulation'. For early agbiotech products undergoing the EU regulatory procedure, safety claims accepted intensive agriculture and its normal hazards as obvious environmental standards. These narrow accounts of harm allowed less scope for national differences in risk assessment, thus facilitating EU-wide authorisation of GM crops. Broader accounts of harm were dismissed as misinterpreting EU law. Thus EU regulatory procedures complemented a policy of agri-industrial intensification. The agricultural environment was modelled as an efficient factory for standard global commodities.

In those ways, Europe was being 'made safe' for the agbiotech innovation trajectory. The EU's policy framework adapted the US neoliberal model in co-producing agbiotech with specific forms of nature and society (cf. Jasanoff, 2004, 2005a). As a precisely improved form of nature, GM crops were 'biotechnological inventions' warranting patent rights. These safe products would protect agriculture from the threats of wild unruly Nature, while also strengthening the European economy against competitive threats. By projecting beneficent natural qualities onto GM products, the market could be represented as the natural regulator; any challenge could be dismissed as illegitimate interference in the natural order (cf. Williams, 1980). This neoliberal project was discursively depoliticised by invoking objective imperatives.

Agbiotech encountered significant opposition since the 1980s in some EU member states, e.g. Germany and Denmark, and then Europe-wide public revolt by the late 1990s. Governments came under pressure to justify or soften their pro-agbiotech commitments. Yet these were deeply embedded in EU institutions and in most national policy frameworks (Chapter 3).

Reforms aimed to mediate the consequent societal conflicts over those commitments. Institutions were created or adapted for incorporating, managing or marginalising dissent. In participatory Technology Assessment exercises too, contentious issues about private control over agricultural innovation were reduced to issues of regulatory control measures. This complemented the similar role of the regulatory procedures that they simulated or informed (Chapter 4).

Centralised EU-level advisory expertise was designed to adjudicate regulatory disagreements, in parallel with a policy shift from 'risk-based regulation' to 'science-based regulation'. In 2003 the European Food Safety Authority (EFSA) was established to provide expert advice on regulatory issues. EFSA's safety claims would be based on authoritative 'science', by implicit contrast to the 'political' basis of government objections.

These reforms operationalised strategies from the 2001 *White Paper on European Governance,* which had proposed greater transparency and citizen involvement in order to enhance public confidence in policy delivery (e.g. CEC, 2001a: 5, 10). This meant instrumentalising civil society in order to implement and legitimise policies already set, according to a critical account (Eriksen, 2001: 63). In that vein, EFSA incorporated a representative of EU-wide consumer groups as social partners.

These EU reforms were soon put on trial. For agbiotech in particular, the GM Food and Feed Regulation established procedures for public access to regulatory documents and public comment. Consultation procedures privileged scientific evidence as the only basis for admissible comments, thus modelling the public as would-be expert risk assessors. Technical comments were dismissed as contributing nothing new, while 'political' or 'ethical' comments were dismissed as irrelevant, thus frustrating those who took part.

'Science-based regulation' depended on credibly 'objective, independent' advice from EFSA, but this official status was disputed. Expert disagreements among member states revealed different accounts of the relevant science. These disputes destabilised any science/policy boundary, EFSA's cognitive authority and thus the Commission's political authority to approve GM products. The EU system once again faced difficulties in finding an expert basis for legitimate regulatory decisions. In 2006 the Commission shifted its policy, now acknowledging that expert advice involves policy and precautionary issues. The Commission put EFSA's advice under scrutiny, even asking EFSA to re-do its risk assessments. Thus EFSA itself became a defendant, not simply a judge of dissenting governments and experts.

Rules on labelling products as GM also remained contentious. Any GM labelling criteria made normative assumptions about product characteristics as legitimate grounds for consumer choice. From a standpoint of economic individualism, any GM label should be based on scientific information relevant to consumer safety. From this standpoint, the Commission initially approved GM products with no mandatory labelling, on several grounds: that such a requirement would have no scientific basis, would jeopardise the internal market and would provoke a US challenge at the WTO.

However, demands for comprehensive GM labelling united a wide range of civil society groups. Governments and companies were put onto the defensive for denying consumers the freedom to choose non-GM versus GM food. Through protest and NGO surveillance of GM 'contamination', GM grain was turned into an instability for conventional agri-food chains. In response, retail chains devised their own labelling rules and then soon excluded GM ingredients from their own-brand products, rather than label them. When put 'on trial' by NGO surveillance and market forces, agbiotech was found guilty and was forced out as a commercial liability. This outcome contradicted promises that agbiotech would enhance the economic competitiveness of the European agri-food industry.

EU labelling rules became progressively more stringent, applying to a wider range of GM products. Moreover, GM labelling was linked with traceability in the 2003 GM Food and Feed Regulation. To justify full labelling and traceability of GM products, advocates cited precautionary reasons: consumers and regulators must have adequate information to deal with uncertain risks. Although this regulatory framework was officially meant to overcome blockages of biotech, the new rules facilitated a different aim: agbiotech could be more readily kept 'on trial' for any uncertain risks, and its commercial prospects could be influenced by consumer decisions. Agbiotech opponents gained extra means to deter food companies from using GM grain (Chapter 7).

The outcome was largely a commercial blockage of GM products, especially for cultivation uses. As an exception which perhaps proves the rule, Spain had Europe's only significant cultivation of GM crops. Bt maize was produced there mainly for a national market in animal feed, thus resembling the US agri-industrial context for which GM crops were designed.

As an extra issue, gene flow from GM to non-GM products became more contentious from the late 1990s onwards. As a Commission response, a 'coexistence' policy framework was meant to manage this problem of inadvertent admixture and thus avoid conflicts among farmers. According to the Commission, any segregation rules must 'allow market forces to operate freely' by allowing farmers a free choice of GM, conventional or organic agriculture. GM crops could justifiably face extra burdens only if necessary to maintain any admixtures below the threshold for GM labelling and thus avoid economic loss to other agricultural systems (Chapter 8). This policy made normative assumptions about the inevitability and desirability of GM crops entering large-scale cultivation. Given these assumptions, various proposals for 'coexistence' rules came under attack rather than overcoming the conflict.

Some regional authorities aimed to minimise the presence of GM material through stringent measures. Uncertain environmental risks were cited as precautionary grounds. Such measures were rejected by the Commission as disproportionate. The Commission's policy framework in turn was criticised for 'coercion', i.e. forcing non-GM farmers to accept GM contamination and thus denying their freedom of choice.

Moreover, some authorities treated 'GM contamination' as an inherent threat to 'quality' alternatives and territorial resources. They drafted 'coexistence' laws which would effectively put GM crops 'on trial'—e.g. by requiring prior evidence that their cultivation would not contaminate other crops, or by strongly penalising any contamination through strict liability. The Commission scrutinised these rules for possible incompatibility with operator freedom in the internal market. From those opposite standpoints, the same draft rules were defended as market freedom or else denounced as coercion, thus undergoing informal trials from two opposite standpoints.

Conflict over segregation led to at least one judicial trial. Under Upper Austria's draft law, 'GM-free agricultural areas' would protect products and natural biodiversity from 'GMO contamination', especially in 'sensitive ecological areas'. Its proposal was declared illegal by the European Court of First Instance. At the same time, many other regional authorities found more subtle ways to undermine Commission policy, especially by blurring its key distinction between environmental and economic harm. Moreover, the Commission's policy came under similar challenges from other EU institutions, e.g. the Parliament and Committee of Regions.

Also on trial has been EU-level expert advice as a basis for Commission policy on the above issues—risk assessment, GM labelling criteria and segregation rules. In the Commission's policy frameworks, 'science' was meant to provide a basis for politically neutral, objective advice from EU expert bodies. For each issue, the Commission discursively conflated 'science' with such advice, as if it lay above policy issues. Yet EU-level expert advice was often disputed along several lines: that it depended on weak research methods, optimistic assumptions and narrow accounts of harm. By posing different questions, precautionary approaches identified ignorance in official risk assessments and so opened up extra uncertainties (cf. Stirling, 1999).

Although these criticisms came from diverse national perspectives, with no coherent alternative view, together they destabilised the putative expert basis of pro-agbiotech policies. Evidential standards for 'science' were turned into contentious policy issues. The conflict stimulated more cautious risk-assessment approaches, as a dual means to manage scientific uncertainty and political dissent. Calls for greater precaution served to open up official assumptions about the expert basis of rules for the internal market. Expert disagreements highlighted the normative basis of supposedly technical judgements (cf. Nowotny et al., 2003).

In sum, returning to the earlier questions about policies on trial: Early EU policies were designed to co-produce agbiotech with a particular socio-natural order. This depoliticised the innovation pathway as an objective imperative, thus marginalising alternative agri-development pathways. As those policies underwent multiple trials, new policies implied a different socio-natural order, now treating biotechnologised nature as suspect and potentially abnormal. The 2003 labelling and traceability regulations, alongside requirements for market-stage monitoring, for example, implied a long-term uncertainty about potential risks, encompassing a broader range of agri-environmental effects than before. Segregation rules for GM material remained contentious; some proposals blurred any boundary between environmental and economic harm, as a basis for lower tolerance limits and greater restrictions on GM crops.

It was hypothetically possible for agbiotech to be co-produced with such a socio-natural order, e.g. by managing broader uncertainties about operator behaviour, agri-environmental effects and 'GM contamination'. But the European outcome was largely a non-production of agbiotech, given the regulatory and commercial blockages, encouraged by numerous stakeholder groups (NGOs, retailers, some farmers, etc.). At the same time, this opposition stimulated interest in alternative futures around quality agri-food production.

3 DEMOCRATIC LEGITIMACY ON TRIAL

Questions

- What difficulties arose for the accountability of representative democracy?
- How has this case tested European integration, its democratic legitimacy, its deficits and possible remedies?

Agbiotech became a focus for contesting claims about European 'democracy'. Its legitimacy was put symbolically on trial, along with agbiotech, its EU decision procedures and institutions. The defendant on trial was expanded from agbiotech to an entire political system. Europe was told that it had no choice but to accept agbiotech, yet this imperative was turned into a test of democratic accountability for societal choices. Agbiotech became an explicit test case of the EU's democratic deficit—its sources, diagnoses and possible remedies. Through the conflict, European integration has been tested and reformed for democratic legitimacy, thus reshaping the 'EU' that was on trial.

In academic as well as political contexts, democracy is often idealised—as connoting particular qualities (e.g. pluralism, freedom, self-government, accountability), or as denoting exemplary institutions (e.g. the Parliamentary system). Rather than adopt a single ideal account of democracy, this book has applied a realist method. It draws out the actors' accounts of democracy, in relation to various theoretical accounts (Gupta, 2007; see our Chapter 1). Thus a story about 'testing European democracy' need not depend upon the authors' normative views about what democracy should be. In this book 'European democracy' refers to a contested, elusive concept—rather than any ideal account.

In early EU policy, a deep commitment to a biotechnological innovation pathway remained beyond state accountability or public deliberation. Only through a fierce conflict did the EU adopt a requirement for prior authorisation of all GM products. These procedures would facilitate approval of 'safe' GM products, which could then circulate freely within an EU-wide internal market. Product safety would be the sole criterion for approval or delay. Technological innovation was idealised as an essential basis for economic survival and even democratic renewal.

Within that policy framework, societal conflicts were channelled into regulatory issues, initially product safety alone. According to a Commission strategy document, regulatory oversight 'is the expression of societal choices': EU rules should ensure that 'market mechanisms function effectively', by providing safe products as a basis to accommodate consumer preferences (CEC, 2002b: 14). That mechanical metaphor denotes impersonal relations beyond political responsibility; the market would somehow depoliticise agbiotech and avoid governmental accountability. However, any market depended upon formal and informal rules, which readily became contentious. These rules were put symbolically on trial for imposing a single future and pre-empting alternatives.

Discursive strategies appealed to ideal concepts of democracy. Some promoted regulatory harmonisation for the internal market as a model of EU democracy. At the same time, regulatory decisions favourable to agbiotech were attacked as 'undemocratic', a phrase connoting democracy in various ways—e.g. as free choice versus coercion, as the popular will versus minority special interests, as national sovereignty versus globalisation, etc. The democratic legitimacy of decisions was challenged on both procedural and substantive grounds. These conflicts can be analysed in terms of input and output legitimacy.

Input legitimacy, also known as 'government by the people', depends upon key instruments of accountability: collective identities, institutional infrastructure and policy discourses. These instruments have been generally absent on a Europe-wide basis, according to a political scientist (e.g. Scharpf, 1999: 187). Although this may be true, their embryonic forms may be in conflict. Moreover, those instruments—collective identities, institutional infrastructure and policy discourses—shape the social and natural order that is co-produced with specific technologies (Jasanoff, 2004b).

Regarding input legitimacy in the agbiotech case, neoliberal agendas both shaped and limited EU-wide instruments of accountability. Dominant policy discourses promoted collective societal identification with agbiotech as a beneficent, modern eco-efficiency tool, essential for economic competitiveness. The public was constructed as unwitting consumers of presumably safe products, as eventual beneficiaries of more efficient production methods and thus as rational supporters of technological progress.

This framework provided a vulnerable target for agbiotech critics. They counterposed different identities and discourses—e.g. consumer freedom to choose non-GM products, the precautionary principle versus safety claims, and democratic sovereignty through national or regional policies. Popularised through protest activity, these oppositional discourses undermined the Commission's earlier basis for input legitimacy.

Facing various trials over agbiotech, state bodies debated how to address their legitimacy problems, generally understood as an issue of 'public confidence'. Official procedures addressed ever-wider issues and involved more stakeholder groups. Policymakers devised various institutional reforms, which became political experiments in European integration. These reforms too have been kept on trial.

State bodies often competed for decision-making authority, while also avoiding responsibility for unpopular decisions. Under EU procedures, each national authority had a weak capacity to influence EU rules but had political-legal responsibility for implementing them, thus undermining state accountability to a national citizenry. Responsibility for decision-making was variously transferred, shared, sought or evaded by state authorities—the European Commission, European Council, member states, regions, etc.

These rivalries weakened the Commission's authority for policies favouring agbiotech. Although internally divided, the Parliament generally went furthest in accommodating demands for greater state accountability to citizens, e.g. regarding new regulatory procedures, GM labelling criteria and segregation measures. Measures to minimise the presence of GM material in other products, as proposed by the autonomous Assembly of European Regions, eventually gained support from the Parliament and the Committee of the Regions, another official EU body; such measures contradicted Commission policies.

Output legitimacy, also known as 'government for the people', means government capacity to solve collective problems for the public good. This presumes a shared account of such problems, yet the European integration project generates contending accounts. 'Completing the internal market' depends upon policy changes that readily become contentious (Scharpf, 1999: 23).

Indeed, the agbiotech case has featured divergent accounts of collective problems. GM crops were promoted as an eco-efficiency tool for sustainably intensifying agriculture, as if this were a common understanding of the public good. Yet opponents framed greater productive efficiency itself as a threat, at least in the agri-food sector, as a basis to block the internal market for such products.

As the central commitment of EU treaties, the internal market project became a contentious reference point for agbiotech policy debate, e.g. regarding patent rights, product safety, GM labelling, segregation measures, etc. Pro-agbiotech policies were officially justified as measures necessary to create, complete or stabilise the EU's internal market. Any restrictions or burdens on GM products had to be officially justified in similar terms.

When EU institutions promoted regulatory harmonisation for agbiotech products to circulate in the internal market, this effort revealed and even stimulated divergent national standards. According to an earlier study of EU agbiotech policy, European integration has undergone a tension between two political models: one seeking to eliminate national divergences in policy framings, the other protecting deep-seated national values that generate them (Jasanoff, 2005a: 71). Since the late 1990s, however, the EU-wide dynamics have become more complex than those two models. Whenever one member state has proposed more stringent criteria or broader framings for risk assessment, their proposals have often gained support from others as a potential European standard. Societal conflicts over agbiotech were continually translated into regulatory standards—e.g. for risk assessment, GM labelling and segregation measures. Outcomes went beyond any prior national standards or values. Consequently, regulatory harmonisation became a dynamic process of disputing, proliferating, combining and raising standards.

According to a prevalent model of European integration, EU regulatory procedures can gain democratic legitimacy through a multi-stage, multi-level governance process. Through the co-decision procedure, legislation can be shaped by national governments and the Parliament, which represent people's concerns more directly than the Commission may do. In implementing EU laws, national representatives deliberate regulatory criteria for specific products through a comitology procedure involving national and EU expertise (Joerges, 1999). Member states have retained their expertise and capacity for evaluating product safety, unlike other EU policy areas that have been subject to market liberalisation (e.g. public utilities). So the comitology procedure for product approval depends upon a positive integration of diverse national standards, as the basis for regulatory harmonisation and thus an internal market, argues Scharpf (1999).

In deliberating standards for GM products, the comitology procedure also became an arena for trials of democratic legitimacy. Since the late 1990s institutional reforms have enhanced the state's accountability for regulatory decisions and broadened their criteria. EU procedures had great scope to deliberate national differences in regulatory norms. Yet the Commission ultimately approved GM products with little support from member states, given the gap between their domestic concerns and official safety assessments. Alongside denunciations by NGOs, some politicians too challenged the democratic legitimacy of the EU's decision procedure. Through such criticisms (or simple abstention), national governments could also avoid responsibility for unpopular decisions.

As this case illustrates, the comitology procedure is a microcosm of the EU's legitimacy problems, rather than a solution. Any regulatory decision favours some norms over others (cf. Weiler, 1999: 349). In the agbiotech case, the EU-wide conflicts were visibly concentrated by the comitology procedure; contending norms generated legitimacy problems for European integration.

This multiple, overlapping trial of European democracy resulted from a specific policy commitment: namely, promoting a contentious innovation as if it were an objective imperative, beyond challenge by different societal choices or public deliberation. By the late 1990s, state bodies faced stronger pressure to justify agbiotech development as a desirable choice for the public good—or else to favour alternatives. Authorities sought governance methods to mediate the consequent societal conflicts. Diagnosing a problem of public distrust, they sought to enhance 'public confidence'. Formal policy debate has focused on the management of a common inevitable future (cf. Pestre, 2008). The EU's democratic deficit was understood as a problem of legitimising EU institutions vis à vis passive citizens, according to a critic (Bonefeld, 2002).

More subtly, state institutions could be legitimised also through forms of public representation. Anticipating or responding to societal conflict over agbiotech, state-sponsored participatory TA exercises generally narrowed the problem-definition to its safe management. Participants deliberated the appropriate expertise for addressing that problem. In some cases, these TA exercises enhanced state accountability for regulatory frameworks—but not for innovation choices. Their reinforcement potentially legitimised state commitments to agbiotech. This pattern turns upside-down the democratic aim for participatory exercises to open up such issues (cf. Stirling, 2006: 5; Stirling, 2005: 229).

By contrast to those exercises, uninvited forms of public participation—e.g. demonstrations, food boycotts, sabotage and public meetings—served to keep open the possible future of European agri-food systems. Antagonistic to the internal market project, these initiatives arose from outside EU institutions and official procedures. Agbiotech opponents popularised alternative visions for a future Europe. The slogan 'Another agriculture is possible' adapted the main slogan of the anti-globalisation movement, 'Another World is Possible'. This also implied other forms of democratic decision-making to realise an alternative 'green, quality' agriculture. Such alternatives futures have been promoted by civil society networks of environmental, consumer, retail and farmer groups, with support from the Assembly of European Regions. They generated a *de facto* subsidiarity under the slogan 'GM free'. But they had no means to transform EU institutions or European integration for such alternatives.

When pro-biotech policies and their EU level decision procedures were denounced as 'undemocratic', this illustrates a paradox of transnational state forms:

> . . . boundaries within which democratic practice can be contained are simply not delineable any longer—they slip away between fragmentation and interdependence . . . this situation, while apparently undermining the operation of democracy, also raises demands for democracy. (Gupta, 2007: 27)

Indeed, in the European agbiotech controversy, opponents invoked and demanded democratic accountability—perhaps a euphemism for political power to counter the dominant political-industrial power. They found national or regional entry points which had greater opportunities for undermining EU-level pro-biotech policies.

These political conflicts warrant caution in idealising the EU as a democratic space. Among such idealisations, Habermas is somewhat ambivalent: EU policies for market liberalisation 'desolidarise' society, so European democracy depends on citizens creating cross-national practices of common opinion and collective will, he argues (Habermas, 2001: 99–100). If his phrases mean a significant popular demand on a trans-European scale, then agbiotech provoked a negative collective will. This was initiated through cross-border links among activists and eventually attracted mainstream organisations of civil society. 'GM-free' became a popular European slogan, implicitly antagonistic to the EU's internal market project for agbiotech, while expressing or stimulating alternative agri-food developments beyond that project. These dynamics contradict scenarios of the EU per se as a democratic space.

In sum, let us return to the earlier questions about democratic legitimacy: In prevalent models of European democracy, the EU is understood as a series of treaty arrangements among member states, i.e. as a supranational polity rather than as a state. This cannot reproduce national forms of state accountability to a citizenry, so it depends on a different basis for democratic legitimacy. Such a basis is meant to derive from the co-decision procedure (among the Commission, Council and Parliament) in the legislative phase. Laws are then implemented through the comitology procedure involving national representatives. This framework aims to create an internal market for the free circulation of 'safe' products.

Through that model of European integration, conflicts over innovation pathways were reduced to ever-complex regulatory issues and expert disagreements. Diverse cultural values and agri-environmental contexts were technicised in the name of 'science'—equated with objective expert advice, supposedly standing above policy. Resulting conflicts were misdiagnosed as 'political interference' by national governments in EU procedures, thus warranting moves towards centralised regulation; this was being proposed more generally as a remedy (e.g. by Majone, 1996, and Bernauer, 2003).

At least in the agbiotech sector, those reforms have aggravated the EU's legitimacy problems. These problems have arisen from a specific model of European integration—a techno-fix for the internal market and economic competitiveness, to be imposed through a technicist form of regulatory harmonisation. Informed by such a model, the outcome has undermined the public accountability of representative democracy.

4 BROADER LESSONS (TO BE) LEARNED

Questions

- What broader lessons can be learned from this case?
- How can 'another GM' be avoided—or perhaps created?

Any broader lessons depend upon how the European agbiotech controversy is explained—itself a contentious matter. Multiple sources have been identified in this book. Genetic modification techniques were appropriated for a specific technological trajectory, corresponding to models of the social and natural order. Agbiotech was initially co-produced by EU policies shaping the social and natural order according to neoliberal policies; in turn, agbiotech was used to promote those policies and institutional changes. This agenda was discursively depoliticised by invoking objective imperatives, as if Europe had no choice. Government could avoid political responsibility and public accountability for innovation choices. That policy framework provided a vulnerable target, linking opponents of both agri-industrial innovation models and neoliberal policies facilitating them. State authorities were put 'on trial' for pre-empting societal choices and concealing their agendas behind a technical-expert facade.

Neoliberal agendas likewise have characterised the design and context of many other technological trajectories in the last two decades, but agbiotech had extra vulnerabilities. Agbiotech was reaching the commercial stage in a period when food hazards were widely attributed to agri-industrial methods, deregulatory policies and official expert ignorance. So attacks on agbiotech for 'tampering with our food' resonated strongly with public anxieties. Agbiotech also acquired public meanings through EU policy frameworks—regulatory harmonisation, 'objective' expert advice, trade liberalisation and marketisation of public goods, especially research funds and plant genetic resources. All these policies became defendants in 'GM on trial'.

Thus the European agbiotech controversy has a special combination of sources which may never arise again. Yet it also has relevance beyond the agbiotech case, even beyond technological controversies.

As a less obvious analogy, let us remember the original proposal for an EU Constitutional Treaty, which mandated further market integration and militarisation. According to its Preamble, Europe 'wishes to deepen the democratic and transparent nature of its public life, and to strive for peace, justice and solidarity throughout the world'. Statutory force was absent for those aims but strong for neoliberal aims: the EU has an exclusive competence 'to establish the competition rules necessary for the functioning of the internal market'. This would offer its citizens 'a single market where competition is free and undistorted'. Thus public services would be forcibly subjected to market competition; the text was pervaded by the language of 'competition' and 'market'. Moreover, 'Member States shall undertake progressively to improve their military capabilities' (IGC, 2004).

Given those emphases in the proposed EU Constitution, it became a vulnerable target for attack by diverse opponents, including Left-wing internationalists and Right-wing nationalists in many countries. Although the proposal easily gained approval in national Parliaments, it was defeated in the three popular referenda held in 2005. What could explain this public response? 'There is almost certainly an information deficit and efforts will have to be made to explain things more clearly to citizens', according to a spokesperson for the European Commission (le Bail, 2005).

The agbiotech controversy likewise attracted diagnoses of public deficits and mis-communication, along with a didactic remedy: to reshape public attitudes and citizen representation along lines accepting technological imperatives of European progress. Comparison to the proposed EU Constitution becomes more plausible if a marketisation agenda is seen as a prime defendant in 'GM on trial', as well as a social order being co-produced with agbiotech. In both conflicts, formally democratic procedures were denounced as undemocratic, e.g. by counterposing popular referenda to Parliamentary votes on the EU Constitution.

From the experience of the European agbiotech controversy, many commentators have drawn wider lessons, including some dubious ones. For example, it is said: We must ensure that nanotechnology does not become 'another GM'. In other words, technophobes may again polarise the issues through public ignorance and irrational fears, which can be countered through better public communication. For example, 'The easiest way for the nanotechnology community to avoid the problems experienced in the deployment of biotechnology is to provide accurate information and encourage critical, informed analyses' (McHugen, 2008: 51).

As another lesson often heard: The next novel technology could become 'another GM' if the public is not adequately consulted at an early stage. Put positively, it is said, greater public involvement could help to avoid societal conflict over technological innovations. For example, 'Given the opportunity to deliberate on such innovations, the public voice can be expected to be measured and moderate' (Gaskell, 2008: 257).

Although those commentators draw different lessons, they decontextualise technology from political-economic agendas. Yet, as the commentator above acknowledges, 'Science is evidently more commercialised, less accountable to the financial markets; and this at a time when many new developments . . . raise many scientific, ethical and moral uncertainties' (ibid.: 256). This book has located the agbiotech conflict within such contexts: at stake are models of society, nature and expert authority.

Finally, let us return to the questions posed above: Prospects for avoiding 'another GM' controversy—or perhaps for creating one—depend upon how an innovation models the social and natural order, and how state bodies attempt to promote that order. If a political-economic choice is represented as an objective imperative, then such an innovation may be successfully naturalised and imposed through a technicist harmonisation process. Or else critics may tap public unease, generate alternative identities, challenge official expert authority, provoke policy conflicts and thus undermine that choice.

Those different prospects pose either a threat or an opportunity, depending on one's aims. So divergent lessons may be (and are) drawn from the EU agbiotech controversy. Those divergent lessons symptomise legitimacy problems inherent in a European integration project driven by the EU's internal market and marketisation policies. This project remains vulnerable to legitimacy problems. Yet more issues may be turned into symbols of democratic deficit, thus testing the forms and limits of European democracy.

References

Note on EU documents: CEC denotes a viewpoint from the Commission of the European Communities. EC (or EEC before 1995) denotes statutory decisions of the European Community as published in the *Official Journal*, series L.

ACRE (2003) Advice on notification C/NL/98/11, oilseed rape genetically modified for tolerance to glyphosate herbicides, line GT73, 10 March, http://www.defra.gov.uk/environment/acre/advice/advice23.htm

ACRE (2004) Advice on Implications of the Farm-Scale Evaluations of GM Herbicide-Tolerant Crops, http://www.defra.gov.uk/environment/acre; and advice on Bayer notification C/BE/96/01.

ACRE/ACP (2001) Advisory Committee on Releases to the Environment and Advisory Committee on Pesticides, 'Issues Raised for the Ecological Risk Assessment of Plant Protection Products from a Broader Biodiversity Approach to the Protection of Biodiversity'.

Actu-Environnement (2008) Le Conseil d'État rejette le recours des producteurs de maïs concernant le MON810, 19 March, http://www.actu-environnement.com

AEBC (2001) *Crops on Trial*, London: Agriculture and Environment Biotechnology Commission, http://www.aebc.gov.uk

AEBC (2003) *GM Crops? Coexistence & Liability*, London: Agriculture and Environment Biotechnology Commission, http://www.aebc.gov.uk/aebc/coexistence_liability.shtml

AER (2006) 2nd Berlin conference of GMO-free regions, http://www.a-e-r.org

AER/FoEE (2005) Safeguarding Sustainable European Agriculture: Coexistence, GMO free zones and the promotion of quality food produce in Europe, 17 May conference, http://www.a-e-r.org

AFSSA (2003) Dossier maïs doux Bt11 au titre du règlement (CE) 258/97, http://www.afssa.fr/ftp/basedoc/biot2003sa0353.pdf

Agence Press Europe (2002) 'Parliament strengthens labelling rules on GM food but avoid upsetting balance', 3 July.

Agence Press Europe (2004) 'Greenpeace demands Environment Council to reject GT73 oilseed rape', 21 June.

Agence Press Europe (2005) 'Twenty European "GMO-free" regions sign charter', 7 February.

Anderson, L. (1999) 'Hot potatoes', *The Guardian*, 17 February: 4.

Anon (1999) 'Trial "in public interest", judge rules', *Chemistry & Industry*, 3 May: 332.

Anon (2003) 'MEPs attack Commission over GM co-existence', *Agra Europe*, 7 November: 8.

Anon (2004a) 'EU ministers sound death knell for GM moratorium', *AgraEurope*, 30 April: 1–2.

Anon (2004b) 'Commission rejects Greens' accusations over GM authorisations', *AgraFood Biotech,* 2 August: 12.

Anon (2005) 'Opposition to Spanish GM coexistence proposal', *Agra Europe,* 24 March: N/6.

Anon (2006a) 'GM coexistence rules could push up production costs', *Agra Europe,* 24 February: 7–8.

Anon (2006b) 'Dimas, Boel clash over GM crops', *Agra Europe,* 7 April: 2.

Anon (2006c) 'Member states press Commission for more open GM regime', *AgraEurope,* 10 March: 1–2.

Austrian Presidency (2006) Co-existence of genetically modified, conventional and organic crops: the freedom of choice, Conclusions of Vienna conference, 4–6 April, http://ec.europa.eu//comm/agriculture/events/vienna2006/index_en.htm

BAC (2004a) Advice of the Biosafety Advisory Council on the notification C/BE/96/01 (transgenic oilseed rape—Bayer CropScience) under Article 35 of Directive 2001/18/EC, 26 January, http://www.biosafety-council.be/docs/BAC_2004_SC_084.pdf

BAC (2004b) Advice of the Biosafety Advisory Countil on the British report 'On the rationale and interpretation of the Farm-Scale Evaluation (FSE) of genetically modified herbicide-tolerant crops', 26 January, http://www.biosafety-council.be/docs/BAC_2004_SC_087.pdf

Bakonyi, G., Szira, F., Kiss, I., Villányi, I., Seres, A. and Székács, A. (2006) 'Preference tests with collembolas on isogenic and Bt-maize', *European Journal of Soil Biology* 42: S132–35.

Barben, D. (1998) 'The political economy of genetic engineering', *Organization & Environment* 11(4): 406–22.

Barns, I. (1995) 'Manufacturing consensus? Reflections on the UK National Consensus Conference on Plant Biotechnology', *Science as Culture* 5(2): 200–16.

Barry, A. (2001) *Political Machines: Governing a Technological Society.* London: Athlone.

Barry, J. and Paterson, M. (2004) 'Globalisation, ecological modernisation and New Labour', *Political Studies* 52: 767–84.

Bartle, I. (1991) *Herbicide-Tolerant Plants: Weed Control with the Environment in Mind,* Haslemere, Surrey: ICI Seeds [later Zeneca].

Bayer Crop Science (2002) Full notification C/BE/96/01 for oilseed rape Ms8xRf3, including Environmental Risk Assessment & Post-Marketing Monitoring Plan, December.

Bayer Crop Science (2003) SNIFs for glufosinate-tolerant oilseed rape Ms8xRf3, http://gmoinfo.jrc.it/csnifs/C-BE-96-01.pdf, and for glufosinate-tolerant oilseed rape Liberator pHoe6/Ac, http://gmoinfo.jrc.it/csnifs/C-DE-98-6.pdf

Bayer Crop Science (2004) 'Bayer Crop Science discontinues further efforts to commercialise GM forage maize in the UK', 31 March, http://www.bayercropscience.com

Belasco, W. J. (1989) *Appetite for Change,* New York: Pantheon.

Belgium-FPS (2004) Assessment report in accordance with Directive 2001/18/EC: notification C/BE/96/01 oilseed rape, Federal Publique Service of Public Health, Food Chain Safety and Environment, 2 February, , http://www.biosafety.be/TP/partC/StatementBE_C_BE_96_01.pdf

Benford, R. D. and Snow, D. A. (2000) 'Framing processes and social movements: an overview and assessment', *Annual Review of Sociology* 26: 611–39.

Bernauer, T. (2003) *Genes, Trade and Regulation: The Seeds of Conflict in Food Biotechnology,* Princeton: Princeton University Press.

BEUC (2003) Position Paper on Coexistence, Brussels: Bureau Européen des Unions de Consommateurs, 31 October.

Binimelis, R. (2008) 'Coexistence of plants and coexistence of farmers: is an individual choice possible?', *Journal of Agricultural and Environmental Ethics* 21: 437–57.

BÖLW (2003) Grüne Gentechnik bedroht die ökologische Lebensmittelwirtschaft, Press release 17 July, Organic Farming and Processing Association, http://boelw.de

Bonefeld, W. (2002) 'European integration: the market, the political and class', *Capital & Class* 77: 117–44.

Bora, A. and Hauseldorf, H. (2006) Participatory science governance revisited: normative expectations versus empirical evidence, *Science and Public Policy* 33 (2): 479-488.

Borrás, S. (2006) 'Legitimate governance of risk at the EU level? The case of genetically modified organisms', *Technological Forecasting & Social Change* 73: 61–75.

Boschert, K. and Gill, B. (2004) *Germany: Precaution for Choice and Alternatives, National Report for PEG Project,* Munich: Institute of Sociology.

Boschert, K. and Gill, B. (2005) 'Germany's agri-biotechnology policy: precaution for choice and alternatives', *Science and Public Policy* 32(4): 285-92.

Bové, J. (1998) Speech at the Agen court, 12 February, translated by Greenpeace France, http://www.greenpeace.org.fr, http://www.cpefarmers.org

Boy, D., Donnet-Kamel, D. and Roqueplo, P. (1998) A Report on the Citizens Conference on Genetically Modified Foods (France, 21–22 June 1998), including the report prepared by the French Lay Panel, http://www.loka.org/pages/Frenchgenefood.htm

Brown, P. (1998) 'Genetic crops move upsets green groups, Brown', *The Guardian,* 22 October.

Brown, P. (1999) 'Crops case puts GM food on trial', *The Guardian,* 21 April.

Burchell, J. and Lightfoot, S. (2001) *The Greening of the European Union: Assessing the EU's Environmental Credentials,* Sheffield: Continuum.

Busch, L. and Lacy, W. B. (1983) *Science, Agriculture, and the Politics of Research,* Boulder, CO: Westview Press.

Buttel, F. (1998) 'Nature's place in the technological transformation of agriculture: some reflections on the recombinant BST controversy in the USA', *Environment and Planning A* 30(12): 1151–63.

Buttel, F. (2000) 'GMOs: the Achilles heel of the globalization regime?', paper for annual meeting of the Rural Sociological Society, Washington, DC.

BWG (1990) *Biotechnology's Bitter Harvest: Herbicide Tolerant Crops and the Threat to Sustainable Agriculture,* New York: Biotechnology Working Group, Environmental Defense Fund.

Byrne, D. (2002) Speech at inaugural meeting of the EFSA Management Board, 18 September, 'EFSA: Excellence, integrity and openness', available at http://www.efsa.europa.eu/en/press_room/speech/60.html

Cantley, M. (1995) The regulation of modern biotechnology: a historical and European perspective. a case study in how societies cope with new knowledge in the last quarter of the 20th century, in D Brauer (ed.), *Legal, Economic and Ethical Dimensions,* vol. 12 of the *Multi-Volume Comprehensive Treatise Biotechnology,* Weinheim: VCH, pp. 503–681.

Castree, N. (2000) 'Marxism and the production of nature', *Capital & Class* 72: 5–37.

CEC (1983a) Third Environmental Action Programme, *Official Journal of the European Communities,* C 46, 17 February. Brussels: Commission of the European Communities.

CEC (1983b) *Community Strategy for Biotechnology in Europe,* FAST Occasional Paper.

CEC (1987) Fourth Environmental Action Programme, *Official Journal of the European Communities,* C 329, 7 December.
CEC (1988) Proposal for a Council Directive on Protection of Biotechnological Inventions.
CEC (1992) 'Research after Maastricht: an assessment, a strategy', Bulletin of the European Communities, supplement 2/92, Brussels: Commission of the European Communities.
CEC (1993a) *Growth, Competitiveness and Employment: The Challenges and Ways Forward into the 21st Century,* Brussels: Commission of the European Communities.
CEC (1993b) *Towards Sustainable Development,* Fifth Environmental Action Programme; also in *Official Journal of the European Communities,* C 138, 17 May: 5–98.
CEC (1993c) *Official Journal of the European Communities,* C 329, 6 December.
CEC (1994) EC Fourth Framework Programme, in *Official Journal of the European Communities,* C 37: 177–87.
CEC (1996) Report on the Review of Directive 90/220/EEC in the Context of the Commission's Communication on 'Biotechnology and the White Paper'.
CEC (1997) Agenda 2000: For a Stronger and Wider Union, *Bulletin of the European Communities,* supplement 5/97, Brussels: Commission of the European Communities.
CEC (1998a) Review of the 5th Environmental Action Programme.
CEC (1998b) Proposal for a European Parliament and Council Directive amending Directive 90/220/EEC, on the Deliberate Release into the Environment of Genetically Modified Organisms, *Official Journal of the European Communities,* C 139, 4 May: 1–23.
CEC (2000) Communication from the Commission on the Precautionary Principle, Brussels: Commission of the European Communities (CEC), http://ec.europa.eu/environment/docum/20001_en.htm
CEC (2001a) *European Governance: A White Paper,* http://ec.europa.eu/comm/governance
CEC (2001b) *A Sustainable Europe for a Better World: A European Union Strategy for Sustainable Development,* COM (2001) 264.
CEC (2001c) Towards a Strategic Vision of Biotechnology and Life Sciences: Consultation Document, http://ec.europa.eu/comm/biotechnology
CEC (2001d) Commission improves rules on tracing and labelling of GMOs in Europe to enable freedom of choice and ensure environmental safety, IP/01/1095, 25 July.
CEC (2001e) Commission of the European Communities, Proposal for a regulation on genetically modified food and feed, COM (2001) 425 final, *Official Journal of the European Communities,* C 304, 30 October: 221–40.
CEC (2001f) Commission of the European Communities, proposal for a regulation concerning the traceability and labelling of GMOs and traceability of food and feed produced from GMOs, COM (2001) 182 final, 25 July, http://ec.europa.eu/comm/food/fs/biotech/biotech09_en.pdf
CEC (2002a) Towards a Strategic Vision of Biotechnology and Life Sciences, http://ec.europa.eu/comm/biotechnology
CEC (2002b) *Life Sciences and Biotechnology—A Strategy for Europe.*
CEC (2002c) Towards a European Food Safety Authority, http://ec.europa.eu/comm/food/fs/efa/index_en.html
CEC (2002d) The operating framework for the European Regulatory Agencies, Communication from the Commission, COM (2002) 718 final, Brussels, 11 December.

CEC (2003a) Commission Recommendation of 23 July on guidelines for the development of national strategies and best practices to ensure coexistence of GM crops with conventional and organic farming.

CEC (2003b) Communication from the Commission, Directive 98/34/EC, Notification 2003/200/A–OGM–Carinthie.

CEC (2003c) *Agriculture and Environment,* Brussels: Directorate-General for Agriculture, http://ec.europa.eu/agriculture/publi/fact/envir/2003_en.pdf

CEC (2003d) Towards a Strategic Vision of Biotechnology and Life Sciences: Progress Report and Future Orientations, http://ec.europa.eu/eur-lex/en/com/cnc/2003/com2003_0096en01.pdf, also http://ec.europa.eu/comm/biotechnology/pdf/com2003-96_en.pdf

CEC (2004) *The Common Agricultural Policy Explained.* http://ec.europa.eu/agriculture/pu bli/capexplained/cap_en.pdf

CEC (2005a) Report from the Commission to the Council and the European Parliament: Development and implications of patent law in the field of biotechnology and genetic engineering, SEC(2005) 943, COM (2005) 312 final, Brussels, 14 July.

CEC (2005b) Commission authorises import of GM oilseed rape for use in animal feed, Press Release, 31 August.

CEC (2006a) EC proposes practical improvements to the way the European GMO legislative framework is implemented, 12 April, http://www.europa-eu-un.org/articles/fi/article_5908_fi.htm

CEC (2006b) Report on the implementation of national measures on the coexistence of GM crops with conventional and organic farming, SEC (2006) 313. Communication from the Commission to the Council and the European Parliament.

CEC-SJ (2004a) 'Oral statement of the European Communities at the first meeting of the Panel with the Parties, European Communities—measures affecting the approval and marketing of biotech products (DS291, DS292, DS293)', 2 June, Commission of the European Communities-Service Juridique.

CEC-SJ (2004b) 'Second written submission by the European Communities: European Communities—measures affecting the approval and marketing of biotech products (DS291, DS292, DS293)', Geneva, 19 July.

CEC-SJ (2005) 'Comments by the European Communities on the Scientific and Technical Advice to the Panel: European Communities—measures affecting the approval and marketing of biotech products (DS291, DS292, DS293)', Geneva, 28 January, Commission of the European Communities-Service Juridique.

Cerny, P. G. (1999) 'Reconstructing the political in a globalizing world: states, institutions, actors and governance', in F. Buelens (ed.), *Globalization and the Nation-State,* Cheltenham: Edward Elgar, pp. 89–137.

CESE (2004) Coexistence between genetically modified crops, and conventional and organic crops, European Economic and Social Committee, CESE 1656/2004, 16 December, http://www.cese.europa.eu

Champion, G. T. et al. (2003) 'Crop management and agronomic context of the Farm Scale Evaluations of genetically modified herbicide-tolerant crops', *Philosophical Transactions:* Biological Sciences, Series B 358(1439): 1801–18.

Charles, D. (2001) *Lords of the Harvest: Biotech, Big Money and the Future of Food,* Cambridge, MA: Perseus.

Chataway, J., Tait, J. and Wield, D. (2004) 'Understanding company R&D strategies in agro-biotechnology: trajectories and blind spots', *Research Policy* 33(6-7): 1041-57.

Christoforou, T. (2003) 'The precautionary principle and democratising expertise: a European legal perspective', *Science and Public Policy* 30(3): 205–13.

Clark, J.R.A., Jones, A., Potter, C.A., Lobley, M. (1997) 'Conceptualising the evolution of the European Union's agri-environmental policy: a discourse approach', *Environment and Planning A* 29: 1869-85.

Coghlan, A. (2002) 'Peace of mind for Europe's shoppers may raise food bills all around', *New Scientist*, 1 June: 14.

Coghlan, A., Concar, D. and Mackenzie, D. (1999) 'Frankenfears', *New Scientist*, 20 February: 4–5.

Consiglio dei Diritti Genetici (2005) 'Analysi delle osservazioni del pubblico alle notifiche presentate secondo la direttiva 2001/18/CE', Roma.

COPA (2002a) Remarks on draft regulations on GM Food & Feed; and on Traceability and Labelling, http://www.copa.be

COPA (2002b) 'Implementation of Compulsory Labelling for GM Seed, as provided by Directive 98/95', Comments on working paper SANCO/1542/2002, for implementing recommendation no. 77 of the 1999 White Paper on Food Safety.

COPA (2003) Comments on the Fischler Communication on Coexistence of GM, Conventional and Organic Crops, Brussels: Committee of Agricultural Organisations of the EU, 19 May.

CoR (1995) Opinion of the Committee of the Regions on the Revision of the Treaty on European Union and of the Treaty Establishing the European Community, 21 April, http://www.cor.europa.eu.

CoR (2006) Opinion of the Committee of the Regions of 6 December 2006 on the Communication from the Commission to the Council and the European Parliament Report on the implementation of national measures on the coexistence of genetically modified crops with conventional and organic farming, COM (2006) 104 final, DEV IV-006, http://www.cor.europa.eu/cms/pages/documents/deve/en/Avis/cdr149-2006_fin_ac_en.pdf

CoR (2007) The Committee of the Regions—an introduction, http://www.cor.europa.eu

Cordis News (2003) 'Coexistence about economic risks, not health or food safety: Fischler', 26 April, http://www.checkbiotech.org

Corner House (1997) 'No Patents on Life!', http://www.thecornerhouse.org.uk

CPE (1999) Agenda 2000 negotiations: the funding of the CAP, Brussels: Coordination Paysanne Européenne, http://www.cpefarmers.org

CPE (2001) To change the CAP, Brussels: Coordination Paysanne Européenne, http://www.cpefarmers.org

CPE (2007) L'agriculture biologique en danger de contamination et d'industrialisation, http://www.cpefarmers.org

Cronon, W. (1991) *Nature's Metropolis: Chicago and the Great West*, New York: W. W. Norton.

Darvas, B., Székács, A., Bakonyi, G., Kiss, I., Biró, B., Villányi, I., Ronkay, L., Peregovits, L., Lauber, É. and Polgár, A. L. (2006) Authors' response to the Statement of the European Food Safety Authority GMO Panel concerning Environmental Analytical and Ecotoxicological Experiments Carried out in Hungary (Az Európai Élelmiszerbiztonsági Hivatal GMO paneljének a magyarországi környezetanalitikai és ökotoxikológiai vizsgálataival kapcsolatos állásfoglalásáról) *Növényvédelem*, 42: 313–25 (in Hungarian).

DEFRA (2004) Statement on GM policy by Secretary of State Margaret Beckett, 9 March, http://www.defra.gov.uk/corporate/ministers/statements/mb040309.htm

De Marchi, B. and Ravetz, J. (1999) 'Risk management and governance: a post-normal science approach', *Futures* 31: 743–57.

de Visser, A.J.C. *et al.* (2002) *Crops of Uncertain Nature: Controversies and Knowledge Gaps Concerning Genetically Modified Crops—An Inventory*, Plant Research International B.V., Commissioned by Greenpeace Netherlands, http://www.plant.wageningen-ur.nl

Decker, F. (2002) 'Governance beyond the nation-state: Reflections on the democratic deficit of the European Union', *Journal of European Public Policy* 9(2): 256–272.

Denys, S. (2005) 'Engager les OGM dans une innovation démocratique et scientifique', typescript, Bruxelles: ULB, sdenys@ulb.ac.be

Detaille, S. (2004) 'Pas dans les champs, pas encore dans l'assiette', *Le Soir,* 3 February.

DETR (1998) 'Government Announces Fuller Evaluations of Growing Genetically Modified Crops', news release, 21 October, Department of the Environment, Transport and the Regions, London.

DETR (1999a) The commercial use of genetically modified crops in the UK: the potential wider impact of farmland wildlife, http://www.environment.detr.gov.uk/acre/wildlife/index/htm.

DETR (1999b) Farm-scale evaluations of the impact of the management of GM herbicide tolerant oil-seed rape and maize on farmland wildlife, http://www.environment.detr.gov.uk/acre/wildlife/index/htm

Deutsche Bank (1999). 'GMOs are dead', 12 July, http://www.biotech-info.net/Deutsche.html

Dewar, A., May, M. J., Woiwood, I. P., Haylock, L. A., Champion, G. T., Garner, B. H., Sands, R. J., Qi, A. and Pidgeon, J. D. (2003) 'A novel approach to the use of genetically modified herbicide tolerant crops for environmental benefit', *Proceedings of the Royal Society of London,* Series B, 270: 33540.

DFNA (2003) Danish Forestry and Nature Agency, letters to DG Environment, 24 March and 30 September.

DFS (2006) 'Educate democracy', Dansk Folkeoplysnings Samråd, http://www.dfs.dk, Danish Adult Education Association.

DG-Envt (2002) Commission Decision of [. . .] establishing guidance notes supplementing Annex VII to European Parliament and Council Directive 2001/18/EC on the deliberate release into the environment of genetically modified organisms and repealing Council Directive 90/220/EEC, ENV/02/07, 30 April draft [finalized as EC, 2002b].

DG-Envt (2003) Working Group on Herbicide-Tolerant Plants: Final Report, doc nr Env/03/23.

DG-Envt (2007) Draft Commission Decision of [ ] concerning the placing on the market, in accordance with Directive 2001/18/EC of the European Parliament and of the Council, of a maize product (Zea mays L., line Bt11) genetically modified for resistance to certain lepidopteran pests and for tolerance to the herbicide glufosinate-ammonium [October proposal, not adopted by the Commission].

DG-JRC (2002) *Coexistence of GM Crops with Conventional and Organic Crops,* Joint Research Centre, European Commission.

DG-JRC (2006) *New Case Studies on the Coexistence of GM and non-GM Crops in European Agriculture.*

DG Research (2002) *Science for Society, Science with Society: How can research on food and agriculture in Europe better respond to citizens' expectations and demand?* Summary of EURAGRI conference, Brussels, 14–15 October.

DG Research (2004) The European Group on Life Sciences.

DG-SANCO (2002) Working document for a draft Commission Directive amending seeds directives, January.

DG XII (1994) *RTD Magazine* 1: 8–9, Brussels: Commission of the European Communities.

Dimas, S. (2006) Speech at conference on 'Co-existence of genetically modified, conventional and organic crops: the freedom of choice', Vienna, 4–6 April, http://ec.europa.eu/comm/agriculture/events/vienna2006/index_en.htm

Dively, G. P., Rose, R., Sears, M. K., Hellmich, R. L. Stanley-Horn, D. E., Calvin, D. D., Russo, J. M. and Anderson, P. L. (2004) 'Effects on monarch butterfly larvae (Lepidoptera: Danaidae) after continuous exposure to Cry1Ab-expressing corn during anthesis', *Environmental Entomology* 33(4): 1116–25.

Dobson, A. (1996) 'Environmental sustainabilities: an analysis and a typology', *Environmental Politics* 5(3): 401–28.

Dratwa, J. (2004) 'Social learning with the precautionary principle at the European Commission and the Codex Alimentarius', in B. Reinalda and B. Verbeek (eds), *Decision Making within International Organizations,* London: Routledge, 215–29.

Dreyer, M. and Gill, B. (2000) 'Germany: continued "elite precaution" alongside continued public opposition', *Journal of Risk Research* 3(3): 219–26.

DTI (2003) Reports on 'GM Nation?', London: Dept of Trade & Industry, www.gmnation.org.uk/

Duesing, J. (1989) 'Plant patenting as seen by a plant breeding professional', in *Patenting Life Forms in Europe*, Proceedings of an international conference at the European Parliament—Brussels, 7-8 February.

EC (1996a) Commission Decision 96/158/EC of 6 February 1996 concerning the placing on the market of a product consisting of a GMO, hybrid herbicide-tolerant swede-rape seeds, *Official Journal of the European Communities,* L 37, 15 February: 30–31.

EC (1996b) Commission Decision 96/281/EC of 3 April 1996 concerning the placing on the market of genetically modified soybeans, *Official Journal of the European Communities,* L 107, 30 April: 10.

EC (1997a) Commission Decision 97/98/EC of 23 January 1997 concerning the placing on the market of genetically modified maize, *Official Journal of the European Communities,* L 31, 1 February: 69–70.

EC (1997b) Regulation 258/97/EC of 27 January 1997 concerning novel foods and novel food ingredients, *Official Journal of the European Communities,* L 43, 14 February: 1–6.

EC (1997c) Commission Decision 97/579 of 23 July 1997 setting up Scientific Committees in the field of consumer heath and food safety, *Official Journal of the European Communities,* L 237, 28 August: 18–23.

EC (1997d) Commission decision concerning the placing on the market of a genetically modified oilseed rape (*B.napus L.oleifera* Metzg. MS1, RF1), in conformity with Council Directive 90/220/EEC, *Official Journal of the European Communities,* L 164, 6 June: 38–39.

EC (1998a) Directive 98/44/EC of the European Parliament and of the Council on Protection of Biotechnological Inventions, *Official Journal of the European Communities,* L 213, 30 July: 13.

EC (1998b) Council Regulation 1139/98 concerning the compulsory indication on the labelling of certain foodstuffs produced from genetically modified organisms, *Official Journal of the European Communities,* L 159, 3 June: 4.

EC (1998c) Directive 98/34/EC of the European Parliament and of the Council of 22 June 1998 laying down a procedure for the provision of information in the field of technical standards and regulation, *Official Journal of the European Communities,* L 204, 21 July: 37.

EC (2000) Commission Regulation 49/2000 amending Council Regulation 1139/98 concerning the compulsory indication on the labelling of certain foodstuffs produced from genetically modified organisms of particulars other than those provided for in Directive 79/112/EEC, *Official Journal of the European Communities,* L 6, 11 January: 13–14.

EC (2001) European Parliament and Council Directive 2001/18/EC of 12 March on the deliberate release into the environment of genetically modified organisms and repealing Council Directive 90/220/EEC, *Official Journal of the European Communities,* L 106, 17 April: 1–38.

EC (2002a) Regulation 178/2002 of 28 January 2002 laying down the general principles and requirements of food law, establishing the European Food Safety Authority, and laying down procedures in matters of food safety, *Official Journal of the European Communities,* L 31, 1 February: 1–23.

EC (2002b) Commission Decision of 24 July establishing guidance notes supplementing Annex II to European Parliament and Council Directive 2001/18/EC on the deliberate release into the environment of genetically modified organisms and repealing Council Directive 90/220/EEC, *Official Journal of the European Communities,* L 200, 30 July: 22–30.

EC (2002c) Commission Decision of 3 October establishing guidance notes supplementing Annex VII to European Parliament and Council Directive 2001/18/EC on the deliberate release into the environment of genetically modified organisms and repealing Council Directive 90/220/EEC, *Official Journal of the European Communities,* L 280, 18 October: 27–36.

EC (2003a) Regulation 1829/2003 of 22 September 2003 on genetically modified food and feed, *Official Journal of the European Communities,* L 268, 18 October: 1–23.

EC (2003b) Regulation 1830/2003 of 22 September 2003 concerning the traceability and labelling of GMOs and traceability of food and feed produced from GMOs and amending Directive 2001/18, *Official Journal of the European Communities,* L 268, 18 October: 24–28.

EC (2003c) Commission Decision of 2 September relating to national provisions on banning the use of GMOs in the region of Upper Austria notified by the Republic of Austria pursuant to Article 95(5) of the Treaty, 2003/653/EC, *Official Journal of the European Community,* 16 September, L 230: 34–43.

EC (2004) Commission Regulation 65/2004 of 14 January 2004 establishing a system for the development and assignment of unique identifiers for GMOs, *Official Journal of the European Communities,* L 10, 16 January: 5–7.

EC (2005) Commission Decision of 31 August 2005 concerning the placing on the market, in accordance with Directive 2001/18/EC of the European Parliament and of the Council, of an oilseed rape product (*Brassica napus* L., GT73 line) genetically modified for tolerance to the herbicide glyphosate, *Official Journal of the European Communities* L 228, 3 September: 11–13.

EC (2007a) Commission Decision 2007/232/EC of 26 March 2007 concerning the placing on the market, in accordance with Directive 2001/18/EC of the European Parliament and of the Council, of oilseed rape products (*Brassica napus* L., lines Ms8, Rf3 and Ms8xRf3) genetically modified for tolerance to the herbicide glufosinate-ammonium (notified under document number C(2007) 1234). *Official Journal of the European Communities.* L 100, 17 April: 20.

EC (2007b) Council Regulation (EC) No 834/2007 of 28 June 2007 on organic production and labelling of organic products and repealing Regulation (EEC) No 2092/91, *Official Journal of the European Communities,* L 189, 20 July: 1–23.

EcoSoc (1998) Opinion of the Economic and Social Committee on 'Genetically modified organisms in agriculture—impact on the CAP', *Official Journal of the European Communities,* C 284: 39–50.

Editorial (1999) 'Health risks of GM foods', *The Lancet* 353: 1811.

EEA (2002b) K. Eastham and J. Sweet, *Genetically Modified Organisms: The Significance of Gene Flow Through Pollen Transfer,* Copenhagen: European Environment Agency, http://www.eea.europa.eu

EEC (1990) Council Directive 90/220 on the Deliberate Release to the Environment of Genetically Modified Organisms, *Official Journal of the European Communities,* L 117, 8 May: 15–27.

EFSA (2006) Final Report of the EFSA Technical Meeting with NGOs on Genetically Modified Organisms, Parma, 22 February.

EFSA GMO Panel (2004a) Opinion related to the Austrian invoke of Article 23, July.
EFSA GMO Panel (2004b) Guidance document of the Scientific Panel on Genetically Modified Organisms for the risk assessment of genetically modified plants and derived food and feed, September.
EFSA GMO Panel (2005a) Scientific Panel on GMOs, PMEM WG, Minutes of general surveillance workshop with environmental organisations, held in December 2004.
EFSA GMO Panel (2005b) Scientific Panel on GMOs: Opinion on notification C/ES/01/01, Bt insect-protected maize 1507, January.
EFSA GMO Panel (2008) Scientific Panel on GMOs: Request from the European Commission related to the safeguard clause invoked by Hungary on maize MON810 according to Article 23 of Directive 2001/18/EC, 2 July.
Einsiedel, E.F., Jelsøe, E. and Breck, T. (2001) 'Publics at the technology table: The consensus conference in Denmark, Canada, and Australia', *Public Understanding of Science* 10: 83–98.
Elert, C. et al. (1991) *Biotechnology at Work in Denmark,* Copenhagen: Danish Board of Technology.
Elmegaard, N. and Bruus Pedersen, M. (2001) *Flora and Fauna in Roundup Tolerant Fodder Beet Fields,* Roskilde: National Environmental Research Institute, NERI Technical Report 349, http://www.dmu.dk/1_viden/2_Publikationer/3_fagrapporter/rapporter/FR349.pdf
Emmott, S. (2001) 'No patents on life! The incredible ten-year campaign against the European Patent Directive', in B. Tokar, *Redesigning Life? The Worldwide Challenge of Genetic Engineering,* London: Zed, pp. 373–84.
ENDS (1998a) 'The spiralling agenda of agricultural biotechnology', *ENDS Report* 283: 18–30.
ENDS (1998b) 'Iceland bans genetically modified ingredients from own-brands', *ENDS Report* 278: 24.
ENDS (1998c) 'The inhospitable climate for genetically modified crops', *ENDS Report* 287: 22–23.
ENDS (1999) 'Unilever, Nestle to drop GM ingredients from food products', *ENDS Report* 291: 30.
ENDS (2001) 'Commission sets out strategy for sustainability policy', *ENDS Report* 316: 47.
ENDS (2003) 'Sound science has its say on GM crops and biodiversity', *ENDS Report* 345: 27–31.
ENDS (2004a) Commission lifts EU's de facto GM moratorium, *ENDS Daily,* no. 1671, 19 May.
ENDS (2004b) 'Limited go-ahead for GM crops', *ENDS Report* 348: 12.
EP (1997) European Parliament, Report by the Temporary Committee of Inquiry Into BSE on Alleged Contraventions or Maladministration in the Implementation of Community Law in Relation to BSE, Without Prejudice to the Jurisdiction of the Community and National Courts, PE 220.544/fin, also in *Official Journal of the European Communities,* C 85: 61.
EP (2003) Report on Coexistence of Genetically Modified Crops and Conventional and Organic Crops. Rapporteur: Friedrich-Wilhelm Graefe zu Baringdorf. Brussels: Committee on Agriculture and Rural Development, European Parliament, 4 December, 2003/2098 (INI).
EP Plenary (2002) Debates of the European Parliament, Plenary sitting of 2 July, http://www.europarl.europa.eu
EP Plenary (2003a) Debates of the European Parliament, Plenary sitting of 1 July, http://www.europarl.europa.eu

EP Plenary (2003b) Debates of the European Parliament, Plenary sitting of 2 July, http://www.europarl.europa.eu

Eriksen, E. O. (2001) 'Democratic or technocratic governance?', in C. Joerges, Y. Mény and J. H. H. Weiler (eds), *Mountain or Molehill? A Critical Appraisal of the Commission White Paper on Governance,* Jean Monnet Working Paper No. 6/01, pp. 61–72, http://www.iue.it/RSC/, http://www.JeanMonnetProgram.org

EU (2000) *Agenda 2000: Reform of the Common Agricultural Policy,* Brussels: European Council.

EU Council (2000a) 'Council Resolution on the Precautionary Principle', Nice European Council.

European Council (2000b) *An Agenda of Economic and Social Renewal for Europe.* Brussels: European Council, DOC/00/7, held in Lisbon.

EU Council (2007) Environment Ministers, minutes of 30 October meeting, http://www.consilium.europa.eu/ueDocs/cms_Data/docs/pressdata/en/envir/96961.pdf

EU Environment Council (2006) Minutes of meeting, 27 June.

EU Food Law News (2000) 'Food law from farm to table: creating a European Food Authority', http://www.foodlaw.rdg.ac.uk/news/eu-00-85.htm

EuroCommerce (2002) Position Paper on GM F&F and T&L draft regulations, Brussels: EuroCommerce (Retail, Wholesalers and International Trade Representative to the EU), http://www.eurocommerce.be

EuroCommerce (2003) Eurocommerce calls for strict purity criteria for GM contamination of seeds, October.

EuroCommerce (2004) Implementation of Traceability in the Commerce Sector.

Euro-Coop (2003) Comments on the draft Recommendation for implementation of EC Regulation 1829/2003.

EuropaBio (2003) 'Coexisting with GM crops: European Commission publishes guidelines', 23 July.

EuropaBio (2004a) Technical Advisory Group (TAG), Plant Biotechnology Unit, EuropaBio, comments on Draft Guidance [from EFSA GMO Panel].

EuropaBio (2004b) Comments on draft Commission decision establishing minimum thresholds for adventitious or technically unavoidable traces of GM seeds in other products, April.

EuropaBio (2005) Coexistence of GM crops with non-GM crops is possible in the EU, 1 February.

EuropaBio (2006a) 'Co-existence—choice or denial?', 13 March.

EuropaBio (2006b) 'Where is the co-existence? Where is the choice?', 5 April.

Everard, M. and Ray, D. (1999) *Genetic Modification and Sustainability,* 2020 Vision Series No. 1, Bristol: Environment Agency/Cheltenham: The Natural Step.

Ewen, S. W. B. and Pusztai, A. (1999) 'Effect of diets containing genetically modified potatoes expressing Galanthus nivalis lectin on rat small intestine', *The Lancet* 354: 1353–54.

Fairbank, L. G. et al. (2003) An introduction to the Farm-Scale Evaluations of genetically modified herbicide-tolerant crops, *Journal of Applied Ecology* 40: 2–16.

FEBC (1997) Views on 'The Directive on the Legal Protection of Biotechnological Inventions', Forum for European Bioindustry Coordination, hosted by EuropaBio, http://www.europabio.be

FEC (2004) *Just Knowledge? Governing Research on Food and Farming,* Brighton: Food Ethics Council.

Fenton, B. and Irwin, A. (1999) 'Vested interests cloud search for truth', *The Daily Telegraph* 15 February: 4.

Ferretti, M. P. (2007) 'What do we expect from public participation? The case of authorising GMO products in the European Union', *Science as Culture* 16(4): 377–95.
FFA (2005) Foundation Future Farming, Assembly of European Regions, 'Berlin Manifesto for GMO-free Regions and Biodiversity in Europe', January, http://www.are-regions-europe.org
Fischer, F. (2005) 'Are scientists irrational? Risk assessment in practical reason', in M. Leach, I. Scoones and B. Wynne (eds), *Science and Citizens: Globalisation and the Challenge of Engagement,* London: Zed, pp. 54–65.
Fischer Boel, M. (2005) Harmonising coexistence rules, *EuroBiotechNews* 4(1–2): 24–25.
Fischer Boel, M. (2006) Speech at conference on 'Co-existence of genetically modified, conventional and organic crops: the freedom of choice', Vienna, 4–6 April, http://ec.europa.eu/comm/agriculture/events/vienna2006/index_en.htm, http://ec.europa.eu/commission_barroso/fischer-boel/speeches/archive_en.htm
Fischler, F. (2003) Communication from Mr Fischler to the Commission, 'Coexistence of Genetically Modified, Conventional and Organic Crops', SEC (2003) 258/2, March.
Flynn, L. and Gillard, M. (1999) 'Pro-GM scientist "threatened editor"', *The Guardian,* 1 November.
FoEE (1996) 'Working Group on risk assessment meets again', *Biotech Mailout* 4: 4-5. Brussels: Friends of the Earth Europe.
FoEE (1997) 'France authorizes cultivation of GM maize', *Biotech Mailout* 3(8), 15 December, Brussels: Friends of the Earth Europe Biotechnology Programme, http://www.foeeurope.org/biotechnology
FoEE (1998) 'France: NGOs take action against Bt maize', *Biotech Mailout* 4(6), 15 September.
FoEE (1999) 'EU Environment Council adopts de facto moratorium on GMOs', *FoEE Biotech Mailout* 5(5), 31 July: 1–4. Brussels: Friends of the Earth Europe, http://www.foeeurope.org
FoEE (2000) 'The European campaign to halt GMO pollution', *FoEE Biotech Mailout* 6(2), 15 March: 4, Brussels: Friends of the Earth Europe.
FoEE (2001) 'GM contamination in seeds too', *FoEE Biotech Mailout* 7(4), August: 6–7.
FoEE (2002a) 'Risk assessment and monitoring of GMOs', *Biotech Mailout,* 1 April.
FoEE (2002b) 'Action on GM: science war continues', 24 June.
FoEE (2002c) 'Coexistence: what is it?', Brussels: Friends of the Earth Europe, http://www.foeeurope.org/GMOs/Coexistence.htm
FoEE (2003a) 'New moves to end the EU moratorium on GMOs; Coexistence is expensive', *FoEE Biotech Mailout* 9(1), February, http://www.foeeurope.org/GMOs/Index.htm
FoEE (2003b) Comments on Part C product files: Monsanto RR oilseed rape, event GT73 (C/NL/98/11); Bt maize Mon 863 and Mon 863 x Mon 810 (C/DE/02/9); Syngenta Bt 11 maize (C/F/96/05/10).
FoEE (2003c) 'The Commission dodges its responsibility', *FoEE Biotech Mailout,* May: 1–3.
FoEE (2003d) 'EU regions call for GM free zones', *FoEE Biotech Mailout,* December: 15.
FoEE (2004a) 'Member states vote against Commission proposals to end national bans', *FoEE Biotech Mailout,* November: 6–7.
FoEE (2004b) *Throwing Caution to the Wind: A review of the European Food Safety Authority and its work on genetically modified foods and crops,* Brussels:

Friends of the Earth Europe Biotechnology Programme, http://www.foeeurope.org/GMOs/publications/EFSAreport.pdf
FoEE (2005a) Table on how the EU member states voted on GMOs, http://www.foeeurope.org/GMOs/
FoEE (2005b) *Biotech Mailout,* July: 1–4, http://www.foeeurope.org/GMOs
FoEE (2005c) 'Legal advice on coexistence', *FoEE Biotech Mailout,* July: 4–6.
FoEE (2005d) 'Italy bans GM crops', *FoEE Biotech Mailout,* March: 1–2.
FoEE (2005e) 'Regions demand "power-sharing" over GMO decisions', *FoEE Biotech Mailout,* July: 15–16.
FoEE (2006a) 'Can We Trust EFSA's Science?', presentation at meeting with EFSA's GMO Panel, 22 February, http://www.foeeurope.org/GMOs/
FoEE (2006b) 'Contaminate or legislate? European Commission policy on "coexistence"'; Position Paper, April.
FoEE/GP/EEB (2003) 'Coexistence of GMOs and non-GM agriculture: the European Commission dodges its responsibility', 3 March, Brussels: Friends of the Earth Europe, European Environmental Bureau, Greenpeace.
FoEE/Greenpeace/Global 2000 (2006) *Hidden Uncertainties: What the European Commission doesn't want us to know about the risks of GMOs.*
Frouws , J. and van Tatenhove, J. (1993) 'Agriculture, environment and the state: the development of agro-environmental policy-making in the Netherlands', *Sociologia Ruralis* 33(2): 240-261.
Gaskell, G. (2006) Rapporteur of workshop on 'Consumer attitude and market response: viewpoints of stakeholders', conference on 'Co-existence of genetically modified, conventional and organic crops: the freedom of choice', Vienna, 4–6 April, http://ec.europa.eu/comm/agriculture/events/vienna2006/index_en.htm
Gaskell, G. (2008) 'Lessons from the bio-decade', in K. David and P. B. Thompson (eds), *What Can Nanotechnology Learn from Biotechnology?: Social and Ethical Lessons for Nanoscience from the Debate over Agrifood Biotechnology and GMOs,* London: Academic Press, pp. 237–58.
Gaskell, G., Allum, N., Bauer, M., Durant, J., Allansdottir, A., Bonfadelli, H., Boy, D., de Cheveigné, S., Fjaestad, B., Gutteling, J. M., Hampel, J., Jelsøe, E., Jesuino, J. C., Kohring, M., Kronberger, N., Midden, C., Nielsen, T. H., Przestalski, A., Rusanen, T., Sakellaris, G., Torgersen, H., Twardowski, T. and Wagner, W. (2000) 'Biotechnology and the European public', *Nature Biotechnology* 18: 935–8.
GEA (1999) 'Five Year Freeze on genetic engineering and patenting in food and farming', London: Genetic Engineering Alliance, http://www.gmfreeze.org
Geitner, P. (2004) 'EU backs GM maize, pushes to end ban', Associated Press, 28 January.
GEN (1997–1998) *Genetix Update,* bi-monthly newsletter from the Genetic Engineering Network.
Genetics Forum (1989) Submission to Royal Commission on Environmental Pollution, typescript.
Genetics Forum (1994) 'Whose consensus?', *The Splice of Life,* December.
Genewatch UK (2000) 'Privatising Knowledge, Patenting Genes: The Race to Control Genetic Information', Briefing Number 11, http://www.genewatch.org
GIBiP (1990) 'The green industry biotechnology platform', *Agro-Industry Hi-Tech* 1(1): 55–57.
Gifford, R. (2000) 'The United States and Europe: Different Reactions to Biotechnology', speech at EU-US Conference, 'Have Allies Become Adversaries?', 6 July, Brussels.
Gill, B. (1993) 'Technology assessment in Germany's biotechnology debate', *Science as Culture* 4(1): 69–84.

Gill, B. (1996) 'Germany: splicing genes, splitting society', *Science and Public Policy* 23(3): 175–79.
Global 2000 (1997) 'No Patents on Life!', http://www.global2000.at/index.php
GM Science Review (2003) *First Report: An open review of the science relevant to GM crops and food based on the interests and concerns of the public*. London: GM Science Review Panel, www.gmsciencedebate.org.uk/report/pdf/gmsci-report1-pt1.pdf
GMO-Free Regions (2007) 3rd International Conference on GMO-Free Regions, Biodiversity, and Rural Development, http://www.gmo-free-regions.org/
Goodman, D. and Redclift, M. (1991) *Refashioning Nature: Food, Ecology & Culture*. London: Routledge.
Goodman, Robert M. (1989) 'Biotechnology and sustainable agriculture: policy alternatives', in J.F.MacDonald, ed., *Agricultural Biotechnology at the Crossroads*, pp.48-57. Ithaca, NY: NABC.
Gottweis, H. (1998) *Governing Molecules: The Discursive Politics of Genetic Engineering in Europe and the United States*, Cambridge, MA: MIT Press.
Gottweis, H. (2005) 'Transnationalizing recombinant-DNA regulation: between Asilomar, EMBO, the OECD, and the European Community', *Science as Culture* 14(4): 325–38.
Goven, J. (2006) 'Dialogue, governance and biotechnology: acknowledging the context of the conversation', *Integrated Assessment Journal* 6(2): 99–116.
Greenpeace (1997) 'From BSE to Genetically Modified Organisms: Science, Uncertainty and the Precautionary Principle', London: Greenpeace.
Greenpeace (2000) *GM on Trial*, http://www.greenpeace.org.uk
Greenpeace (2004) 'EFSA: Failing Consumers and the Environment', comments on draft guidance.
Greenpeace (2006) *Impossible Coexistence*, translation of Greenpeace España report, *La Imposible Coexistencia*, http://www.greenpeace.eu/issues/gmo.html
Greenpeace España (2004) Los cultivos transgénicos contaminan: casos conocidos en España. Greenpeace España.
Greenpeace Germany (2006) 'Failure of EFSA in GMO risk assessment—case studies', presented by Christoph Then at meeting with EFSA on 22 February.
Greenpeace Research Laboratories (2003) 'Greenpeace comments on Bt 11 maize, C/FR/96/05/10', 23 July.
Greens/EFA (2001) *Agroecology: Toward a New Agriculture for Europe*, Brussels: Greens/EFA in the European Parliament.
Grundahl, J. (1995) 'The Danish consensus conference model', in S. Joss and J. Durant (eds), *Public Participation in Science: the Role of Consensus Conferences in Europe*, London: Science Museum, pp. 3–40.
Gupta, S. (2007) *The Theory and Reality of Democracy: A Case Study in Iraq*, London: Continuum.
Habermas, J. (2001) *The Postnational Constellation: Political Essays*, Oxford: Polity.
Haerlin, B. (1990) 'Genetic engineering in Europe', in P. Wheale and R. McNally, (eds), *The BioRevolution: Cornucopia or Pandora's Box?* London: Pluto, pp. 253–61.
Hajer, M. (1995) *The Politics of Environmental Discourse: Ecological Modernisation and the Policy Process*, Oxford: Oxford University Press.
Hajer, M. (2005) 'Setting the stage: a dramaturgy of policy deliberation', *Administration & Society* 36(6): 624-47.
Hajer, M. and Versteeg, W. (2005) 'A decade of discourse analysis of environmental politics: achievements, challenges, perspectives', *Journal of Environmental Policy & Planning* 7(3): 175–84.

Hajer, M. and Wagenaar, H. (2003) *Deliberative Policy Analysis: Understanding Governance in the Network Society,* Cambridge: Cambridge University Press.

Halffman, W. and Hoppe, R. (2005) 'Science/policy boundaries: a changing division of labour in Dutch expert policy advice', in S. Maasen and P .Weingart (eds), *Democratization of Expertise? Exploring Novel Forms of Scientific Advice in Political Decision-Making,* Dordrecht: Springer, pp. 135–41.

Halliday, J. (2007) EU has opened door to GM in organics, say activists, 15 June, Decision News Media SAS, http://www.checkbiotech.org/green_News_Genetics.asp

Hamlett, P. (2003) 'Technology theory and deliberative theory', *Science, Technology and Human Values* 28(1): 122-40.

Hanf, K. (1996) 'Implementing environmental policies', in A. Blowers and P. Glasbergen (eds), *Environmental Policy in an International Context—Prospects for Environmental Change,* London: Arnold, pp. 197–221.

Hansen, L. et al. (1992) 'Consensus Conferences', Copenhagen: Danish Board of Technology.

Hansen, J. (2006) 'Operationalising the public in participatory technology assessment', *Science and Public Policy* 33(8): 571-584.

Haskins, Lord (2002) *The Future of European Rural Communities,* London: Foreign Policy Centre.

Heller, C. (2002) 'From scientific risk to *paysan savoir-faire*: peasant expertise in the French and global debate over GM crops', *Science as Culture* 11(1): 5–37.

Hennessy, P. (2004) 'GM firm pulls out of scheme to grow maize', *Evening Standard,* 31 March: 18.

Hilbeck, A. et al. (1998) 'Effects of transgenic Bt corn-fed prey on mortality and development time of immature Chrysoperla carnea', *Environmental Entomology* 27(2): 480-87.

Hilbeck, A. et al. (1998) 'Toxicity of Bt Cry1Ab toxin to the predator Chrysoperla carnea', *Environmental Entomology* 27(5): 1255-63.

Hilgartner, S. and Bosk, C.L. (1988) 'The rise and fall of social problems: A public arenas model', *American Journal of Sociology* 94(1): 53-78.

Hinsliff, G. (2002) 'Alert to threat over organic farms', *The Observer,* 26 May.

Hobbelink, H. (1991) *Biotechnology and the Future of World Agriculture,* London: Zed.

Hoppichler, J. (1999) *Concepts of GMO-free Environmentally Sensitive Areas,* Vienna: Federal Institute for Less-Favoured and Mountainous Areas, http://www.bergbauern.com

Horlick-Jones, T. et al (2004) *A Deliberative Future? An Independent Evaluation of the GM Nation? Public Debate about the Possible Commercialisation of Transgenic Crops in Britain,* Understanding Risk Working Paper 04-02, Norwich: University of East Anglia.

Horlick-Jones, T., Walls, J., Rowe, G., Pidgeon, N., Poortinga, W. and O'Riordan, T. (2006) 'On evaluating the GM Nation? Public debate about the commercialisation of transgenic crops in Britain', *New Genetics and Society* 25(3): 265–88.

House of Lords (1993) Regulation of the United Kingdom Biotechnology Industry and Global Competitiveness, House of Lords, Select Committee on Science and Technology, 7th report, London: HMSO.

Huffman, W. and Just, R. (1999) 'The organization of agricultural research in Western developed countries', *Agricultural Economics* 21: 1–18.

Iceland (1998) 'Genetic modification of food and how it affects you', May.

IGC (2004) Inter-Governmental Conference, Treaty establishing a Constitution for Europe, CIG/87/2/04, *Official Journal of the European Communities,* C 310, 16 December.

Imhof, H. (1998) 'Challenges for sustainability in the crop protection and seeds industry', talk by Novartis officer at Rabobank International's Global Conference.

Irwin, A. (1995) *Citizen Science: A Study of People, Expertise and Sustainable Development. London*: Routledge.

Italy (2007) European Policy on GMOs and the role of EFSA, Italian delegation to EU Environment Council, 18 October, document 13919/07, unofficial.

James, P., Kemper, F. and Pascal, G. (1999) *The Future of Scientific Advice in the EU*, http://ec.europa.eu/comm/food/fs/sc/future_en.html

Jank, B., Rath, J. and Gaugitsch, H. (2006) 'Co-existence of agricultural production systems', *Trends in Biotechnology* 24(5): 198–201.

Jasanoff, S. (1986) *Risk Management and Political Culture*. London: Russell Sage.

Jasanoff, S. (1987) 'Contested boundaries in policy-relevant science', *Social Studies of Science* 17(2): 231–56.

Jasanoff, S. (1995) 'Product, process, or programme: three cultures and the regulation of biotechnology', in M. Bauer (ed.), *Resistance to New Technology*, pp. 311-31. Cambridge: Cambridge UP.

Jasanoff, S. (1997) 'Civilisation and madness: the great BSE scare of 1996', *Public Understanding of Science* 6: 221-32.

Jasanoff, S. (2004) 'Ordering knowledge, ordering society', in S. Jasanoff (ed.), *States of Knowledge: The Co-production of Science and Social Order*, London: Routledge, pp.13–45.

Jasanoff, S. (2005a) *Designs on Nature: Science and Democracy in Europe and the United States*, Princeton: Princeton University Press.

Jasanoff, S. (2005b) 'Judgment under siege: the three-body problem of expert legitimacy', in S.Maasen and P.Weingart, eds, *Democratization of Expertise? Exploring Novel Forms of Scientific Advice in Political Decision-Making*, pp.209-24, Springer, Sociology of the Sciences Yearbook.

JNCC (1997) Response from the Conservation Agencies [to MAFF, 1997], Joint Nature Conservancy Committee, including English Nature, Scottish Natural Heritage.

Joerges, C. (1997) 'Scientific expertise in social regulation and the European Court of Justice', in C. Joerges, K.-H. Ladeur and E. Vos (eds), *Integrating Scientific Expertise into Regulatory Decision-Making: National Traditions and European Innovations*, Baden-Baden: Nomos, pp. 295–323.

Joerges, C. (1999) '"Good governance" through comitology', in C. Joerges and E. Vos (eds), *EU Committees: Social Regulation, Law and Politics*, Portland, OR: Hart Publications, 311–38.

Joerges, C. (2006) *Free Trade with Hazardous Products? The Emergence of Transnational Governance with Eroding State Government*, EUI Working Paper LAW No. 2006/5, http://www.iue.it

Joly, P.-B.; Marris, C., Hermitte, M.A. (2003) 'A la recherche d'une 'démocratie technique'. Enseignements de la conférence citoyenne sur les OGM en France', *Natures, Sciences et Sociétés* 11(1): 3-15.

Joly, P.-B. and Assouline, G. (2001) *Assessing Debate and Participative Technology Assessment in Europe*, Final report, Grenoble: INRA Economie et Sociologie rurales; Teys: QAP Decision, Chapter 2, http://www.inra.fr/sed/science-gouvernance/

Jordan, T. (2002) 'Introduction: European Union environmental policy—actors, institutions and policy processes', in A. Jordan (ed.), *Environmental Policy in the European Union*, London: Earthscan, pp. 1–12.

Joss, S. 1998. 'The Danish consensus conferences as model of participatory technology assessment', *Science & Public Policy* 25 (1): 2-22.

Joss, S. (2005a) 'Lost in translation? Challenges for participatory governance of science and technology', in H. Torgersen and A. Bogner, eds, *Wozu Experten/ Why Experts?*, pp.197-219. Wiesbaden: Verlag für Sozialwissenschaften.

Joss, S. (2005b) 'Between policy and politics', in S. Maasen and P. Weingart (eds), *Democratization of Expertise? Exploring Novel Forms of Scientific Advice in Political Decision-Making*, Dordrecht: Springer, Sociology of the Sciences Yearbook, pp. 171–88.

Joss, S. and Durant, J. (eds) (1995) *Public Participation in Science: The Role of Consensus Conferences in Europe*, London: Science Museum.

Judge, D., ed. (2003) *A Green Dimension for the European Community*. London: Frank Cass.

Kay, L. (1992) *The Molecular Vision of Life: Caltech, the Rockefeller Foundation and the Rise of the New Biology*, New York/Oxford: Oxford University Press.

Kenney, M. (1986) *Biotechnology: The University Industrial Complex*, New Haven, CT: Yale University Press.

Klintman, M. (2002) 'The GM food labelling controversy: ideological and epistemic crossovers', *Social Studies of Science* 32(1): 71–91.

Kloppenburg, J. (1988) *First the Seed: The Political Economy of Plant Biotechnology, 1492–2000*, Cambridge: Cambridge University Press.

Klüver, L. (1995) 'Consensus conferences at the Danish Board of Technology', in S. Joss and J. Durant (eds), *Public Participation in Science: The Role of Consensus Conferences in Europe*, London: Science Museum, pp. 41–49.

Krimsky, S. (1991) *Biotechnics and Society*, New York: Praeger.

Krimsky, S. and Plough, A. (1988) *Environmental Hazards: Communicating Risks as a Social Process*, Dover, MA: Auburn House.

Krimsky, S. and Wrubel, R. (1996) *Agricultural Biotechnology and the Environment: Science, Policy and Social Issues*, Chicago: University of Illinois Press.

Krohn, W. and Weyer, J. (1994) 'Society as a laboratory: the social risks of experimental research', *Science and Public Policy* 21(3): 173–83.

Lagnado, L. (1999) 'Strained peace: Gerber baby food, grilled by Greenpeace, plans swift overhaul. Gene-modified corn and soy will go', *Wall Street Journal*, 30 July: A1.

Lake, G. (1991) 'Scientific uncertainty and political regulation: European legislation on the contained use and deliberate release of genetically modified (micro) organisms', *Project Appraisal* 6(1): 7-15.

Landfried, C. (1999) 'The European regulation of biotechnology by polycratic governance', in C. Joerges and E. Vos (eds), *EU Committees: Social Regulation, Law and Politics*, Portland, OR: Hart Publications, pp. 173–94.

Lang, A. (2004) 'Monitoring the impact of Bt maize on butterflies in the field: estimation of required sample sizes', *Environmental Biosafety Research* 3: 55–66.

Lasok, K. and Haynes, R. (2005) Advice: In the matter of coexistence, traceability and labelling of GMOs, 21 January.

Latour, B. (1993) *We Have Never Been Modern*, Cambridge, MA: Harvard University Press.

Lawrence, R. (1988) 'New applications of biotechnology in the food industry', in *Biotechnology and the Food Supply*, Washington, DC: National Academy Press, pp. 19–45.

Lean, G. (2005) 'Mandelson wants to fast-track GM', *Independent on Sunday*, 6 November.

le Bail, F. (2005) quoted in *The Guardian*, 31 May, also at http://www.referendum.org/uk/main/euindenial.html

Le Tourbillon Masqué (2008) MON 810, Monsanto échoue au Conseil d'État, 19 March.

Levidow, L. (2000) 'Pollution metaphors in the UK biotechnology controversy', *Science as Culture* 9(3): 325–51.

Levidow, L. (2001) 'Precautionary uncertainty: regulating GM crops in Europe', *Social Studies of Science* 31(6): 845–78.

Levidow, L. (2002) 'Ignorance-based risk assessment? Scientific controversy over GM food safety', *Science as Culture* 11(1): 61–67.

Levidow, L. (2003) 'Policing the Scientific Debate on GM Crops: the Royal Society meeting of 11th Feb 2003', http://www.gmsciencedebate.org.uk/topics/forum/0070.htm

Levidow, L. (2005) 'Governing conflicts over sustainability: agricultural biotechnology in Europe', in V. Higgins and G. Lawrence, eds, *Agricultural Governance: Globalization and the New Politics of Regulation,* London: Routledge, pp. 98–117.

Levidow, L. (2007) 'European public participation as risk governance: enhancing democratic accountability for agbiotech policy?', *East Asian Science, Technology and Society: An International Journal* 1(1): 19–51, special issue on Public Participation in Science, http://www.springerlink.com/content/7wv9764lw42r7867/fulltext.pdf

Levidow, L. and Bijman, J. (2002) 'Farm inputs under pressure from the European food industry', *Food Policy* 27(1): 31–45.

Levidow, L. and Boschert, K. (2008) Coexistence or contradiction? GM crops versus alternative agricultures in Europe, *Geoforum* 39: 174–90.

Levidow, L. and Carr, S. (1996) 'UK: disputing boundaries of biotechnology regulation', *Science and Public Policy* 23(3): 164–70.

Levidow, L. and Carr, S. (2000) 'Normalizing novelty: regulating biotechnological risk at the US EPA', *Risk—Health, Safety and Environment* 11(1): 61–86.

Levidow, L. and Carr, S. (2007a) 'Europeanising advisory expertise: the role of "independent, objective and transparent" scientific advice in agri-biotech regulation', *Environment & Planning C: Government & Politics* 25(6): 880–95.

Levidow, L. and Carr, S. (2007b) 'GM crops on trial: technological development as a real-world experiment', *Futures* 39(4): 408–31.

Levidow, L., Carr, S., von Schomberg, R. and Wield, D. (1996) 'Regulating agricultural biotechnology in Europe: harmonization difficulties, opportunities, dilemmas', *Science and Public Policy* 23(3): 135–57.

Levidow, L., Carr, S. and Wield, D. (2000) 'Genetically modified crops in the European Union: regulatory conflicts as precautionary opportunities', *Journal of Risk Research* 3(3): 189–208.

Levidow, L., Carr, S. and Wield, D. (2005) 'EU regulation of agri-biotechnology: precautionary links between science, expertise and policy', *Science and Public Policy* 32(4): 261–76, http://technology.open.ac.uk/cts/peg/sppaug2005eu%20fin.pdf

Levidow, L. and Marris, C. (2001) 'Science and governance in Europe: lessons from the case of agbiotech', *Science and Public Policy* 28(5): 345–60.

Levidow, L., Murphy, J. and Carr, S. (2007) 'Recasting "Substantial Equivalence": transatlantic governance of GM food', *Science, Technology and Human Values* 32(1): 26–64.

Levidow, L., Søgaard, V. and Carr, S. (2002) 'Agricultural PSREs in Western Europe: research priorities in conflict', *Science and Public Policy* 29(4): 287–95.

Lewis, S. (2004a) 'EU sends mixed signals on Roundup Ready corn', *Food Chemical News,* 26 July: 7.

Lewis, S. (2004b) 'Delayed studies stall EU seeds thresholds proposal until spring', *Food Chemical News*, 13 December: 6–7.

Lewis, S. (2006a) 'EU Ministers blast biotech approval scheme', *Food Chemical News,* 13 March: 5–6.

Lewis, S. (2006b) 'European Commission rules out harmonized coexistence law', *Food Chemical News,* 20 March: 5–6.
Lewis, S. (2006c) 'EU environment chief resists "adventitious presence" for seeds', *Food Chemical News,* 15 May: 7–8.
Lewis, S. (2006d) 'EFSA yields to pressure', *Food Chemical News,* 22 May: 6.
Lezaun, J. (2006) 'Creating a new object of government: making genetically modified organisms traceable', *Social Studies of Science* 36(4): 499–531.
Lezaun, J. and Millo, Y. (2006) 'Regulatory experiments: GM crops and financial derivatives on trial', *Science and Public Policy* 33(3): 179–90.
Lezaun, J. and Soneryd, L. (2006) *Government by Elicitation: Engaging Stakeholders or Listening to the Idiots?* London: LSE, CARR Discussion Paper no. 34, http://www.lse.ac.uk/carr
LGC/DTI (1991) *Biotechnology: A Plain Man's Guide to the Support and Regulations in the UK,* Cambridge: Laboratory of the Government Chemist, UK Department of Trade & Industry.
Liberatore, A., rapporteur (2001) 'Democratising Expertise and Establishing Scientific Reference Systems', Report of the Working Group 1b, Broadening and enriching the public debate on European matters, *White Paper on European Governance,* http://ec.europa.eu/comm/governance/areas/group2/index_en.htm
Liberatore, A. and Funtowicz, S. (2003) 'Democratising expertise, expertising democracy: what does this mean, and why bother?', *Science and Public Policy* 30: 146-50.
Lord, C. (2001) 'Democracy and democratization in the European Union', in S. Bromley (ed.), *Governing the European Union,* London: Sage, pp.165-91.
Lowe, P., Ward, N., Seymour, S., Clark, J. (1996) 'Farm pollution as environmental crime', *Science as Culture* 5(4): 588-612.
Lowe, P., Clark, J., Seymour, S., Ward, N. (1997) *Moralising the Environment: Countryside Change, Farming and Pollution.* London: UCL Press.
Luján, J. L., Mirabal, O., Borrillo, D., Santesmases, M. J. and Muñoz, E. (1996) 'Spain: transposing EC regulation through negotiation', *Science and Public Policy* 23(3): 181–84.
McAfee, K. (2003) 'Neoliberalism on the molecular scale: Economic and genetic reductionism in biotechnology battles', *GeoForum* 34: 203-19.
MAFF (1998) Scientific Review of the Impact of Herbicide Use on Genetically Modified Crops.
Magnaghi, A. (2005) 'Local self-sustainable development: subjects of transformation', *Tailoring Biotechnologies* 1(1): 79–102, http://www.tailoringbiotechnologies.com
Magnien, E. and de Nettancourt, D. (1993) 'What drives European biotechnological research?', in E. J. Blakelely and K. W. Willoughby (eds), *Biotechnology Review no.1: The Management and Economic Potential of Biotechnology,* Brussels: Commission of the European Communities, pp. 47–48.
Magretta, J. (1997) 'Growth through global sustainability: an interview with Monsanto's CEO, Robert Shapiro', *Harvard Business Review,* January–February: 79–88.
Majone, G. (1996) 'Regulatory legitimacy', in G. Majone (ed.), *Regulating Europe,* London: Routledge, pp. 284–301.
Majone, G. (1997) 'From the positive to the regulatory state: causes and consequences of changes in the mode of governance', *Journal of Public Policy* 17: 139–67.
Marris, C. (2000) 'Swings and roundabouts: French public policy on agricultural GMOs since 1996', *Politeia* 60: 22–37.
Marris, C. and Joly, P.-B. (1999) 'Between consensus and citizens: public participation in technology assessment in France', *Science Studies,* 12(2): 3–32.

Marris, C., Joly, P.-B., Ronda, S. and Bonneuil, C. (2005) 'How the GM controversy in France led to the reciprocal emancipation of scientific expertise and policy making', *Science and Public Policy* 32(4): 301–8.

Marris, C., Ronda, S., Bonneuil, C. and Joly, P.-B. (2004) 'France: battling with expertise', national report for 'Precautionary Expertise for GM Crops' project, Paris: INRA-TSV.

Marsden, T. K. and Smith, E. (2005) 'Ecological entrepreneurship: sustainable development in local communities through quality food production and local branding', *Geoforum* 36: 440–51.

Marsden, T. and Sonnino, R. (2005) Rural food and agri-food governance in Europe: tracing the development of alternatives, in V. Higgins and G. Lawrence (eds), *Agricultural Governance: Globalization and the New Politics of Regulation,* London: Routledge, pp. 50–68.

Mayer, S. (1999) 'Science's secret garden', *Financial Times,* 5 August.

McGiffen, S. (2005) *Biotechnology: Corporate Power versus the Public Interest,* London: Pluto.

McHugen, A. (2008) 'Learning from mistakes', in K. David and P. B. Thompson (eds), *What Can Nanotechnology Learn from Biotechnology?: Social and Ethical Lessons for Nanoscience from the Debate over Agrifood Biotechnology and GMOs,* London: Academic Press, pp. 33–53.

McKechnie, S. (1999) 'Food fright', *The Guardian,* 10 February [director of the Consumers Association].

McMichael, P. (1998) 'Global food politics', *Monthly Review* 50: 97–111.

Mellon, M. (1991) 'Biotechnology and the environmental vision', in J. F. MacDonald (ed.), *Agricultural Biotechnology at the Crossroads,* Ithaca, NY: NABC, pp. 65–70, http://www.cals.cornell.edu/extension/nabc

Millstone, E., Brunner, E. and Mayer, S. (1999) 'Beyond "substantial equivalence"', *Nature* 401(7): 525–26.

Millstone, E. and van Zwanenberg, P. (2001) 'Politics of expert advice: lessons from the early history of the BSE saga', *Science and Public Policy* 28(2): 99–112.

Miraglia, M. et al. (2004) 'Detection and traceability of genetically modified organisms in the food production chain', *Food and Chemical Toxicology* 42: 1157–80, http://www.elsevier.com/locate/foodchemtox

Mol, A. P. J. (1996) 'Ecological modernisation and institutional reflexivity: environmental reform in the late modern age', *Environmental Politics* 5(2): 302–23

Monitoraggio normative OGM (2006) Email service, 13 November.

Monsanto (1984) *Genetic Engineering: A Natural Science.*

Monsanto (1997) *Report on Sustainable Development.* St Louis, MO: Monsanto Company.

Monsanto (2003) Summary Notification Information Form (SNIF) for GT73 OSR.

Murphy, J. and Levidow, L. (2006) *Governing the Transatlantic Conflict over Agricultural Biotechnology: Contending Coalitions, Trade Liberalisation and Standard Setting,* London: Routledge.

Nguyen, H. T. and Jehle, J. A. (2007) 'Quantitative analysis of the seasonal and tissue-specific expression of Cry1Ab in transgenic maize MON810', *Journal of Plant Diseases and Protection* 114(2): 820–87.

Niespolo, F. (2005) The main differences between European Community law and Italian national law concerning GMOs, http://www.codex-politics.com

Novartis (1998) *Novartis Bt-11 Maize,* Basel: Novartis, http://www.novartis.com

Nowotny, H., Scott, P. and Gibbons, M. (2003) '"Mode 2" revisited: the new production of knowledge', *Minerva* 41: 179–94.

OECD (1986) *Recombinant DNA Safety Considerations.* Paris: OECD.

Økonomi- og Erhvervsministeriet (2001) Government response to the CEC Consultation on 'A Strategic Vision for Life Sciences and Biotechnology'. Copenhagen: Economy and Business Ministry.

Öko-Institut (2002a) *Genetic Engineering and Organic Farming,* Berlin: Umweltbundesamt.

Öko-Institut (2002b) *Forschungsvielfalt für die Agrarwende.*

Öko-Institut (2004) *Developing Agrobiodiversity! Strategies for action and impulses for sustainable animal and plant breeding,* http://www.oeko.de, http://www.agrobiodiversitaet.net

OPECST (1998a) Letter from Steering Committee to participants in the preparatory weekends, 16 April, Paris: L'Office Parlementaire d'Évaluation des Choix Scientifiques et Technologiques (OPECST), Assemblée Nationale, http://www.senat.fr/opecst/

OPECST (1998b) Conférence de Citoyens Sur l'Utilisation des Organismes Génétiquement Modifiés en Agriculture et dans l'Alimentation. Paris: L'Office Parlementaire d'Évaluation des Choix Scientifiques et Technologiques (OPECST), Assemblée Nationale, http://www.senat.fr/opecst/

Oreszczyn, S. (2005) GM crops in the UK: precaution as process, *Science & Public Policy* 32(4): 317-24.

Pelkmans, J. (1987) 'The new approach to technical harmonization and standardization', *Journal of Common Market Studies* 25(3): 249–69.

Pellizoni, L. (2003) 'Uncertainty and participatory democracy', *Environmental Values* 12: 195-224.

Pestre, D. (2008) 'Challenges for the democratic management of technoscience: governance, participation and the political today', *Science as Culture* 17(2): 101–20.

Peterson, T.R. (1997) *Sharing the Earth: The Rhetoric of Sustainable Development.* Columbia, SC: University of South Carolina Press.

Pettauer, D. (2002) 'Interpretation of substantial equivalence in the EU', in H. Gaugitsch and A. Spök (eds), *Evaluating Substantial Equivalence: A Step Towards Improving the \Risk/Safety Evaluation of GMOs,* Vienna: Federal Environment Agency, pp. 15–24.

Prasifka, P. L., Hellmich, R. L., Prasifka, J. R. and Lewis, L. C. (2007) 'Effects of Cry1Ab-expressing corn anthers on the movement of monarch butterfly larvae', *Environmental Entomology* 36(1): 228–33.

Purdue, D. (1995) 'Whose knowledge counts? "Experts", "counter-experts" and the "lay" public', *The Ecologist* 25(5), September–October: 170–72.

Purdue, D. (1996) 'Contested expertise: plant biotechnology and social movements', *Science as Culture* 5(4): 526–45.

Purdue, D. (2000) *Anti-GenetiX: The Emergence of the Anti-GM Movement,* Aldershot, UK: Ashgate.

Rainbow Group (1989) Deliberate Release into the Environment of Genetically Engineered Organisms [brochure], Brussels: Rainbow Group of European Parliament.

Rayner, S. (2003) 'Democracy in the age of assessment: reflections on the roles of expertise and democracy in public-sector decision making', *Science and Public Policy* 30(3): 163–81.

Redclift, M. (1987) *Sustainable Development.* London: Routledge

Rein, M. and Schön, D. (1991) *Frame Reflection: Toward the Resolution of Intractable Policy Controversies,* New York: Basic Books.

Renn, O. (1992) 'The social arena concept of risk debates', in S. Krimsky and D. Golding (eds), *Social Theories of Risk,* London: Praeger, pp. 179–96.

Renn, O. (1995) 'Style of using scientific expertise: a comparative analysis', *Science and Public Policy* 22(3): 147–56.

Reynolds, L. and Szerszynski, B. (2006) 'Representing GM Nation', PATH conference paper, http://www.macaulay.ac.uk/pathconference

Rich, A. (1997) 'Après la vache folle, récidive sur le maïs transgénique', p. 1; 'Pourquoi ce maïs transgénique et quelles garanties sanitaires', p. 8; '"Une décision réfléchie" ou "Une décision dans l'urgence"?', p. 8, *Le Soir,* 27 January, Brussels.

Rigby, D. and Bown, S. (2007) 'Whatever happened to organic? Food, nature and the market for "sustainable" food', *Capitalism Nature Socialism* 18(3): 81–102.

Ritsema, G. (2002) Presentation by GMO Campaign Coordinator FoEE at European Round Table on GMO Safety Research, 18 April.

Roqueplo, P. (1996) *Entre savoir et decision, l'expertise scientifique.* Paris: INRA.

Rosi-Marshall, E. J., Tank, J. L., Royer, T. V., Whiles, M. R., Evans-White, M., Chambers, C., Griffiths, N. A., Pokelsek, J. and Stephen, M. L. (2007) 'Toxins in transgenic crop byproducts may affect headwater stream ecosystems', *PNAS* 104: 16204–208.

Roy, A. and Joly, P.-B. (2000) 'France: broadening precautionary expertise?', *Journal of Risk Research* 3(3): 247–54.

Royal Society (1999) 'Review of data on possible toxicity of GM potatoes', June, http://www.royalsoc.ac.uk, see under Reports.

SAGB (1990) *Community Policy for Biotechnology: Priorities and Actions,* Brussels: CEFIC.

Sagoff, M. (1991) 'On making nature safe for biotechnology', in L. Ginzburg (ed.), *Assessing Ecological Risks of Biotechnology,* Stoneham, MA: Butterworth-Heineman, pp. 341–65.

Sagoff, M. (2001) 'Genetic engineering and the concept of the natural', in *Genetically Modified Food and the Consumer,* Ithaca, NY: NABC, pp. 137–40, http://www.cals.cornell.edu/extension/nabc

SBC (2001) *Stakeholder Dialogue on Environmental Risks and Safety of GM Plants,* Leiden, 28 February–1 March, workshop report by Schenkelaars Biotechnology Consultancy on behalf of the EFB Task Group on Public Perceptions of Biotechnology, http://www.sbcbiotech.nl

SBC (2002) *Report of a European Workshop: Public Information and Public Participation in the Context of EU Directives 90/220 and 2001/18,* Leiden, 30–31 May, Schenkelaars Biotechnology Consultancy.

SCF (2002) Opinion of the Scientific Committee on Food on the safety assessment of GM sweet maize line Bt 11, 17 April, http://ec.europa.eu/comm/food/fs/sc/scf/out129_en.pdf

SCFCAH (2003) Standing Committee on the Food Chain and Animal Health, Minutes of 8 December meeting, http://ec.europa.eu/comm/food/fs/rc/scfcah/general/out07_en.pdf

Scharpf, F. W. (1999) *Governing in Europe: Effective and Democratic?* Oxford: Oxford University Press.

Schmitter, P. (2001) 'What is there to legitimate in the European Union . . . and how might this be accomplished?', in C. Joerges, Y. Mény and J. H. H. Weiler (eds), *Mountain or Molehill? A Critical Appraisal of the Commission White Paper on Governance,* Jean Monnet Working Paper No. 6/01, http://www.iue.it/RSC/, http://www.JeanMonnetProgram.org

Schurman, R. (2004) 'Fighting "Frankenfoods": industry opportunity structures and the efficacy of the anti-biotech movement in Western Europe', *Social Problems* 51(2): 243–268.

Schütte, G. (2002) 'Prospects of biodiversity in herbicide-resistant crops', *Outlook on Agriculture* 31(3): 193–98.

Schwarz, M. and Thompson, M. (1990) *Divided We Stand: Redefining Politics, Technology and Social Choice*, London: Harvester.
Schweiger, T. (2001) 'Europe: hostile lands for GMOs', in B. Tokar (ed.), *Redesigning Life? The Worldwide Challenge of Genetic Engineering*, London: Zed, pp. 361–72.
Science Museum/BBSRC (1994) *UK National Consensus Conference on Plant Biotechnology: Final Report.*
SCIMAC (1999) 'Industry stewardship programme launched to promote responsible introduction of GM crops', 21 May, Supply Chain Initiative on Modified Agricultural Crops.
SCP (1998a) Opinion of the Scientific Committee on Plants Regarding AgroEvo's glufosinate-tolerant oilseed rape, 10 February.
SCP (1998b) Opinion of the Scientific Committee on Plants Regarding Pioneer's MON9 Bt, glyphosate-tolerant maize, 19 May.
SCP (2001a) Scientific Committee on Plants, 'Opinion on adventitious presence of GM seeds in conventional seeds', http://ec.europa.eu/comm/food/fs/sc/scp/out93_gmo_en.pdf
SCP (2001b) Scientific Committee on Plants, Opinion on the invocation by the UK of Article 16 regarding genetically modified maize line T25 notified by Agrevo, 8 November, http://ec.europa.eu/comm/food/fs/sc/scp/out110_gmo_en.pdf
SCSP (2003) Standing Committee on Seeds and Propagating Material for Agriculture, Horticulture and Forestry, short report of meeting held on 27–28 October, http://ec.europa.eu/comm/food/fs/rc/scsp/rap47_en.pdf
Skogstad, G. (2001) 'The WTO and food safety regulatory policy innovation in the European Union', *Journal of Common Market Studies* 39(3): 485–505.
Skov-og Naturstyrelsen (2003) Notat til Folketingets europaudvalg vedr. ansøgning om godkendelse til markedsføring i EU af genetisk modificeret majs (C/ES/01/01) i henhold til Europa-parlamentets og Rådets direktiv 2001/18/EF, 8 October. Forestry and Nature Protection Board, Copenhagen.
Smith, E., Marsden, T. and Flynn, A. (2004) 'Regulating food risks: rebuilding confidence in Europe's food?', *Environmental and Planning C: Government and Policy* 22: 543–67.
Smith, J. (2006) 'EU Commissioners split on GM food', 6 April, http://www.checkbiotech.org
Soil Association (2003) Comment on Broom's Barn GM Sugar Beet Research, January, http://www.soilassociation.org
Spinart, D. (2003) 'GMO law proposal sparks deep division', *European Voice*, 2 October: 16.
SSC (2003) Scientific Steering Committee, Report on the Future of Risk Assessment in the European Union: Second Report on the Harmonisation of Risk Assessment Procedures, http://ec.europa.eu/comm/food/fs/sc/ssc/out361_en.pdf
StCF (2000) 'Summary record of 78th meeting', Standing Committee on Foodstuffs, 18–19 October, http://ec.europa.eu/comm/food/fs/rc/scfs/rap02_en.html
Stirling, A. (1999) *On Science and Precaution in the Management of Technological Risk,* Sussex: SPRU (based on contributions from O. Renn, A. Rip and A. Salo), Final Report for EC Forward Studies Unit, ftp://ftp.jrc.es/pub/EURdoc/eur19056en.pdf
Stirling, A. (2003) 'Risk, uncertainty and precaution: some instrumental implications from the social sciences', in F. Berkhout et al. (eds), *Negotiating Environmental Change,* Cheltenham: Edward Elgar, pp. 33–76.
Stirling, A. (2005) 'Opening up or closing down? Analysis, participation and power in the social appraisal of technology', in M. Leach, I. Scoones and B. Wynne (eds), *Science and Citizens: Globalisation and the Challenge of Engagement,* London: Zed, pp. 218–31.

Stirling, A. (2006) 'From science and society to science in society: Towards a framework for "co-operative research"', Report of a European Commission Workshop, 24–25 November 2005, http://cipast.org, http://eurosfaire.prd.fr/7pc/bibliotheque/consulter.php?id=308

Strandberg, B. and Bruus Pedersen, M. (2002) *Biodiversity in glyphosate-tolerant fodder beet fields: timing of herbicide application,* National Environmental Research Institute, Silkeborg, NERI Technical Report no. 410, http://www.dmu.dk/1_viden/2_Publikationer/3_fagrapporter/rapporter/FR410.pdf

Surel, Y. (2000) 'The role of cognitive and normative frames in policy-making', *Journal of European Public Policy* 7: 495–612.

Székács, A., Juracsek, J., Polgar, L. A. and Darvas, B. (2005) 'Levels of expressed Cry1Ab toxin in genetically modified corn DK-440-BTY (Yieldgard) and stubble', *FEBS Journal* 272(Suppl. 1): 508.

Tàbara, J. D., Polo, D. and Lemkow, L. (2004) 'Spain: Caution versus Precaution', national report for PEG project, Barcelona: Institute of Environmental Sciences and Technology (IEST), Autonomous University of Barcelona, http://technology.open.ac.uk/cts/bpg

Taeymans, D. (2001) 'An unworkable package', *Parliament Magazine* 123, 10 September: 35 [Director of Scientific & Regulatory Affairs, Confederation de l'Industries Agro-Alimentaires].

Tait, J. and Levidow, L. (1992) 'Proactive and reactive approaches to risk regulation: the case of biotechnology', *Futures* 24: 219-31.

Tait, J., Chataway, J. and Wield, D. (2002) 'The life science industry sector: evolution of agro-biotechnology in Europe', *Science and Public Policy* 29(4): 253–58.

Teknologinævnet (1987) *Genteknologi i industri og landbrug.* Teknologinævnets Rapporter 1987/2 og 1987/4. Copenhagen: Danish Board of Technology, www.tekno.dk

Teknologinævnet (1999) Final document of the consensus conference on genetically-modified foods, Copenhagen: Danish Board of Technology

Terragni, F. and Recchia, E. (1999) 'Italy: Precaution for Environmental Diversity?', report for 'Safety Regulation of Transgenic Crops: Completing the Internal Market', DGXII RTD project coordinated by the Open University, http://technology.open.ac.uk//cts/srtc/index.html

Thomas, J. (2001) 'Princes, aliens, superheroes and snowballs: the playful world of UK genetic engineering resistance', in B. Tokar (ed.), *Redesigning Life? The Worldwide Challenge of Genetic Engineering,* London: Zed, pp. 337–50.

Tierney, K. (2006) 'Implementing EU legislation on GMOs—is there room for precaution?', talk at conference on 'GMOs and Precaution', Vienna, April; in *The Role of Precaution in GMO Policy,* Vienna: BMGF, pp. 17–20, http://www.bmgf.gv.at

Toft, J. (1996) 'Denmark: seeking a broad-based consensus on gene technology', *Science and Public Policy* 23(3): 171–74.

Toft, J. (2000) 'Denmark, potential polarization or consensus?', *Journal of Risk Research* 3(3): 227–36.

Toft, J. (2005) 'Denmark's regulation of agribiotechnology: coexistence bypassing risk issues', *Science and Public Policy* 32(4): 293–300.

Toke, D. (2002) 'Ecological modernisation and GM food', *Environmental Politics* 11(3): 145–63.

Toke, D. (2004) *The Politics of GM Food,* London: Routledge.

Toke, D. and McGough, S. (2005) 'Farmers stage a comeback: but have the greens sold out?', *Political Quarterly* 76: 505-15.

Tolstrup, K. et al. (2003) Report from the Working Group on the co-existence of genetically modified crops with conventional and organic crops. Copenhagen: Fødevareminsteriet (Ministry for Food, Agriculture and Fisheries), January.

Torgersen, H. and Bogner, A. (2005) 'Austria's agri-biotechnology regulation: political consensus despite divergent concepts of precaution', *Science and Public Policy* 32(4): 277–84.
Torgersen, H. and Seifert, F. (2000) 'Austria: precautionary blockage of agricultural biotechnology', *Journal of Risk Research* 3(3): 209–17.
UK DEFRA (2003a) UK responses to marketing notifications under the Deliberate Release Directive, http://www.defra.gov.uk/environment/gm/regulation/euconsent.htm
UK DEFRA (2003b) GM Crops: Effects on Farmland Wildlife, http://www.defra.gov.uk/environment/gm/fse/results/fse-summary.pdf
UK SSC (1999) GM Crop Farm-Scale Evaluations, minutes of the Scientific Steering Committee, 14 June, http://www.environment.detr.gov.uk/fse/steering/jun14.htm
UK SSC (2000) GM Crop Farm-Scale Evaluations, minutes of the Scientific Steering Committee, 10 March, http://www.environment.detr.gov.uk/fse/steering/00mar10.htm
UK SSC (2003) Scientific Steering Committee for the GM crop farm-scale evaluations: final advice to Ministers, http://www.defra.gov.uk/environment/gm/fse/results/ssc-advice.htm
Unilever (2000) 'Genetically Modified Organisms', http://www.unilever.com
USDA (1996) 'Sustainable Development', Secretary's Memorandum 9500-6, http://www.usda.gov/agency/oce/oce/sustainable-development/secmemo.htm
van Apeldoorn, B. (2000) 'Transnational class agency and European governance: the case of the European Roundtable of Industrialists', *New Political Economy* 5(2): 157-81.
van Apeldoorn, B. (2002) *Transnational Capitalism and the Struggle over European Integration*. London: Routledge.
van den Daele, W. (1994) *Technology Assessment as a Political Experiment*, Berlin: Wissenschaftszentrum.
van den Daele, W. (1995) 'Technology assessment as a political experiment', in R. von Schomberg (ed.), *Contested Technology: Ethics, Risk and Public Debate*, Tilburg: International Centre for Human and Public Affairs, pp. 63–89.
van der Straaten, J. (1993) 'A sound European environmental policy: challenges, possibilities and barriers', in D. Judge (ed.), *A Green Dimension for the European Community*, London: Frank Cass, pp. 65–83.
Vogel, D. (1986) *National Styles of Regulation: Environmental Policy in Great Britain and the United States*. Ithaca: Cornell University Press.
von Schomberg, R. (1996) 'Netherlands: deliberating biotechnology regulation', *Science and Public Policy* 23(3): 158-63.
Vidal, J. (2004a) 'Belgian authorities accused of making political decision on GM rape', *The Guardian*, 3 February.
Vidal, J. (2004b) 'Firm drops plan to grow GM maize', 1 April, http://www.guardian.co.uk/gm
Vos, E. (1997) 'Market building, social regulation and scientific expertise: an introduction', in C. Joerges, K.-H. Ladeur and E. Vos (eds), *Integrating Scientific Expertise into Regulatory Decision-Making*, Baden-Baden: Nomos Verlagsgesellschaft, pp. 127–39.
Walls, J., Horlick-Jones, T., Niewöhner, J., O'Riordan, T. (2005)'The meta-governance of risk and new technologies: GM crops and mobile telephones', *Jnl of Risk Research* 8: 635-61.
Waterton, C. and Wynne, B. (2004) 'Knowledge and political order in the European Environment Agency', in S. Jasanoff (ed.), *States of Knowledge: The Co-production of Science and Social Order*, London: Routledge, pp. 87–108.

Weale, A. (1992) *The New Politics of Pollution*, Manchester: Manchester University Press.
Weale, A. (1999) *Democracy*, Basingstoke: Macmillan.
Weale, A. (2000) *Environmental Governance in Europe: An Ever Closer Ecological Union?* Oxford: Oxford University Press.
Weale, A. and Williams, A. (1993) 'Between economy and ecology? The single market and the integration of environmental policy', in D. Judge (ed.), *A Green Dimension for the European Community*, London: Frank Cass, pp. 45–64.
Weiler, J. (1999) 'Epilogue: "comitology" as revolution—infranationalism, constitutionalism and democracy', in C. Joerges and E. Vos (eds), *EU Committees: Social Regulation, Law and Politics*, Portland, OR: Hart Publications, pp. 339–51.
WEN (1998) 'Test Tube Harvest: Genetically Engineered Soybeans', leaflet, London: Women's Environmental Network.
Wield, D., Tait, J. and Chataway, J. (2004) 'Understanding company R&D strategies in agro-biotechnology: trajectories and blind spots', *Research Policy* 33: 1041–57.
Williams, R. (1980) 'Ideas of nature', in R. Williams, *Problems in Materialism and Culture*, London: Verso, pp. 67–85.
Woodhouse, P. (2000) 'Environmental degradation and sustainability', in T. Allen and A. Thomas (eds), *Poverty and Development into the 21st Century*, Oxford: Oxford University Press, pp. 141–62, in association with the Open University, Milton Keynes.
Wrong, M. (1999) 'Fodder for the GM debate', *Financial Times*, 18 November.
Wynne, B. (2005) 'Risk as globalising "democratic" discourse? Framing subjects and citizens', in M. Leach, I. Scoones and B. Wynne (eds), *Science and Citizens: Globalisation and the Challenge of Engagement*, London: Zed, pp. 66–82.
Wynne, B. (1995) 'Technology Assessment and reflexive social learning: observations from the risk field', in A. Rip et al., eds, *Managing Technology in Society: The Approach of Constructive Technology Assessment*, pp.19-36, London: Pinter.
Wynne, B. (2006) 'GMO Risk Assessment Under Conditions of Biological (and Social) Complexity: How European politics and science mutually construct—and maybe destruct', in *The Role of Precaution in GMO Policy*, pp.30-46, Vienna: BMGF, www.bmgf.gv.at
Wynne, B. et al. (2001) 'Public Attitudes towards Agricultural Biotechnologies in Europe (PABE)', final report of project with five partner country teams (Spain, Italy, Germany, France UK), funded by DG-Research, Brussels. Centre for the Study of Environmental Change (CSEC), Lancaster University, http://www.inra.fr/internet/Directions/SED/science-gouvernance, or http://csec.lancs.ac.uk/pabe/docs/summary.doc
Young, O. R. (1994) *International Governance: Protecting the Environment in a Stateless Society*, Ithaca, NY: Cornell University Press.
Yoxen, E. (1981) 'Life as a productive force: capitalizing the science and technology of molecular biology', in L. Levidow and R. M. Young (eds), *Science, Technology and the Labour Process*, London: CSE Books & Atlantic Highlands, NJ: Humanities Press, vol. 1, pp. 66–122; reissued by Free Association Books, 1983.
Ziltener, P. (2004) 'The economic effects of the European single market: projections, simulations and the reality', *Review of International Political Economy* 11(5): 953–79.
Zolo, D. (1992) *Democracy and Complexity: A Realist Approach*, Cambridge: Polity.

ZS-L (2003) The EU's planned Directive regarding the adventitious presence of GMOs in seeds, Berlin: Zukunftsstiftung Landwirtschaft/Foundation on Future Farming, http://www.saveourseeds.org

Zwahlen, C., Hilbeck, A., Gugerli, P. and Nentwig, W. (2003a) 'Degradation of the Cry1Ab protein within transgenic *Bacillus thuringiensis* corn tissue in the field', *Molecular Ecology* 12: 765–75.

Zwahlen, C. Hilbeck, A., Howald, R. and Nentwig, W. (2003b) 'Effects of transgenic *Bt* corn litter on the earthworm *Lumbricus terrestris*', *Molecular Ecology* 12: 1077–86.

Index

D

H

T

Printed in the United States
by Baker & Taylor Publisher Services